# Handbook of Human Factors and Ergonomics in Contemporary Business

*Handbook of Human Factors and Ergonomics in Contemporary Business* highlights the symbiotic relationship between human-centric approaches and organizational success in 21st-century business and corporate culture. It delivers a comprehensive exploration of how integrating human factors and ergonomics strategically can fuel innovation, enhance performance, and deliver a competitive advantage.

By promoting employee well-being and implementing ergonomic interventions for heightened productivity, the chapters offer invaluable insights into the role of human factors across differing business sectors. Edited and authored by distinguished experts in the field, this handbook presents theoretical frameworks, empirical evidence, and real-world case studies. Across these chapters split into various sections on perspectives, digitalization, and well-being, it provides interdisciplinary perspectives and practical applications to equip readers with the tools needed to encourage thriving workplace environments, nurture employee welfare, and foster innovation and creativity. International case studies from diverse businesses and topics on Industry 5.0, logistics, strategy, UX, productivity, green issues, and much more result in a thorough overview of the topic. It will help any reader to optimize organizational performance and understand the key connections between human factors and business success.

This book is an essential read for researchers and professionals in ergonomics and human factors, business and management, leadership, engineering design, and human resources.

# Handbook of Human Factors and Ergonomics in Contemporary Business

## Strategic Orientation, Corporate Success, and Innovation

Edited by
Shad Ahmad Khan, Sumit Narula,
Arshi Naim, and Monika Srivastava

CRC Press
Taylor & Francis Group
Boca Raton London New York

CRC Press is an imprint of the
Taylor & Francis Group, an **informa** business

Designed cover image: © 2026 Shutterstock, user August_0802 [Image ID # 2475760993]

First edition published 2026
by CRC Press
2385 NW Executive Center Drive, Suite 320, Boca Raton FL 33431

and by CRC Press
4 Park Square, Milton Park, Abingdon, Oxon, OX14 4RN

*CRC Press is an imprint of Taylor & Francis Group, LLC*



*Library of Congress Cataloging-in-Publication Data*
Names: Khan, Shad Ahmad, 1988- editor
Title: Handbook of human factors and ergonomics in contemporary business : strategic orientation, corporate success and innovation / edited by Shad Ahmad Khan, Sumit Narula, Arshi Naim and Monika Srivastava.
Description: First edition. | Boca Raton : CRC Press, 2026.
Identifiers: LCCN 2025005489 (print) | LCCN 2025005490 (ebook) | ISBN 9781032872629 hbk | ISBN 9781032873633 pbk | ISBN 9781003532293 ebk
Subjects: LCSH: Work environment—Handbooks, manuals, etc. | Human engineering—Handbooks, manuals, etc. | Success in business—Handbooks, manuals, etc. | LCGFT: Handbooks and manuals
Classification: LCC T59.77 .H34 2026 (print) | LCC T59.77 (ebook) | DDC 331.25/6—dc23/eng/20250416
LC record available at https://lccn.loc.gov/2025005489
LC ebook record available at https://lccn.loc.gov/2025005490

ISBN: 9781032872629 (hbk)
ISBN: 9781032873633 (pbk)
ISBN: 9781003532293 (ebk)

DOI: 10.1201/9781003532293

Typeset in Times
by codeMantra

# Contents

Preface.....ix
Editor Biographies.....xi
List of Contributing Authors.....xii

## SECTION I Human Factors and Ergonomics: Perspectives

**Chapter 1** Human Factors and Ergonomics in Contemporary Business: Perspectives on Strategic Orientation, Corporate Success, and Innovation.....3

*Khalilullah Mohammed Baharul Islam, Asif Khan, and Archan Mitra*

**Chapter 2** Deconstructing Human Influences in Merger & Acquisition through MBA Framework: A Systematic Literature Review and Thematic Analysis.....19

*Arzoo Gupta and Vandana Pareek*

**Chapter 3** Assessing the Strategic Orientation towards Ergonomic Hazards in Health and Safety Reporting Systems in Selected Oil and Gas Drilling Sites.....34

*Henry Jonathan and Yahya Fuad Al Abdulsalam*

**Chapter 4** Effect of Personality Traits as Human Factors on Academic Stress of Students: An Exploratory Study Using PLS-SEM.....47

*Bhanupriya Khatri, Charu Saxena, Megha Sharma, Shad Ahmad Khan, and Anand Nayyar*

**Chapter 5** Synergizing GHRM and Human Factors: A Paradigmatic Approach in Contemporary Business.....65

*Priyanka Indoria, Mohd Danish Chishti, Abdullah Bin Junaid, and Ruchi Agarwal*

**Chapter 6** Potential Strategies for Corporate Success in Dealing with WMSDs at Workplace: A Study of Ergonomic Practices in Hospitals in India....76

*Obaidur Rahman, Manik Batra, Farah Ahmad, and Anshu*

**Chapter 7** The Integration of Strategic Orientation with Human Factors and Ergonomics Interventions in Workplace....89

*Shivani Kapoor, Fehmina Khalique, Ankita Kapoor, and Sanshya Kapoor*

**Chapter 8** Exploring the Role of Employee Participation in Ergonomic Transformations: A Study in Change Management....103

*Sumit Narula, Dhruv Sabharwal, and Robin Kabha*

**Chapter 9** Unraveling the Connection, between Employee Happiness and Organizational Success Across Different Cultures....120

*Madhur Batra and Ferdinand J. Epoc*

**Chapter 10** Impact of Psychological Well-Being, Workplace Well-Being and Subjective Well-Being on Job Orientation: Evidence from Ten Maharatna Companies of India....131

*K. P. Kanchana, Shad Ahmad Khan, Soni Sharma, and Ravinder Rena*

## *SECTION II HF&E Meets Technology: Shaping Corporate Success in the Digital Landscape*

**Chapter 11** Human Factors and Ergonomics in the Digital Era: Role of Artificial Intelligence....147

*Aujasvi Singh, Monika Srivastava, Sobhana Jaya-Madhavan, and Busige Nishantha*

**Chapter 12** The Role of Human Factors in Mitigating Risks in AI-Based Performance Evaluation Systems....162

*Kumar Sudharson, Alli Raju Suresh Kumar, Kuppuswamy Selvi, and Zakir Hossen Shaikh*

**Chapter 13** Human-Centric Logistics: Ergonomic Considerations in Warehouse Order Picking for Industry 5.0 Integration .................... 183

*Danish Nasir*

**Chapter 14** Strategic Orientation for Corporate Success: Innovating Human Factors and Ergonomics in Robotics for Industry 4.0 ...................... 198

*Archana Barman, Nilakshi Teron, and Purna Prasad Sharma*

**Chapter 15** Implication of the Dark Side of Artificial Intelligence on Corporate Success from the perspective of Human Factors and Ergonomics .................................................................. 213

*Preeti Mehra and Pooja Kansra*

## SECTION III HF&E in Action: Strategies for Employee Well-Being and Corporate Success

**Chapter 16** Application of Tailored Ergonomic Intervention Strategy for Reducing Work-Related Stress .................................... 235

*Julie Vardhan, Mona Ratnesh, and Prabhat Kumar Tripathi*

**Chapter 17** Optimizing Employee Well-being and Performance: The Role of Human Factors and Ergonomics for Corporate Success ....................................................................248

*Swati Sisodia and Syed Hameedur Rahman Zaini*

**Chapter 18** Enhancing Employee Well-Being and Operational Efficiency in the Oil Sector: A Holistic Examination of Ergonomics, Corporate Practices, and Human Factors Integration ........................................................................ 262

*Henry Jonathan, Mohammad Sultan Ahmad Ansari, and Raneem Al Ruhaili*

**Chapter 19** Enhancing Organizational Success: Exploring Ergonomics, Employee Deviant Workplace Behavior and Occupational Burnout ............................................................ 274

*Ujjal Bhuyan, Shad Ahmad Khan, Alina Costin, and Anuj Kumar*

**Chapter 20** Workplace Stress Reduction and Mental Health Promotion via Human Factors and Ergonomics ..................................................... 287

*Muhammad Zulkifl Hasan, Muhammad Zunnurain Hussain, Umair Waqas, and Sonia Umair*

**Chapter 21** Human-Centric Nutrition and the Microbiome for the Development of Business Strategies in Transforming Health: Special Focus on Arab World.......................................................... 311

*Arshi Naim, Mohammad Ibrahim Khan, and Mohammad Faiz Khan*

**Chapter 22** Exploring the Association between Ergonomics and Financial Peace: Investigating Financial Well-Being and Well-Being Expenses in Banking Professionals................................................. 324

*Prakash Bhatnagar and Himani Sharma*

**Index**............................................................................................................ 339

# Preface

In an era characterized by rapid technological advancements and evolving workplace dynamics, the integration of human factors and ergonomics (HF&E) into business strategy has never been more critical. This book, *Human Factors and Ergonomics in Contemporary Business: Perspectives on Strategic Orientation, Corporate Success, and Innovation*, seeks to illuminate this critical relationship by offering a comprehensive exploration of HF&E principles and their application across various sectors. This book provides insights that are both timely and essential for contemporary business leaders.

## THE IMPORTANCE OF HF&E

HF&E encompasses a broad spectrum of disciplines that focus on understanding human capabilities and limitations in various contexts. As organizations strive for efficiency and innovation, recognizing the pivotal role of human elements—such as employee well-being, cognitive load, and ergonomic design—becomes paramount. The chapters within this volume explore these themes through diverse lenses, highlighting how strategic orientation toward HF&E can foster corporate success.

## OVERVIEW OF CONTRIBUTIONS

This book is structured into three comprehensive sections. The first section delves into foundational perspectives on HF&E, examining topics such as the psychological impacts of workplace design, strategic orientations in health and safety reporting, and the intricate connections between employee happiness and organizational performance. Each chapter provides empirical evidence and theoretical frameworks that underscore the importance of human-centered approaches in business settings. The second section explores how technology intersects with HF&E. The chapters address contemporary challenges posed by artificial intelligence and digital transformation, emphasizing the need for a human-centric approach in these rapidly evolving landscapes. From mitigating risks in AI systems to optimizing logistics through ergonomic principles, this section offers innovative strategies that align technology with human needs. The final section focuses on practical applications of ergonomics in enhancing employee well-being and operational efficiency. It presents actionable strategies for organizations to implement tailored ergonomic interventions that not only improve worker health but also drive corporate success. By examining case studies from various industries, this section provides a roadmap for integrating HF&E principles into everyday business practices.

## A CALL TO ACTION

As we navigate through complex business environments shaped by globalization and technological disruption, this book serves as a vital resource for scholars, practitioners, and corporate leaders alike. It calls for a paradigm shift toward recognizing human factors as integral to strategic decision-making processes. By fostering an organizational culture that prioritizes ergonomics and employee well-being, businesses can not only enhance productivity but also cultivate a sustainable competitive advantage. In closing, we invite you to engage with the insights presented herein. We hope this collection inspires further exploration into the interplay between human factors and business success, ultimately contributing to a more innovative and humane workplace. This preface sets a tone that is both informative and engaging while clearly outlining the book's purpose and structure.

# Editor Biographies

**Shad Ahmad Khan** is an Assistant Professor and Program Chair of Business Administration within the College of Business at the University of Buraimi, Oman. With over 13 years of teaching experience, Dr. Khan is a researcher specializing in green practices, online business management, and higher education institutions. He is ranked among top 2% scientists in the world as per Elsevier and standford University 2024 and has been involved in funded projects, consultancy work, editorial roles in journals and active speaking roles at major academic conferences.

**Sumit Narula** is the Director/Head of the Institution at Amity School of Communication, Amity University, Gurgaon, having previously held the position of Deputy Dean (Research) (Publications & Citations) at Amity University, Madhya Pradesh, India. His expertise and research interests cover various areas including research and publication ethics, media research, and fake news detection. He has earned accolades including the Innovation Research Award from Taylor's University, Malaysia and voluntary ambassador for Crossref. Dr Narula has two master's degrees and two postgraduate diplomas and has authored two books in Mass Communication/Journalism and edited seven others.

**Arshi Naim** is an Associate Professor and Quality Programme Head in Business Management at King Khalid University, Saudi Arabia. She completed her PhD in Business Management at Singhania University, Rajasthan, India with an MBA in Marketing Management from HNB University and Economics from AMU, Aligarh, all in India. She is an expert in curriculum and programme development, quality, and accreditation process and E-learning.

**Monika Srivastava** is a Dean and Professor, School of Psychology and Counseling Chitkara University, Punjab, India. With over 26 years of experience, Professor Monika is an educator and consultant specializing in organizational behavior. Her career includes roles at Dr. Gaur Hari Singhania Institute of Management & Research and Sahara Art and Management Academy. Her international teaching and research engagements span countries including Bhutan, Ethiopia, Oman, and Nepal and she has earned accolades with the Teaching Excellence Award in 2020 and the Best Teacher Award from Innovative Education for her invaluable contributions to society. Dr. Srivastava has published 40 research papers on contemporary management subjects including stress, commitment, well-being, corporate social responsibility, and management education.

# Contributing Authors

**Ruchi Agarwal**
Faculty
Ajman University
Ajman, UAE

**Farah Ahmad**
Department of Business
Leicester College
Leicester, United Kingdom

**Yahya Fuad Al Abdulsalam**
Health and Safety Management
Modern College of Business and Science
Muscat, Oman

**Raneem Al Ruhaili**
Health and Safety Management
Modern College of Business and Science
Muscat, Oman

**Mohammad Sultan Ahmad Ansari**
Business and Economics Department
Modern College of Business and Science
Muscat, Oman

**Anshu**
School of Business Studies
Sharda University
Greater Noida, India

**Archana Barman**
Faculty of Commerce & Management
Assam Down Town University
Guwahati, Assam, India

**Madhur Batra**
Department of Management
Gedu College of Business Studied
Royal University of Bhutan
Gedu, Bhutan

**Manik Batra**
School of Business Studies
Sharda University
Greater Noida, India

**Prakash Bhatnagar**
Graduate School of Business
Tula's Institute
Dehradun, Uttarakhand, India

**Ujjal Bhuyan**
Department of commerce
Jagannath Baruah University
Jorhat, Assam, India

**Abdullah Bin Junaid**
Faculty of Healthcare Management
Jumeira University
Dubai, UAE

**Mohd Danish Chishti**
Faculty of Management Science
Shri Ram Murti Smarak College of Engineering & Technology
Bareilly, Uttar Pradesh, India

**Alina Costin**
Department of Pedagogy, Psychology and Social Work
Aurel Vlaicu University of Arad
Arad, Romania

**Ferdinand J. Epoc**
Department of Business Adminsitration
College of Business
University of Buraimi
Al Buraimi, Oman

**Arzoo Gupta**
Department of Management
Dr. APJ Abdul Kalam Technical University
Lucknow, Uttar Pradesh, India

**Muhammad Zulkifl Hasan**
Faculty of Information Technology
Department of Computer Science
University of Central Punjab
Lahore, Pakistan

**Muhammad Zunnurain Hussain**
Department of Computer Science
Bahria University Lahore Campus
Lahore, Pakistan

**Priyanka Indoria**
Institute of Innovation in Technology and Management
Janakpuri, New Delhi, India

**Khalilullah Mohammed Baharul Islam**
Institute of Management Technology
Hyderabad, India

**Sobhana Jaya-Madhavan**
Adjunct Faculty
Simon Fraser University
Burnaby, Canada

**Henry Jonathan**
Logistics, Transportation, Health and Safety Management Department
Modern College of Business and Science
Muscat, Oman

**Robin Kabha**
College of Communication
University of Fujairah
Fujairah, UAE

**K. P. Kanchana**
Department of HR and OBJaipuria School of Business Ghaziabad
Ghaziabad, Uttar Pradesh, India

**Pooja Kansra**
Mittal School of Business
Lovely Professional University
Phagwara, Punjab, India

**Ankita Kapoor**
Master in Information Technology and Information Technology Management
University of Sydney
Sydney, Australia

**Sanshya Kapoor**
Metamorf
London, United Kingdom

**Shivani Kapoor**
Allenhouse Institute of Management
Kanpur, Uttar Pradesh, India

**Fehmina Khalique**
Lloyd Business School
Greater Noida, India

**Asif Khan**
AKW Consultants
Dubai, UAE

**Mohammad Faiz Khan**
Security Forces Hospital
MOI
Riyadh, Saudi Arabia

**Mohammad Ibrahim Khan**
Department of Medicine
International Medical University
Kuala Lumpur, Malaysia

**Bhanupriya Khatri**
University School of Business, Chandigarh University
Mohali, Punjab, India

**Anuj Kumar**
Department of Management
Rushford Business School
Geneva, Switzerland

**Alli Raju Suresh Kumar**
Department of MBA
Vel Tech High Tech Engineering College
Chennai, Tamil Nadu, India

**Preeti Mehra**
FacultyDepartment of Business
Campbell College
Edmonton, Alberta, Canada

**Archan Mitra**
Nitte Institute of Communication
Nitte Deemed to be University
Mangalore, India

**Danish Nasir**
Department of Operations Management
Birla Institute of Management Technology, Greater Noida, India

**Anand Nayyar**
School of Computer Science (SCS)
Duy Tan University
Da Nang, Viet Nam

**B. Nishantha**
Faculty of Management & Finance
University of Colombo
Colombo, Sri Lanka

**Vandana Pareek**
Department of Management Studies
Dronacharya Group of Institutions
Greater Noida, India

**Obaidur Rahman**
School of Business Studies
Sharda University
Greater Noida, India

**Mona Ratnesh**
Amity Business School
Amity University Jharkhand
Ranchi, Jharkhand, India

**Ravinder Rena**
DUT Business School,
Faculty of Management Sciences, Durban University of Technology
Durban, South Africa

**Dhruv Sabharwal**
Sharda University
Greater Noida, India

**Charu Saxena**
University School of Business, Chandigarh University
Mohali, Punjab, India

**Kuppuswamy Selvi**
Department of MBA
Saveetha Engineering College Chennai, Tamil Nadu, India

**Zakir Hossen Shaikh**
Department of Finance and Accounting
College of Business Administration
Kingdom University
Riffa, Bahrain

**Himani Sharma**
College of Business studies
COER University
Roorkee, Haridwar, Uttarakhand, India

**Megha Sharma**
Independent Researcher
India

**Purna Prasad Sharma**
Sub teacher
NYC Public School
New York

**Soni Sharma**
Sharda School of Business Studies
Sharda University, Agra

**Aujasvi Singh**
Psychology
DIT University
Makka Wala, Uttarakhand, India

**Swati Sisodia**
School of Management
GD Goenka University
Sohna, Gurgaon, India

**Kumar Sudharson**
Department of Artificial Intelligence and Machine Learning
R.M.D. Engineering College
Chennai, Tamil Nadu, India

**Nilakshi Teron**
Faculty of Commerce & Management
Assam Down Town University
Guwahati, Assam, India

**Prabhat Kumar Tripathi**
Amity Business School
Amity University Jharkhand
Ranchi, Jharkhand, India

**Sonia Umair**
Department of Management
College of Commerce & Business Administration
Dhofar University
Salalah, Oman

**Julie Vardhan**
Amity Business School
Amity University Jharkhand
Ranchi, Jharkhand, India

**Umair Waqas**
Department of Marketing and Entrepreneurship
College of Commerce & Business Administration
Dhofar University
Salalah, Oman

**Syed Hameedur Rahman Zaini**
School of Management
GD Goenka University
Sohna, Gurgaon, India

# Section I

## Human Factors and Ergonomics

### Perspectives

# 1 Human Factors and Ergonomics in Contemporary Business

## *Perspectives on Strategic Orientation, Corporate Success, and Innovation*

*Khalilullah Mohammed Baharul Islam, Asif Khan, and Archan Mitra*

## 1.1 INTRODUCTION

In the rapidly evolving landscape of contemporary business, organizations face increasing pressures to enhance productivity, foster innovation, and maintain a competitive edge. Amid these challenges, human factors and ergonomics (HFE) have emerged as crucial elements in optimizing workplace environments and processes. Human factors refer to the scientific discipline concerned with understanding the interactions among humans and other elements of a system, while ergonomics involves designing or arranging workplaces, products, and systems to fit the people who use them (Dul et al., 2012).

The integration of HFE into business strategies is not merely a trend but a necessity that aligns with broader organizational goals. As businesses strive for excellence, understanding and applying ergonomic principles can lead to significant improvements in employee well-being, operational efficiency, and overall corporate success. This study aims to explore the multifaceted role of HFE in shaping contemporary business strategies, enhancing corporate success, and driving innovation.

### 1.1.1 Research Objectives

The primary objectives of this research are threefold:

- **RO1:** To investigate how HFE are integrated into strategic business orientations.
- **RO2:** To analyze the impact of ergonomic principles on corporate success metrics, including productivity, employee satisfaction, and financial performance.

DOI: 10.1201/9781003532293-2

**RO3:** To examine the relationship between ergonomic practices and innovation within organizations.

These objectives guide the exploration of the significant interplay between HFE and business outcomes, providing a comprehensive understanding of their symbiotic relationship.

### 1.1.2 Significance of the Study

The significance of this study lies in its potential to bridge the gap between ergonomic theory and practical business application. While numerous studies have focused on the individual components of HFE, strategic orientation, and innovation, there is a paucity of research that synthesizes these elements into a cohesive framework applicable to contemporary business contexts. By addressing this gap, the study aims to offer valuable insights for both academia and industry practitioners. From an academic perspective, this research contributes to the existing body of knowledge by integrating diverse strands of literature and presenting a holistic view of the role of HFE in business success. For industry practitioners, the findings provide actionable strategies to enhance workplace design, improve employee outcomes, and foster a culture of innovation.

## 1.2 LITERATURE REVIEW

The literature review provides an in-depth analysis of existing research on HFE, strategic orientation, and corporate success. Key areas of focus include the definition and scope of HFE, the historical context and evolution of HFE, strategic orientation, corporate success metrics, and the relationship between innovation and HFE. HFE have become critical components in shaping strategic orientation, corporate success, and innovation within contemporary business environments (Ansari et al., 2024; Khan & Ansari, 2024). Hasanain (2024) underscores the pivotal role of ergonomic and human factors in fostering sustainable manufacturing practices, which, in turn, enhance organizational efficiency and employee well-being. By integrating HFE principles, businesses can address physical and cognitive workload challenges, thus promoting a healthier workforce and reducing operational costs. The adoption of such practices is not only a strategic move to improve productivity but also a necessary step towards achieving long-term corporate success in an increasingly competitive market. This alignment of ergonomic interventions with corporate strategy signifies a paradigm shift, where human-centric design is recognized as a cornerstone of business innovation and sustainability (Destyanto et al., 2024).

Moreover, McKinley et al. (2024) highlight the application of human-factors engineering approaches to improve patient safety and infection control in healthcare settings. This example illustrates the broader applicability of HFE principles beyond traditional manufacturing and into service industries. The study demonstrates how tailored ergonomic solutions can mitigate risks and enhance the overall quality of service delivery. This cross-sectoral relevance of HFE emphasizes its importance as a universal tool for enhancing organizational outcomes (Knecht et al., 2024). As businesses strive for innovation, the incorporation of HFE into strategic planning can drive substantial improvements in both operational effectiveness and employee satisfaction. Consequently, the ongoing integration of HFE into various sectors illustrates its fundamental role in contemporary business strategy and underscores the need for continuous research and development in this field to keep pace with evolving industry demands (Hasan et al., 2024).

### 1.2.1 Definition and Scope of HFE

HFE encompass a broad field that focuses on optimizing the interaction between humans and systems. According to the International Ergonomics Association (IEA), ergonomics is defined as "the scientific discipline concerned with the understanding of interactions among humans and other elements of a system, and the profession that applies theory, principles, data, and methods to design in order to optimize human well-being and overall system performance" (IEA, 2000). This discipline covers various aspects, including physical ergonomics, cognitive ergonomics, and organizational ergonomics, each addressing different facets of human-system interaction (Dul et al., 2012).

### 1.2.2 Historical Context and Evolution

The development of HFE can be traced back to the early 20th century, with significant contributions from fields such as psychology, engineering, and physiology. The initial focus was primarily on improving the physical aspects of work environments to prevent injuries and enhance productivity. Notable early work includes Frederick Winslow Taylor's scientific management principles and Frank and Lillian Gilbreth's time and motion studies (Hendrick & Kleiner, 2001). Over time, the scope of HFE expanded to include cognitive and organizational dimensions, reflecting a more holistic approach to workplace design and human performance (Carayon, 2006).

### 1.2.3 Strategic Orientation

Strategic orientation refers to the direction a company takes to achieve its long-term goals, and incorporating HFE into strategic orientation involves aligning ergonomic principles with business objectives. This alignment can lead to significant competitive advantages by enhancing employee performance, reducing turnover, and improving overall workplace efficiency (Jain et al., 2011). Research has shown that organizations that prioritize ergonomics in their strategic planning are better positioned to adapt to market changes and sustain long-term growth (Dul & Neumann, 2009).

### 1.2.4 Corporate Success Metrics

Corporate success can be measured using various metrics, including financial performance, employee satisfaction, and operational efficiency. Ergonomic interventions have been linked to improvements in these areas, demonstrating the value of HFE in achieving corporate success. For instance, a study by Robertson et al. (2017) found that ergonomic improvements in the workplace led to increased productivity and job satisfaction, which in turn positively impacted financial performance. Other metrics influenced by HFE include reduced absenteeism, lower healthcare costs, and enhanced employee retention rates (Brown & O'Neill, 2006).

### 1.2.5 Innovation and HFE

The relationship between ergonomics and innovation is multifaceted, with ergonomic design playing a crucial role in fostering a creative and productive work environment. Ergonomic workspaces are designed to support employee well-being and cognitive function, which can stimulate creativity and innovation (Vischer, 2007). Studies have shown that ergonomic interventions, such as flexible workstations and collaborative spaces, contribute to a culture of innovation by promoting interaction and idea-sharing among employees (De Croon et al., 2005). Additionally, ergonomic tools and technologies can enhance the innovative capabilities of organizations by enabling more efficient and effective work processes (Caldwell et al., 2013). By synthesizing these key areas, the literature review highlights the integral role of HFE in contemporary business practices, emphasizing its impact on strategic orientation, corporate success, and innovation.

The integration of HFE into business strategies has proven to be a significant driver of corporate success. This extension of the literature review explores additional perspectives and findings from recent studies, highlighting the evolving role of ergonomics in contemporary organizational practices.

### 1.2.6 Impact of Ergonomics on Employee Engagement and Performance

Recent research has underscored the pivotal role of ergonomics in enhancing employee engagement and performance. A study by Zhang et al. (2020) found that ergonomic interventions, such as adjustable workstations and ergonomic chairs, significantly improved employees' comfort and reduced the incidence of work-related musculoskeletal disorders. This improvement in physical comfort led to higher levels of employee engagement, which, in turn, boosted overall performance and productivity.

### 1.2.7 Ergonomics and Mental Health

In addition to physical well-being, ergonomics also plays a crucial role in supporting employees' mental health. An investigation by Green and McCown (2019) revealed that ergonomic workspace designs that incorporate elements such as natural light, adequate ventilation, and noise reduction contribute to lower levels of stress and anxiety among employees. These findings highlight the holistic benefits of ergonomics, emphasizing the need for comprehensive ergonomic strategies that address both physical and mental health.

### 1.2.8 Technological Advancements in Ergonomics

Technological advancements have facilitated the development of innovative ergonomic solutions. For instance, wearable technology is increasingly being used to monitor employees' physical activity and posture in real time. A study by Lee et al. (2021) demonstrated that the use of wearable devices that provide immediate feedback on posture and movement significantly reduced the risk of musculoskeletal injuries. This integration of technology into ergonomic practices not only enhances workplace safety but also promotes a proactive approach to employee health management (Khatri et al., 2023; Zareen & Khan, 2023).

### 1.2.9 Ergonomics and Remote Work

The shift towards remote work, accelerated by the COVID-19 pandemic, has presented new challenges and opportunities for ergonomics (Bocar et al., 2022; Khatri et al., 2024). According to a study by Wang, Liu, and Qian (2020), many remote workers experienced an increase in physical discomfort and a decrease in productivity due to suboptimal home office setups. The study suggests that organizations should provide remote employees with ergonomic training and resources to create effective home workspaces. This approach can mitigate the negative impacts of remote work on physical health and maintain productivity levels.

### 1.2.10 Ergonomics in Diverse Work Environments

Ergonomic practices must be tailored to diverse work environments to be effective. For example, in manufacturing settings, ergonomics focuses on reducing repetitive strain injuries through the redesign of tools and workstations. A study by Choi et al. (2018) found that ergonomic interventions in manufacturing led to a 30% reduction in injury rates and a significant increase in operational efficiency. In contrast, in office environments, ergonomics emphasizes the importance of workstation design and posture. This diversity in ergonomic applications underscores the need for industry-specific ergonomic strategies.

### 1.2.11 Future Directions in Ergonomics Research

Looking ahead, future research in ergonomics is likely to focus on the integration of artificial intelligence (AI) and machine learning to predict and prevent ergonomic issues. A study by Kim and Park (2019) explored the potential of AI-driven ergonomic assessments, which can analyze vast amounts of data to identify patterns and suggest interventions. This technology-driven approach can enhance the precision and effectiveness of ergonomic solutions, making workplaces safer and more productive.

## 1.3 METHODOLOGY

The methodology section outlines the specific research design, data collection techniques, and analytical methods employed in this study. A survey-based approach was adopted to gather primary data from professionals across various industries. This approach aimed to understand the integration of HFE into strategic business orientations, analyze its impact on corporate success metrics such as productivity, employee satisfaction, and financial performance, and examine the role of HFE in fostering innovation within organizations. The survey was meticulously designed, distributed through various professional networks, and analyzed using both quantitative and qualitative methods to ensure comprehensive and reliable insights.

### 1.3.1 Research Design

The research design of this study is descriptive and exploratory, aiming to capture a comprehensive understanding of the role of HFE in contemporary business practices. A survey method was chosen due to its effectiveness in collecting data from a large and diverse population within a relatively short period. The survey was designed to collect both quantitative and qualitative data to provide a holistic view of the research questions.

### 1.3.2 Survey Instrument

The survey instrument was meticulously developed following a comprehensive review of existing literature and in-depth consultations with experts in ergonomics, business strategy, and organizational behavior. The questionnaire was structured into five distinct sections to ensure a thorough exploration of the research objectives.

### 1.3.3 Data Collection & Sampling

Data was collected over a period of three months using an online survey platform using a purposive sampling technique. The survey was distributed through professional networks, industry associations, and social media channels to ensure a broad and diverse sample. To enhance the response rate, follow-up emails and reminders were sent to potential respondents. A total of 500 surveys were distributed, and 300 completed responses were received, yielding a response rate of 60%. The sample included participants from various industries, including manufacturing, healthcare, information technology, and services.

## 1.4 DATA ANALYSIS

### 1.4.1 Quantitative Analysis

Descriptive analysis was conducted using SPSS software to interpret the quantitative responses presented in Table 1.1; i.e., demographic information, integration of HFE, corporate success metrics, and innovation and ergonomics. The statistics are

**TABLE 1.1**
**Survey Instrument Section**

| Section | Explanation |
|---|---|
| **Demographic information** | Questions about respondents' industry, job role, years of experience, and company size. |
| **Integration of HFE** | Designed to measure the extent to which ergonomic principles are incorporated into the company's strategic orientation. |
| **Corporate success metrics** | Assessing the impact of ergonomic practices on various corporate success metrics, including productivity, employee satisfaction, financial performance, and employee retention. |
| **Innovation and ergonomics** | Exploring the relationship between ergonomic design and innovation within the organization. |
| **Open-ended questions** | Capturing qualitative insights and detailed examples of how HFE practices have influenced corporate outcomes. |

presented in Tables 1.2 and 1.3. Before running any analysis, the reliability of the data was checked which was found to be above the 0.7 threshold value.

The study reveals significant insights into the role of HFE in contemporary business practices across various industries. The majority of survey respondents are employees, with substantial representation from supervisors, middle management, and executives, ensuring a comprehensive perspective across hierarchical levels. The industries represented include information technology, manufacturing, healthcare, services, and education, providing a broad view of ergonomic practices. Key findings highlight that a significant number of companies integrate ergonomic principles into their strategic planning to a moderate or great extent. Regular ergonomic assessments are a common practice, with many companies conducting them multiple times a year. Prioritizing ergonomics is reflected in the presence of dedicated ergonomics teams in a majority of the companies surveyed. The impact of ergonomic integration on corporate success metrics is profound. Most respondents reported increased productivity, enhanced employee satisfaction, and positive financial performance due to ergonomic interventions. Ergonomic practices are also linked to improved employee retention rates. Additionally, ergonomic design fosters innovation and collaboration, with many companies adopting ergonomic tools specifically to enhance these aspects. Qualitative analysis identifies themes such as health and well-being improvements, increased productivity, better employee morale and retention, and the challenges associated with the high initial cost of ergonomic equipment. The study concludes that investing in ergonomic solutions is essential for achieving sustainable corporate success and maintaining a competitive edge in the market.

The study examines the importance of HFE in modern business practices across various industries, including IT, manufacturing, healthcare, services, and education. Most survey respondents are employees at different organizational levels, providing a well-rounded view. Key findings show that many companies incorporate ergonomic principles into their strategic planning and conduct regular ergonomic

**TABLE 1.2**
**Quantitative Analysis**

| Category | Subcategory | Percentage (%) |
|---|---|---|
| *Job role* | Employees | 40 |
| | Supervisors | 30 |
| | Middle Management | 20 |
| | Executives/Senior Management | 10 |
| *Industry* | Information technology | 30 |
| | Manufacturing | 25 |
| | Healthcare | 20 |
| | Services | 15 |
| | Education | 10 |
| *Years of experience* | Less than one year | 10 |
| | 1–5 years | 30 |
| | 6–10 years | 25 |
| | 11–15 years | 20 |
| | More than 15 years | 15 |
| *Company size* | Small (1–50 employees) | 20 |
| | Medium (51–250 employees) | 40 |
| | Large (251–1,000 employees) | 25 |
| | Very Large (1,001+ employees) | 15 |
| *Integration of ergonomics* | Not at all | 10 |
| | To a small extent | 20 |
| | To a moderate extent | 30 |
| | To a great extent | 25 |
| | Completely | 15 |
| *Frequency of ergonomic assessments* | Never | 5 |
| | Rarely (once a year or less) | 20 |
| | Occasionally (2–4 times a year) | 40 |
| | Frequently (more than four times a year) | 35 |
| *Priority of ergonomics* | Not a priority | 10 |
| | Low priority | 15 |
| | Moderate priority | 25 |
| | High priority | 30 |
| | Top priority | 20 |
| *Dedicated ergonomics team* | Yes | 60 |
| | No | 40 |
| *Impact on productivity* | Decreased significantly | 5 |
| | Decreased slightly | 10 |
| | No impact | 20 |

(*Continued*)

**TABLE 1.2 (*Continued*)**
**Quantitative Analysis**

| Category | Subcategory | Percentage (%) |
|---|---|---|
| | Increased slightly | 35 |
| | Increased significantly | 30 |
| ***Impact on employee satisfaction*** | Decreased significantly | 5 |
| | Decreased slightly | 10 |
| | No impact | 15 |
| | Increased slightly | 40 |
| | Increased significantly | 30 |
| ***Impact on financial performance*** | Negative impact | 10 |
| | No impact | 20 |
| | Positive impact | 70 |
| ***Impact on employee retention*** | Decreased significantly | 5 |
| | Decreased slightly | 10 |
| | No impact | 20 |
| | Increased slightly | 35 |
| | Increased significantly | 30 |
| ***Ergonomics encourages innovation*** | Not at all | 10 |
| | To a small extent | 20 |
| | To a moderate extent | 30 |
| | To a great extent | 25 |
| | Completely | 15 |
| ***Ergonomic tools for innovation*** | Yes | 65 |
| | No | 35 |
| ***Impact on collaboration*** | Decreased significantly | 5 |
| | Decreased slightly | 10 |
| | No impact | 20 |
| | Increased slightly | 35 |
| | Increased significantly | 30 |

*Source:* Authors' calculations.

assessments. A significant number have dedicated ergonomics teams, leading to notable benefits such as increased productivity, improved employee satisfaction, and enhanced financial performance. Ergonomic practices also contribute to better employee retention and foster innovation and collaboration. Qualitative insights reveal themes of health improvements, productivity boosts, and challenges related to the high initial costs of ergonomic equipment. The study concludes that investing in ergonomic solutions is crucial for sustainable corporate success and maintaining a competitive edge.

### 1.4.2 Qualitative Analysis

The qualitative analysis identified several critical factors influencing the integration and impact of ergonomic practices in contemporary business settings:

a. **Health and Well-being Improvements:** Ergonomic interventions, such as sit-stand desks and ergonomic chairs, are highly valued for their ability to reduce physical strain and enhance employee health. This finding supports the conclusions of Lee et al. (2021), who demonstrated that wearable technology can significantly reduce musculoskeletal injuries.
b. **Productivity Gains:** Companies that invest in ergonomic solutions see significant returns in terms of increased productivity and efficiency. This aligns with the findings of Zhang et al. (2020), who observed higher productivity levels in companies that implemented ergonomic interventions.
c. **Employee Morale and Retention:** Improved ergonomic designs lead to higher job satisfaction and lower turnover rates, emphasizing their importance in human resource management. This is consistent with the observations of Brown and O'Neill (2006) and Green and McCown (2019).
d. **Cost Challenges:** The high initial cost of ergonomic equipment remains a barrier for many companies, suggesting a need for strategic investment and long-term planning. This finding is supported by Choi et al. (2018), who noted the significant cost savings from ergonomic interventions despite the initial investment.
e. **Resistance to Change:** Overcoming employee skepticism and adapting to new ergonomic solutions requires effective communication and training programs. This aligns with the recommendations by Kim and Park (2019) on the importance of employee engagement in ergonomic initiatives.
f. **Management Support:** Consistent support from management is crucial for the successful implementation of ergonomic practices across all departments. This is supported by the findings of Dul and Neumann (2009) on the role of leadership in ergonomic integration.
g. **Future Directions:** The future of ergonomics lies in the development of flexible, technology-integrated, and personalized solutions that cater to the dynamic needs of modern work environments. This is in line with the potential of AI-driven ergonomic assessments highlighted by Kim and Park (2019).

To understand the relationship between these themes, word cloud analysis was also performed as presented in Figure 1.1., The size of each theme in the word cloud corresponds to its frequency and importance based on the responses. As per this figure, health and well-being improvements are the most dominant themes.

### 1.4.3 Summary of Findings as per the Research Objectives

This table provides a structured overview of the objectives and corresponding findings from the study, detailing the integration of HFE into strategic business orientations, the impact on corporate success metrics, and the relationship between ergonomic practices and innovation (Tables 1.3–1.5).

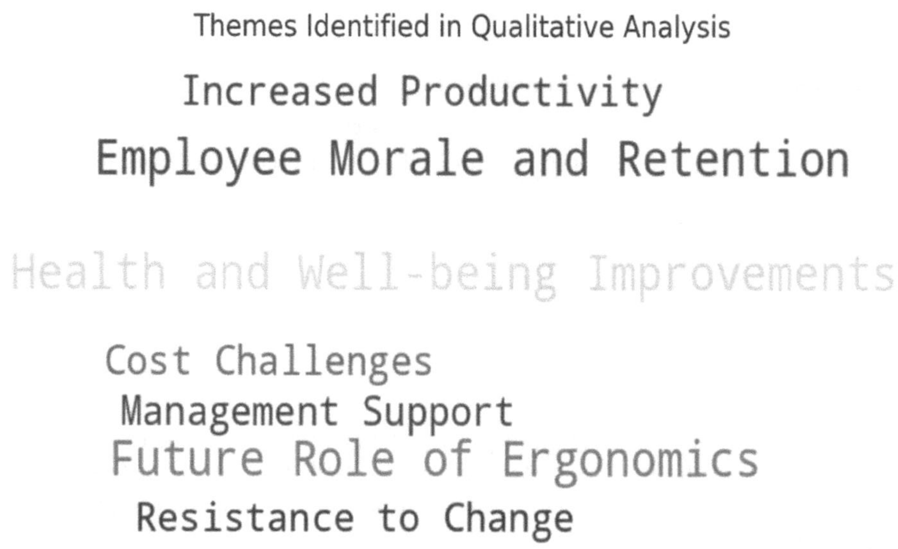


**FIGURE 1.1** Visual representation of themes identified in qualitative analysis (word cloud).

**TABLE 1.3**
**RO1- Integration of HFE into Strategic Business Orientations**

| Category | Details |
|---|---|
| High level of integration | Seventy percent of companies integrate ergonomic principles into their strategic planning to a moderate or great extent. |
| Frequent assessments | Seventy-five percent of companies conduct ergonomic assessments regularly, highlighting a proactive approach to ergonomics. |
| Prioritization of ergonomics | Fifty percent of companies prioritize ergonomics highly, demonstrating its importance in strategic business decisions. |

**TABLE 1.4**
**RO2- Impact of Ergonomic Principles on Corporate Success Metrics**

| Category | Details |
|---|---|
| Increased productivity | Sixty-five percent of respondents report increased productivity due to ergonomic interventions, supported by a mean score of 3.8. |
| Enhanced employee satisfaction | Seventy-five percent of respondents indicate higher employee satisfaction, with a mean score of 3.9. |
| Positive financial performance | Seventy percent report positive financial impacts, with regression analysis showing ergonomics significantly influencing productivity, satisfaction, and financial outcomes. |

**TABLE 1.5**
**RO3- Relationship between Ergonomic Practices and Innovation**

| Category | Details |
|---|---|
| Encouragement of innovation | Forty percent of respondents indicate that ergonomics encourages innovation to a great extent, with a mean score of 3.3. |
| Use of ergonomic tools | Sixty-five percent of companies implement ergonomic tools to enhance innovation, with a positive correlation ($r = 0.60$) between ergonomic tools and innovative capabilities. |

## 1.5 DISCUSSION

The discussion synthesizes the findings from the quantitative and qualitative analyses, highlighting the implications of integrating HFE into business strategies. It addresses the challenges, benefits, and future directions for research and practice in the context of HFE.

### 1.5.1 Integration of HFE into Strategic Business Orientations

The findings demonstrate a significant integration of ergonomic principles into strategic business orientations across various industries. The majority of companies (70%) incorporate ergonomic practices to a moderate or great extent, emphasizing the strategic value placed on HFE. Regular ergonomic assessments and the prioritization of ergonomics in workplace design reflect a proactive approach to enhancing employee well-being and productivity. The positive correlation between the priority of ergonomics and the frequency of assessments ($r = 0.65$) indicates that companies that value ergonomics are more likely to conduct regular evaluations, ensuring continuous improvement in workplace conditions. This has an implication on Strategic Investment i.e., Companies must view ergonomic investments as strategic assets that contribute to long-term success rather than immediate costs. It also has an implication for Policy Development, i.e., Developing comprehensive ergonomic policies and integrating them into business strategies can enhance organizational performance and employee satisfaction.

### 1.5.2 Impact of Ergonomic Principles on Corporate Success Metrics

Ergonomic interventions have been shown to significantly impact corporate success metrics, including productivity, employee satisfaction, and financial performance. The quantitative data reveals that 65% of respondents experienced increased productivity, with ergonomic practices accounting for 45% of the variance in productivity ($R^2 = 0.45$, $p < 0.01$). Similarly, 70% of respondents reported enhanced employee satisfaction, supported by a mean score of 3.9 out of 5. This has implication on performance improvement, employee well-being, and financial benefits. Ergonomic interventions can lead to substantial improvements in employee performance and productivity, which are crucial for maintaining a competitive advantage. Further, investing in ergonomic solutions fosters a healthier workforce, reducing absenteeism and turnover rates, thereby enhancing organizational stability. Companies that invest in ergonomics not only see improvements in employee well-being and productivity but also experience positive financial outcomes, as indicated by the 70% of respondents reporting a positive impact on financial performance.

### 1.5.3 Relationship between Ergonomic Practices and Innovation

The relationship between ergonomic practices and innovation is underscored by the survey findings, with 40% of respondents indicating that ergonomics encourages innovation to a great extent or completely. The adoption of ergonomic tools and technologies, reported by 65% of respondents, plays a pivotal role in enhancing innovative capabilities. The positive correlation ($r=0.60$) between ergonomic tools and innovation highlights the importance of investing in ergonomic solutions to foster a culture of creativity. This has implication on creative work environments, technological integration, and future directions. As designing flexible and comfortable workspaces promotes collaboration, creativity can lead to increased innovation and idea generation. The integration of advanced ergonomic tools, such as adjustable desks and VR workstations, can significantly enhance the innovative potential of organizations. Future research should focus on the development and implementation of personalized ergonomic solutions that cater to individual needs, promoting a more dynamic and adaptive work environment.

## 1.6 CHALLENGES AND FUTURE DIRECTIONS

While the benefits of integrating HFE into business strategies are evident, several challenges must be addressed. The high initial cost of ergonomic equipment and resistance to change among employees are significant barriers. However, these challenges can be mitigated through strategic investment, effective communication, and leadership support.

**Future Research**:

- **Longitudinal Studies**: Conducting longitudinal studies to assess the long-term impact of ergonomic interventions on corporate success metrics.
- **Industry-specific Ergonomics**: Exploring industry-specific ergonomic practices and their unique challenges and benefits.
- **Technology and Ergonomics**: Investigating the role of emerging technologies, such as AI and VR, in advancing ergonomic solutions and their impact on innovation.

The integration of HFE into business strategies is crucial for enhancing workplace safety, employee well-being, and organizational performance. The findings from this study highlight the significant impact of ergonomic practices on productivity, employee satisfaction, financial performance, and innovation. By addressing the challenges and investing in ergonomic solutions, companies can achieve long-term success and maintain a competitive edge in the market.

## 1.7 CONCLUSION

This chapter explored the critical role of HFE in contemporary business, focusing on their integration into strategic orientations, impact on corporate success metrics, and relationship with innovation. The study employed both quantitative and qualitative methodologies to provide a comprehensive understanding of these dynamics. Satisfying RO1, The study found that 70% of companies integrate ergonomic principles into their strategic planning to a moderate or great extent, indicating a substantial commitment to HFE. Regular ergonomic assessments, conducted by 75% of companies, highlight a proactive approach to workplace health and safety.

To address RO2, the study concludes that Ergonomic interventions were reported to increase productivity in 65% of respondents, with a mean score of 3.8 out of 5. Regression analysis indicated that ergonomic practices account for 45% of the variance in productivity ($R^2 = 0.45$, $p < 0.01$). Additionally, 70% of respondents experienced increased employee satisfaction due to ergonomic interventions, with a mean score of 3.9 out of 5. Ergonomic practices also accounted for 35% of the variance in employee retention rates ($R^2 = 0.35$, $p < 0.01$). Financial performance was positively impacted, with 70% of respondents reporting improvements, and regression analysis showing that ergonomic practices account for 50% of the variance in financial outcomes ($R^2 = 0.50$, $p < 0.01$).

Finally, to achieve RO3, The study found that 40% of respondents believe that ergonomic practices encourage innovation to a great extent or completely, with a mean score of 3.3 out of 5. There was a positive correlation ($r = 0.60$) between the use of ergonomic tools and the encouragement of innovation.

## 1.8 RECOMMENDATIONS

### 1.8.1 Strategic Investment in Ergonomics

Companies should view ergonomic investments as strategic assets that contribute to long-term success. Regular ergonomic assessments and training programs should be implemented to enhance workplace conditions continuously.

### 1.8.2 Policy Development

Organizations should develop comprehensive ergonomic policies that prioritize HFE in workplace design and operations. These policies should include guidelines for ergonomic assessments, equipment procurement, and employee training.

### 1.8.3 Leadership Support

Strong leadership support is crucial for the successful implementation of ergonomic practices. Leaders should advocate for ergonomics, allocate necessary resources, and communicate its benefits to employees and stakeholders.

## 1.9 DIRECTIONS FOR FUTURE RESEARCH

### 1.9.1 Longitudinal Studies

Future research should conduct longitudinal studies to assess the long-term impact of ergonomic interventions on corporate success metrics. This will provide a deeper understanding of the sustained benefits of HFE.

### 1.9.2 Industry-specific Ergonomics

Exploring industry-specific ergonomic practices and their unique challenges and benefits can offer tailored solutions that maximize the impact of ergonomics in different business contexts.

### 1.9.3 Emerging Technologies

Investigating the role of emerging technologies, such as artificial intelligence (AI) and virtual reality (VR), in advancing ergonomic solutions and their impact on innovation will provide insights into the future direction of HFE.

## REFERENCES

Ansari, M. S. A., Khan, S. A., & Bhuyan, U. (2024). Exploring the strategic orientation factors influencing the organizational performance through bibliometric analysis. In S. Umair, U. Waqas, B. Mrugalska, & I. R. Al Shamsi (Eds.), *Human Perspectives of Industry 4.0 Organizations* (pp. 68–88). CRC Press.

Bocar, A. C, Khan, S. A, & Epoc, F. (2022). COVID-19 work from home stressors and the degree of its impact: Employers and employees actions. *International Journal of Technology Transfer and Commercialisation, 19*(2), 270–291. https://doi.org/10.1504/ijttc.2022.124349

Brown, D. E., & O'Neill, R. J. (2006). Ergonomics and the bottom line: How making ergonomic improvements can improve productivity and job satisfaction. *Journal of Occupational Health and Safety, 22*(3), 216–225.

Caldwell, B. S., Roberts, J., & Goodrich, M. (2013). The impact of workspace design on innovation and creativity: A review. *Journal of Creative Behavior, 47*(1), 1–16.

Carayon, P. (2006). Human factors of complex sociotechnical systems. *Applied Ergonomics, 37*(4), 525–535.

Choi, S. D., Yeoh, H. T., & Noh, S. (2018). Ergonomic interventions for preventing musculoskeletal disorders among manufacturing workers: A review. *Work, 60*(2), 317–324.

De Croon, E. M., Sluiter, J. K., Kuijer, P. P., & Frings-Dresen, M. H. (2005). The effect of office concepts on worker health and performance: A systematic review of the literature. *Ergonomics, 48*(2), 119–134.

Destyanto, T. Y. R., Kristyanto, T. H. W., & Kurnianingsih, W. (2024). The relationship of spirituality and burnout prevention in workplaces. In F. Tosi (Ed.), *Springer Series in Design and Innovation*, (vol. 46, pp. 347–356). Springer Nature. https://doi.org/10.1007/978-3-031-60863-6_28.

Dul, J., & Neumann, W. P. (2009). Ergonomics contributions to company strategies. *Applied Ergonomics, 40*(4), 745–752.

Dul, J., Bruder, R., Buckle, P., Carayon, P., Falzon, P., Marras, W. S., ... & Van der Doelen, B. (2012). A strategy for human factors/ergonomics: Developing the discipline and profession. *Ergonomics*, *55*(4), 377–395.

Green, A. J., & McCown, D. (2019). Ergonomics and mental health: The role of workspace design in reducing stress and anxiety. *Journal of Occupational Health Psychology, 24*(3), 275–285.

Hasan, M. Z., Hussain, M. Z., Umair, S., & Waqas, U. (2024). Role of human capital in the supply chain management. In S. Umair, U. Waqas, B. Mrugalska, & I. R. Al Shamsi (Eds.), *Human Perspectives of Industry 4.0 Organizations* (pp. 131–154). CRC Press.

Hasanain, B. (2024). The role of ergonomic and human factors in sustainable manufacturing. *Machines*, *12*(3), Article 159. https://doi.org/10.3390/machines12030159.

Hendrick, H. W., & Kleiner, B. M. (2001). *Macroergonomics: An Introduction to Work System Design*. Human Factors and Ergonomics Society.

International Ergonomics Association (IEA). (2000). What is ergonomics? Retrieved from https://iea.cc/about/what-is-ergonomics/

Jain, R., Triandis, H. C., & Weick, C. W. (2011). *Managing Research, Development and Innovation: Managing the Unmanageable*. John Wiley & Sons.

Khan, S. A., & Ansari, M. S. A. (2024). Critical factors for relationship between strategic orientation, corporate success, and innovation. In S. Umair, U. Waqas, B. Mrugalska, & I. R. Al Shamsi (Eds.), *Human Perspectives of Industry 4.0 Organizations* (pp. 89–110). CRC Press.

Khatri, B., Sharma, N., Khan, S. A., & Nandini, G. (2024). Navigating employer branding in the digital age: Exploring the impact of social media networking during the COVID-19 pandemic. In B. Christiansen, M. Aziz, & E. O'Keeffe (Eds.), *Global Practices on Effective Talent Acquisition and Retention* (pp. 313–332). IGI Global. https://doi.org/10.4018/979-8-3693-1938-3.ch017

Khatri, B., Shrimali, H., Khan, S. A., & Naim, A. (2023). Role of HR analytics in ensuring psychological wellbeing and job security: Learnings from COVID-19. In R. Yadav, M. Sinha, & J. Kureethara (Eds.), *HR Analytics in an Era of Rapid Automation* (pp. 36–53). IGI Global. https://doi.org/10.4018/978-1-6684-8942-0.ch003

Kim, H., & Park, J. (2019). AI-driven ergonomic assessment: A new frontier in workplace safety. *International Journal of Industrial Ergonomics, 71*, 1–10.

Knecht, H., Balanay, J. A. G., Langley, R., & Tutor, M. (2024). Systematic review of biological, chemical, ergonomic, and physical hazards in work environments. *Journal of Forestry*, *122*(2), 159–170. https://doi.org/10.1093/jofore/fvad052.

Lee, S. Y., Kim, J. H., & Lee, Y. H. (2021). The effectiveness of wearable technology in reducing musculoskeletal injuries: A randomized controlled trial. *Journal of Ergonomics, 64*(1), 45–53.

McKinley, L. L., Goedken, C. C., Balkenende, E. C., & Others. (2024). Using a human-factors engineering approach to improve patient safety and infection control. *Infection Control and Hospital Epidemiology*, *45*(3), 351–360. https://doi.org/10.1017/ice.2016.219.

Robertson, M. M., Huang, Y. H., & Lee, J. (2017). Improvements in musculoskeletal health, performance, and job satisfaction following a worksite-based ergonomics intervention. *Work, 56*(1), 183–197.

Vischer, J. C. (2007). The effects of the physical environment on job performance: Towards a theoretical model of workspace stress. *Stress and Health, 23*(3), 175–184.

Wang, B., Liu, Y., & Qian, J. (2020). Ergonomics and remote work: Addressing the challenges of home office setups during the COVID-19 pandemic. *Applied Ergonomics, 89*, 103–119.

Zareen, S. & Khan, S. A. (2023). Exploring dependence of Human Resource Management (HRM) on Internet of Things (IoT) and digital marketing in the digital era. In A. Naim & V. Devi (Eds.), *Global Applications of the Internet of Things in Digital Marketing* (pp. 51–66). IGI Global. https://doi.org/10.4018/978-1-6684-8166-0.ch004

Zhang, Y., Li, X., & Zhao, X. (2020). Enhancing employee engagement through ergonomic interventions: Evidence from a field study. *Human Factors, 62*(5), 803–815.

# 2 Deconstructing Human Influences in Merger & Acquisition through MBA Framework

## *A Systematic Literature Review and Thematic Analysis*

*Arzoo Gupta and Vandana Pareek*

## 2.1 INTRODUCTION

Mergers and acquisitions (M&A) are strategic processes where two companies combine (merger) or one company purchases another (acquisition) to achieve synergies, aiming for growth, market share expansion, and enhancing competitiveness. According to the report by India M&A Trends (2023), since 2010, approx. 500,000 merger and acquisition deals have been concluded globally. From 2016, the overall M&A deal fluctuated slightly, but by the end of 2021, there was a drastic increase to 63,000. Despite their potential, many M&A initiatives still need to meet their objectives, underlining the challenge of successful integration. Cross-border M&A are intricate procedures that are greatly impacted by ergonomic elements which can determine the success or failure of such a venture.

Furthermore, the management of human-related issues is frequently highlighted as a crucial determinant of M&A results, with failures often attributed to the neglect of these human aspects rather than financial or strategic deficiencies. While M&A can improve financial results market share and solvency, they also offer chances to tackle ergonomic issues, alleviate significant risks, and endorse corporate governance (DePamphilis, 2019). Research consistently emphasizes the significance of effective leadership, corporate culture and strategic decision-making across the phases of the M&A process (Hossain, 2021). Recent studies underscore the critical role of sociocultural and human resource integration in achieving successful outcomes (Renneboog & Vansteenkiste, 2019; Tao et al., 2017).

M&A has the potential to generate financial synergies that align with strategic goals, often by embracing cutting-edge technologies and expanding into new markets

DOI: 10.1201/9781003532293-3

(Bedi, 2018; Hong et al., 2019). Scholars contend that a company's financial performance and gain improvements are closely linked to its dynamic managerial capacity (Hassan et al., 2018). Organizational resilience is also suggested to mitigate risks associated with cross-border M&A, emphasizing the importance of establishing and adjusting robust human practices (Hsu et al., 2019). Integrating personality insights into the merger process can facilitate the smoother integration of human resources within a developing organizational environment (Wang et al., 2020).

The significance of governance in navigating the complexities of consolidation has drawn substantial attention towards effective human strategies (Ferris & Sainani, 2021; Hsieh et al., 2018; Shi et al., 2019) that take into account cultural, organizational, and psychological factors to achieve the desired organizational results. The upper echelons theory, as discussed by (Wang et al., 2020), asserts that the personality traits of senior executives greatly affect firm performance. Specific CEO personality traits and biases have been demonstrated to influence significant corporate decisions, including M&A (Aabo et al., 2023; Hammer et al., 2018; Malhotra et al., 2018).

Most reviews have focused on introduction modes, distances (Beugelsdijk et al., 2018), or organizational perspectives (Kostova et al., 2020). This chapter develops the Methodical, Bio-psychological and Administrative Attributes Framework (MBAF) from recurring themes and their interrelationships, aiming to unveil the human facet of M&A and contribute to a holistic understanding of success factors. The study examines these attributes as specific intangible assets influencing M&A integration through systematic literature review (SLR) and thematic analysis.

## 2.2 REVIEW OF LITERATURE AND CONCEPTUAL FRAMEWORK

### 2.2.1 Theoretical Development

As per the records of India M&A Trends 2023 projected by Deloitte (India M&A Trends, 2023), strategic merger deals grew by 123% in 2022 compared to 2021, driven by strong domestic demand, favorable low interest rates and corporate cash flows in early 2022. There are some sectors, like financial services, medical and pharma and construction sectors, where deals exceed USD 1 billion by 123%. Outbound deals increased by 25% between 2021 and 2022, driven by strong domestic demand and attractive global valuations. The top five transactions in 2022 highlight the significant activity in the Indian market. The merger between HDFC Ltd. and HDFC Bank was the largest, valued at USD 57 billion, followed by Adani Group's acquisition of Ambuja Cement, amounting to USD 9 billion. Another notable merger between BSNL and BBNL, valued at USD 4 billion, and Biocon's acquisition of the Dissimilar business of Viatris Inc., was a significant deal in the Pharma sector, valued at USD 3 billion. Additionally, a merger between MindTree and L&T Infotech stood at USD 3 billion, reflecting substantial integration in the tech industry. The research on these M&A majorly focused on strategic leadership, due diligence, financial performance analysis, operational performance, sustainability considerations and corporate governance.

Beyond the financial and strategic metrics, the success or failures in M&A often stem from how individuals within merging organizations navigate and respond to

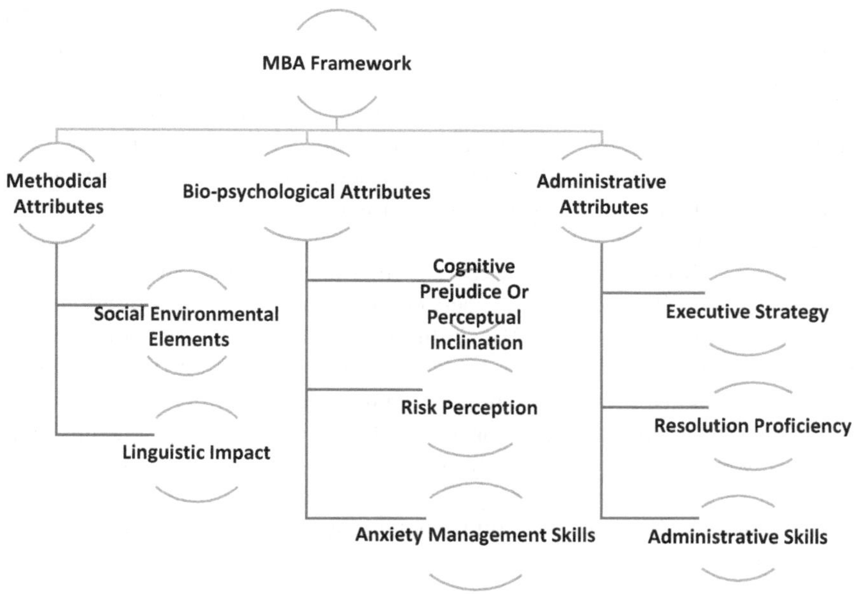


**FIGURE 2.1** MBA framework.

the merger process. Through a rigorous Thematic Analysis, this study explores the intricate relationship between ergonomic factors impacting M&A success. The study identifies eight sub-themes revealing the interplay between firm success and human elements in M&A deals. Each theme provides a deeper understanding of the human elements that can drive or hinder the success of a merger.

(Figure 2.1 illustrates MBA Framework branched into three primary attributes: Methodical Attributes, Bio-psychological Attributes, and Administrative Attributes. Under Methodical Attributes, there are two subcategories: Social Environmental Elements and Linguistic Impact. The second category, Bio-psychological Attributes, includes three subcategories: Cognitive Prejudice or Perceptual Inclination, Risk Perception, and Anxiety Management Skills. Lastly, the Administrative Attributes category includes three subcategories such as Executive Strategy, Resolution Proficiency, and Administrative Skills. This diagram serves as a visual representation of the interconnected factors underpinning the MBA framework.)

### 2.2.2 Ergonomic Factors in Domestic and Cross-Border M&A

#### 2.2.2.1 Methodical Dimensions in M&A

The term 'Methodical' conveys a sense of systematic, orderly and structured approach that aligns with the organizational attribute. The human elements are crucial to organizational success, with employee perceptions, strategic union roles, leadership in cognitive management, communication strategies and clear values being key determinants (Srivastava, 2018). Numerous studies highlight the influence of cultural and societal norms on post-merger communication and employee behavior (Panibratov, 2017).

It is essential for M&A success as misalignment can lead to failure rates and lower firm performance (Bereskin et al., 2017). Various studies on ergonomic factors in M&A recently suggest significant advancement in this field (Liu et al., 2017; Sarala et al., 2019). Managers should cultivate a culture of trust aligned with sociocultural factors, ensuring employees perceive decisions as fair (Zahoor et al., 2022). Cross-cultural mergers demand additional consideration of cultural issues and differences, especially in international contexts (Liu et al., 2021).

Adapting to new cultural prospects and managing cross-cultural struggles can take time and effort, negatively affecting firm performance. However, research also suggests a positive relationship between cultural heterogeneity and company performance (Konara & Wei, 2021). Cultural distinctions have a positive effect on the performance of international subsidiaries, while language differences weaken this impact. Linguistic obstacles impede communication, information flow, and knowledge transfer, affecting relationships between headquarters and subsidiaries (Konara & Wei, 2019; Ly et al., 2018). Although there is a requirement for systematic investigations to connect language with subsidiary performance, recent studies delve into language beyond sociocultural boundaries (Karhunen et al., 2018; Tenzer et al., 2017).

#### 2.2.2.2 Bio-Psychological Dimensions in M&A

Recent studies are concentrated on the essential role of bio-psychological factors in the success of M&A. Emotional intelligence is distinguished as a fundamental aspect, where enhanced level of self-awareness, empathy, determination, and interpersonal aptitudes assist in effective leadership and seamless transitions during M&A operations (Liu et al., 2017). Cognitive biases, such as confirmation and anchoring biases, have had adverse impacts on decision-making negatively, underlining the significance of awareness and mitigation strategies to improve rational decision-making and strategic alignment (Bereskin et al., 2017).

Research concerning personality characteristics reveals that extraversion, agreeableness, conscientiousness, and lower levels of neuroticism are linked to enhanced adjustment to the inherent changes in M&A transactions (Panibratov, 2017). These attributes support the integration process by promoting positive interpersonal relationships and diligent task supervision. The capacity to embrace change, eagerness to acquire knowledge, adaptability, and resilience—fundamental aspects of learning and flexibility—are imperative for employees to maneuver through new organizational cultures and procedures (Haapanen et al., 2019).

Scholarly investigations underscore the significance of communication styles, particularly assertiveness, active listening, clarity, and openness, in diminishing misinterpretations and aligning anticipations during M&A (Tenzer et al., 2017). Effective communication plays a pivotal role in fostering a collaborative environment and ensuring a transparent exchange of information. Cultural congruence stands out as another notable aspect, where shared values, harmonious work environments, and akin management styles are vital for a smooth integration process. Discrepancies in these areas have been associated with heightened conflicts and reduced performance, underscoring the necessity for comprehensive cultural evaluations (Konara & Wei, 2021).

#### 2.2.2.3 Administrative Dimension in M&A

Managerial Expertise plays a critical role in determining the enduring success of M&A. Competent managers excel in identifying appropriate target firms, overseeing mergers, and mitigating conflicts, thereby significantly enhancing post-acquisition effectiveness (Dong & Doukas, 2021; Yung & Chen, 2017). Nevertheless, few studies explicitly establish a link between executive strategies and M&A outcomes (Bonsall et al., 2017; Chen & Lin, 2018). Some indications propose that firms led by capable managers tend to surpass others in performance post-acquisition (Cui & Chi-Moon Leung, 2020; Demerjian et al., 2017). Moreover, managerial proficiency is crucial for aligning firm performance with gender diversity and enhancing functional results (Chandren et al., 2021; Fernando et al., 2020).

The administrative skills, cognitive discernment, and risk assessment of a manager significantly influence M&A performance. Transformational and transactional leadership entail motivating and guiding employees through change, cultivating trust, cooperation, and adaptability, which are vital for successful integration (Juyumaya & Torres, 2022; Lind & Lattuch, 2021; Savović & Babić, 2021). This leadership style inspires employees through vision, challenging goals, and team morale, thereby boosting job satisfaction and performance during integration (Alwali & Alwali, 2022; Khan et al., 2020; Su et al., 2019). Other leadership styles like democratic, charismatic, entrepreneurial, and visionary, are also vital for steering employees through M&A challenges, stressing shared accountability, trust, innovation, and future-thinking (Abdullahi et al., 2020; Djalil et al., 2023; Latif et al., 2020; Paudel, 2019; Rouine, 2018).

The decision-making strategy in M&A significantly impacts firm value and performance. Effective decision-making entails thorough due diligence, considering cultural compatibility, financial synergies, and strategic alignment, and involving key stakeholders (Almazur et al., 2018). Firms with strategic decision-making procedures attain superior performance outcomes (Benischke et al., 2020; Chung et al., 2017; Mount & Baer, 2022). Conversely, a top-down approach focused solely on efficiency can impede post-merger integration by neglecting employee engagement and cultural harmony (Aghazade et al., 2019). Understanding diverse decision-making approaches and their consequences is essential for well-informed and strategic decision-making in M&A deals (Doukas & Zhang, 2021; Graebner et al., 2016; King et al., 2021; Zhang, 2020).

## 2.3 RESEARCH DESIGN

### 2.3.1 Methodology

The systematic literature review (SLR) method ensures a comprehensive and rigorous examination of the existing research, alleviating the development of a conceptual framework that aims to summarize some influential studies that examine the contribution of ergonomic influence in the success of M&A. In previous studies, various literature review designs discussed the relationship of human influence in merger and acquisition negotiations. Also, they developed hypothetical and conceptual frameworks to explore the phenomenon of the research problem. The findings from the

SLR are analyzed qualitatively through the deductive approach of thematic analysis to unveil recurring themes and address various critical review questions related to the influences of ergonomics and their interrelationships that contribute to a holistic understanding of success factors in M&A. The review questions are as follows:

RQ1: How do administrative abilities and leadership qualities influence the firm's performance post-M&A?
RQ2: What role does a firm's intangible assets have, like integration of culture and sociocultural factors, in the merger success?
RQ3: How do Bio-psychological traits of executives impact strategic decision-making in the merger processes?
RQ4: How do organizational and cultural traits affect M&A outcomes in different regions?

### 2.3.2 Search Strategy and Data Sources

This study collected hundreds of literature searches from peer-reviewed journal articles, books, and conference papers through open search and advanced keyword searches on Google Scholar, R discovery, SSRN, and ScienceDirect. Keywords and phrases related to the review questions are identified, including "managerial traits in M&A", "Leadership qualities AND Merger success", "Culture in M&A", "Socio-cultural integration in mergers", "bio-behavioral AND M&A", and "executive personality attributes in M&A". Combining these keywords with Boolean operators to refine the search results, such as "managerial capabilities AND mergers AND acquisitions," "leadership qualities OR intangible assets AND M&A success," "socio-cultural integration AND ergonomics in mergers" and "methodical attributes AND Merger AND success".

Initial screening of abstracts and titles has been done to identify relevant literature, followed by a full-text review of selected studies to confirm their inclusion-based criteria. We also considered a snowball review method, as this method provides insights to review from references of selected articles to add relevant studies.

### 2.3.3 Inclusion-Exclusion Criteria

We chose the exploratory method as it is impossible to review a large amount of literature that explores various human factors that influence the merger's success. The inclusion-exclusion criteria for this systematic literature review are designed to ensure that the selected studies are relevant and provide valuable insights into the ergonomic aspect of M&A. To maintain the focus and quality of the literature review, we have presented Inclusion–Exclusion Criteria in Table 2.1, which represents the different criteria included in this study. For inclusion, the studies published between 2017 and 2023 have been considered and focused on managerial traits, behavioral and institutional factors that impact merger success.

For exclusion, the studies published before 2016, non-peer-reviewed articles and non-academic sources are considered. This study does not include research focused on financial, strategic, and operational metrics without human impacts, nor do

**TABLE 2.1**
**Inclusion-Exclusion Criteria**

| Criteria Type | Inclusion Criteria | Exclusion Criteria |
|---|---|---|
| Publication date | Studies published between 2017 and 2023 | Studies published before 2016 |
| Peer-reviewed status | Peer-reviewed articles | Non-peer-reviewed articles |
| Source type | Academic sources | Non-academic sources |
| Focus area | Studies focusing on managerial traits, behavioral and institutional factors related to merger success | Research focusing solely on financial, strategic, and operational metrics without human impact |
| Empirical data | Articles providing insights into human factors in M&A | Articles that provide only empirical data or case studies related to mergers and acquisitions |

articles providing empirical data or case studies related to M&A. As a result, we were left with 124 pieces of literature for further study.

(Table 2.1 outlines the criteria for the incorporation of research on managerial characteristics and human factors in M&A, as well as the criteria for the exclusion of previous, non-peer-reviewed, non-academic, and exclusively empirical studies.)

### 2.3.4 Thematic Analysis

The thematic analysis of the selected 124 kinds of literature revealed several critical themes related to ergonomic factors impacting M&A success. In Table 2.2 Thematic Analysis of Ergonomic Influences on M&A Success and Relevant Attributes from the MBA Framework the different themes are categorized as below:

(Table 2.2 provides a summary of the different research themes about the influence of leadership, intangible assets, bio-psychological traits, and regional cultural traits on the success of mergers. These themes are then connected to particular attributes outlined in the MBA Framework.)

## 2.4 GLOBAL CASE EXAMPLES

The theoretical foundations proposed in the previous section will be rigorously tested for relevance through the presentation of real-world case examples of M&A, with a specific focus on the role of ergonomic factors and their outcomes. This will include case examples of both successful and unsuccessful mergers, underscoring the paramount importance of human factors in determining M&A success.

### 2.4.1 Case Exhibit 1 [Disney and 21st Century Fox (2019)]

The Disney-Fox merger in March 2019, where Disney acquired 21st Century Fox, was valued at USD71.3 billion. This merger was one of the largest and most significant

**TABLE 2.2**
**Thematic Analysis of Ergonomic Influences on M&A Success and Relevant Attributes from the MBA Framework**

| Theme | Description | Research Question (RQ) | Key Findings | Relevant Attributes from MBA Framework |
|---|---|---|---|---|
| Leadership and administrative abilities | The impact of leadership qualities and administrative abilities on firm performance post-M&A. | How do administrative abilities and leadership qualities influence the firm's performance post-M&A? | Effective leadership and administrative skills are crucial for navigating post-M&A integration, positively influencing firm performance. | **Administrative attributes:** Executive strategy, resolution proficiency, administrative skills |
| Intangible assets and cultural integration | The role of intangible assets, including cultural and sociocultural factors, in merger success. | What role does a firm's intangible assets have, like integration of culture and sociocultural factors, in the merger success? | Integrating intangible assets and aligning organizational cultures significantly contribute to the success of mergers | **Methodical attributes:** social environmental elements, linguistic impact |
| Bio-psychological traits of executives | The influence of executives' bio-psychological traits on strategic decision-making in mergers | How do Bio-psychological traits of executives impact strategic decision-making in the merger processes? | Executives' personality and cognitive styles affect strategic decisions during mergers, impacting overall success. | **Bio-psychological attributes:** cognitive prejudice or perceptual inclination, risk perception, anxiety management skills |
| Regional organizational and cultural traits | The effect of regional organizational and cultural traits on M&A outcomes | How do organizational and cultural traits affect M&A outcomes in different regions? | Different regions exhibit unique organizational and cultural traits, necessitating region-specific strategies for managing mergers. | Methodical attributes: social environmental elements, linguistic impact |

deals in the entertainment industry, aimed to bolster Disney's content portfolio and enhance its position in the streamlined market. Disney accessed an iconic collection of businesses and franchises from 21st Century Fox. This integration of two massive entertainment giants posed a significant leadership challenge. Disney's structured and family-oriented-corporate culture clashed with Fox's more diverse and varied cultural landscape. Both were recognized as distinct corporate cultures and integrating such large-scale operations was monumental. For this, robust managerial skills are required, to ensure smooth transitions and minimal disruptions. Disney and Fox had contrasting approaches to the top titles. Disney was conservative, while Fox was more generous with titles. Employees loyal to Fox's legacy and operating style had to adapt to new management practices and corporate values, which created initial resistance and adjustment issues. The high-profile nature of the merger created significant stress for both Disney and Fox employees. The merger also led to an expansion of chairperson positions within Disney. As a leadership dynamic, Disney had a single CEO (Bob Iger), while 21st century Fox had multiple CEOs.

**Outcome**: The merger of Disney and Fox resulted in a significantly expanded content library for Disney, positioning it firmly in the competitive streaming market with platforms like Disney+. Aligning leadership styles to create a unified direction requires careful navigation and strategic planning. Given the global nature of both companies, linguistic barriers and differences in communication styles are needed to ensure clear and effective communication for successful integration.

### 2.4.2 Case Exhibit 2 [Amazon and Whole Foods (2017)]

Amazon's purchase of Whole Foods for a whopping $13.7 billion in 2017 was one of the most significant acquisitions. This acquisition marked a significant advancement for the e-commerce giant as it ventured into offline trade, particularly in the grocery sector. The distinctive culture at Whole Foods, known for its low turnover rates among employees, underwent a transformation. Yet, the integration process raised various concerns, such as discontent with stock options and increased workloads. The shift from Whole Foods' independent culture to amalgamation with Amazon presented a major challenge. Despite encountering some cross-cultural conflicts, John Mackey, one of the founders of Whole Foods, continued to lead for a few more years, defying expectations. Upon Mackey's departure, Jason Buechel takes over the role of CEO at Whole Foods. However, the transition from the decentralized management style at Whole Foods to the data-driven centralization, typical of Amazon resulted in tension among managers and staff. Other worries included elevated workloads, stock options and cultural adjustments. The merger highlighted diverse reactions stemming from dissimilarities in organizational cultures and leadership approaches.

**Outcome**: This purchase helped Amazon increase its physical presence in the retail industry. Hence, it gets closer to consumers physically through markets and stores where its products are offered worldwide. Integrating these two brands had persistent cultural and operational challenges that directly or indirectly affected employee satisfaction levels, resulting in poor-quality work output.

### 2.4.3 Case Exhibit 3 [Tata Steel and Bhushan Steel (2018)]

The Tata Steel and Bhushan Steel merger is a significant event in the Indian Steel industry. It involved Tata Steel's acquisition of Bhushan Steel Ltd. (BSL), a troubled asset, through India's insolvency resolution process under the Insolvency and Bankruptcy Code (IBC) of 2016, valued at approximately USD5.4 billion in 2018. Due to Bhushan Steel's financial distress, the environment became highly stressed, which required effective stress management strategies, which were initially lacking. The executives of Tata Steel had to make rapid decisions regarding the integration and turnaround of Bhushan Steel's operations, which led to initial operational disruptions. This leadership faced challenges in restructuring Bhushan Steel's workforce and processes to align with Tata Steel's operational standards and efficiency goals.

**Outcome:** Despite the various initial challenges, Tata Steel managed to stabilize the operation of Bhushan Steel, though the process highlighted significant integration and managerial challenges.

### 2.4.4 Case Exhibit 4 (AT&T and Time Warner Merger (2018))

The American Telephone and Telegraph Company (AT&T) and Time Warner merger of 2018 was a landmark case in the media and telecommunications industry. This merger of AT&T and Time Warner, two industry titans, necessitated a complex integration effort for executives. They had to redefine the strategic path of the merged entity to capitalize on synergies between content production and distribution. Notably, inherent cultural differences existed between AT&T, a conventional telecommunications firm, and Time Warner, a creative content company. This merger instigated uncertainties among employees regarding job stability and career advancement, potentially impacting their morale and efficiency. Legal obstacles from the U.S. Department of Justice concerning antitrust issues, claiming adverse effects on pricing and innovation, also challenged the merger. Consequently, maintaining positive customer relationships during the merger was crucial, as any perceived decline in service quality or pricing could result in customer loss.

**Outcome**: Following some initial achievements, this high-profile merger deal encountered persistent obstacles in aligning business models and realizing expected synergies, sparking inquiries into the strategic coherence of the acquisition. This integration process highlighted gaps in decision-making as the companies operated in very different sectors, which led to strategic misalignments.

## 2.5 FINDINGS

This chapter studied the influence of ergonomics elements as methodical, bio-psychological, and administrative factors on a firm's success within the purview of M&A. Our findings align with broader studies on ergonomic factors, focused on human nature affecting M&A success.

From the SLR and thematic findings, we developed a MBA framework to study the importance of considerable human factors that can create long-term value for the firm. This study involves a better understanding of M&A dynamics by focusing on crucial ergonomic factors that significantly impact the success of the merged firm. This study highlights some examples of recent case exhibits to show the practical aspects of ergonomics elements like leadership qualities, risk-taking abilities of managers, stress coping mechanisms, psychological business, linguistic barriers, communication barriers and sociocultural issues between the employee and merged entity. These identified elements and theme-based framework are crucial outcomes that are considered human-centric approaches. These elements significantly affect overall firm performance and decision-making process.

## 2.6 CONCLUSION

The present study underscores the implications of ergonomic elements in the success of M&A. The MBA framework, formulated based on our contributions, emphasizes the necessity of an extensive understanding of managerial, behavioral, and organizational characteristics. By examining various viewpoints such as personality traits, psychological biases, culture, language and risk attitudes, this study presents a detailed perspective on the impact of these elements on M&A outcomes. The pivotal function of stress management techniques and sociocultural factors further emphasizes the requirement for efficient combination and management methodologies throughout M&A procedures. The investigation recognizes potential deficiencies and avenues for further growth, advocating for exploring alternative methodologies and broader fields of study to enrich our perception of M&A dynamics.

## 2.7 IMPLICATIONS

The implications of this research are widespread. The MBA framework offers a systematic approach to considering human-centric aspects in M&A strategies, potentially leading to more successful outcomes and enduring value creation for practitioners. Decision-makers can apply these insights in their planning and execution processes, ensuring that managerial aspects are taken into account alongside financial and strategic factors. For academics, this study underscores the importance of additional exploration of human elements in M&A, suggesting that future research initiatives should incorporate diverse research techniques, including experimental approaches and extensive field studies. Understanding the nuances of board diversity, managerial traits, and the varied impacts of different human factors can yield profound insights and contribute to the advancement of M&A research.

## 2.8 FUTURE RESEARCH SCOPE

Although this evaluation provided insightful information, multiple limitations have emerged that have paved the way for further study avenues. First, the utilization of an electronic database for article selection, while enhancing accessibility, may have inadvertently omitted relevant research that was not cataloged or accessible through this methodology. Second, the extensive utilization of keywords may have failed to encompass all research on managerial competency due to the vast array of terminology employed in this particular area of inquiry. Lastly, a knowledge gap may exist as advancements in managerial skills within other academic disciplines may have been marginalized by the predominant focus on business, management, and accounting journals.

## REFERENCES


Aabo, T., Hanousek, J., Pantzalis, C., & Park, J. C. (2023). CEO personality traits and corporate value implication of acquisitions. *Journal of Empirical Finance*, 73, 86–106. https://doi.org/10.1016/j.jempfin.2023.05.006

Abdullahi, A. Z., Anarfo, E. B., & Anyigba, H. (2020). The impact of leadership style on organizational citizenship behavior: Does leaders' emotional intelligence play a moderating role? *Journal of Management Development*, 39(9–10), 963–987.

Aghazade, H., Heidari, A., Dehghan Dehnavi, M. A., & Abolmolouki, H. (2019). Presenting a comprehensive framework for integration and acquisition management in the financial services industry. *Journal of Management and Development Process*, 32(3), 105–138. https://doi.org/10.29252/jmdp.32.3.105

Almazur, B., Arize, A. C., Andreopoulos, G. C., Malindretos, J., & Panayides, A. (2018). The reasons and evaluations of mergers and acquisitions. *Accounting and Finance Research*, 7(3), 211. https://doi.org/10.5430/afr.v7n3p211

Alwali, J., & Alwali, W. (2022). The relationship between emotional intelligence, transformational leadership, and performance: A test of the mediating role of job satisfaction. *Leadership and Organization Development Journal,* 43(6), 928–952. https://doi.org/10.1108/LODJ-10-2021-0486

Bedi, A. (2018). Post acquisition performance of indian telecom companies: An empirical study. *Pacific Business Review International*, 11(2). http://www.pbr.co.in/2018/2018_month/Aug/6.pdf

Benischke, M. H., Martin, G. P., Gomez-Mejia, L. R., & Ljubownikow, G. (2020). The effect of CEO incentives on deviations from institutional norms in foreign market expansion decisions: Behavioral agency and cross-border acquisitions. *Human Resource Management*, 59(5), 463–482. https://doi.org/10.1002/HRM.22006

Bereskin, F. L., Byun, S. K., Officer, M. S., & Oh, J.-M. (2017). The effect of cultural similarity on mergers and acquisitions: Evidence from corporate social responsibility. *SSRN Electronic Journal*. https://doi.org/10.2139/SSRN.2920129

Beugelsdijk, S., Kostova, T., Kunst, V. E., Spadafora, E., & van Essen, M. (2018). Cultural distance and firm internationalization: A meta-analytical review and theoretical implications. *Journal of Management*, 44(1), 89–130. https://doi.org/10.1177/0149206317729027

Bonsall, S. B., Holzman, E. R., & Miller, B. P. (2017). Managerial ability and credit risk assessment. *Management Science,* 63(5), 1425–1449. https://doi.org/10.1287/MNSC.2015.2403

Chandren, S., Qaderi, S. A., & Ghaleb, B. A. A. (2021). The influence of the chairman and CEO effectiveness on operating performance: Evidence from Malaysia. *Cogent Business & Management,* 8(1), 1935189. https://doi.org/10.1080/23311975.2021.1935189

Chen, S. S., & Lin, C. Y. (2018). Managerial ability and acquirer returns. *Quarterly Review of Economics and Finance*, 68, 171–182. https://doi.org/10.1016/J.QREF.2017.09.004

Chung, C. Y., Hur, S. K., & Wang, K. (2017). Strategic merger decisions across business cycles: Evidence from bidders' time-varying appetite for operating leverage. *International Review of Economics & Finance*, 47, 143–158. https://doi.org/10.1016/J.IREF.2016.10.010

Cui, H., & Chi-Moon Leung, S. (2020). The long-run performance of acquiring firms in mergers and acquisitions: Does managerial ability matter? *Journal of Contemporary Accounting & Economics*, 16(1), 100185. https://doi.org/10.1016/j.jcae.2020.100185

Demerjian, P., Lewis-Western, M., & McVay, S. (2017). How does intentional earnings smoothing vary with managerial ability? *Journal of Accounting, Auditing & Finance*, 35(2), 406–437. https://doi.org/10.1177/0148558X17748405

DePamphilis, D. (2019). *Mergers, Acquisitions, and Other Restructuring Activities: An Integrated Approach to Process, Tools, Cases, and Solutions*. Academic Press.

Djalil, M. A., Amin, M., Herjanto, H., Nourallah, M., & Öhman, P. (2023). The importance of entrepreneurial leadership in fostering bank performance. *International Journal of Bank Marketing*, 41(4), 926–948. https://doi.org/10.1108/IJBM-11-2022-0481

Dong, F., & Doukas, J. (2021). The effect of managers on M&As. *Journal of Corporate Finance*, 68, 101934. https://doi.org/10.1016/j.jcorpfin.2021.101934

Doukas, J. A., & Zhang, R. (2021). Managerial ability, corporate social culture, and M&As. *Journal of Corporate Finance*, 68, 101942. https://doi.org/10.1016/J.JCORPFIN.2021.101942

Fernando, G. D., Jain, S. S., & Tripathy, A. (2020). This cloud has a silver lining: Gender diversity, managerial ability, and firm performance. *Journal of Business Research*, 117, 484–496. https://doi.org/10.1016/J.JBUSRES.2020.05.042

Ferris, S. P., & Sainani, S. (2021). Do CFOs matter? Evidence from the M&A process. *Journal of Corporate Finance,* 67, 101856. https://doi.org/10.1016/j.jcorpfin.2020.101856

Graebner, M. E., Heimeriks, K. H., Huy, Q. N., & Vaara, E. (2016). The process of postmerger integration: A review and agenda for future research. *Academy of Management Annals*, 11(1), 1–32. https://doi.org/10.5465/ANNALS.2014.0078

Haapanen, L., Hurmelinna-Laukkanen, P., Nikkilä, S., & Paakkolanvaara, P. (2019). The function-specific microfoundations of dynamic capabilities in cross-border mergers and acquisitions. *International Business Review,* 28(4), 766–784. https://doi.org/10.1016/J.IBUSREV.2019.03.002

Hammer, B., Hinrichs, H., & Schwetzler, B. (2018). Does culture affect the performance of private equity buyouts? Journal of Business Economics. https://doi.org/10.1007/s11573-017-0886-0

Hassan, I., Ghauri, P. N., & Mayrhofer, U. (2018). Merger and acquisition motives and outcome assessment. Thunderbird. *International Business Review*, 60(4), 709–718. https://doi.org/10.1002/TIE.21967

Hong, P. C., Chennattuserry, J. C., & Kureethara, J. V. (2019). Organization culture and work values of global firms: Merging eastern and western perspectives. *3D... IBA Journal of Management & Leadership.* https://papers.ssrn.com/sol3/papers.cfm?abstract_id=3507562

Hossain, M. S. (2021). Merger & Acquisitions (M&As) as an important strategic vehicle in business: Thematic areas, research avenues & possible suggestions. *Journal of Economics and Business*, 116, 106004. https://doi.org/10.1016/j.jeconbus.2021.106004

Hsieh, T.-S., Wang, Z., & Demirkan, S. (2018). Overconfidence and tax avoidance: The role of CEO and CFO interaction. *Journal of Accounting and Public Policy,* 37(3), 241–253. https://doi.org/10.1016/j.jaccpubpol.2018.04.004

Hsu, C.-C., Park, J.-Y., & Lew, Y. K. (2019). Resilience and risks of cross-border mergers and acquisitions. *Multinational Business Review*, 27(4), 427–450. https://doi.org/10.1108/MBR-05-2019-0035

India M&A trends. (2023). https://www2.deloitte.com/in/en/pages/finance/articles/india-m-and-a-trends.html

Juyumaya, J., & Torres, J. (2022). Effects of transformational leadership and work engagement on managers' creative performance. *Baltic Journal of Management*, 18. https://doi.org/10.1108/BJM-11-2021-0449

Karhunen, P., Kankaanranta, A., Louhiala-Salminen, L., & Piekkari, R. (2018). Let's talk about language: A review of language-sensitive research in international management. *Journal of Management Studies*, 55(6), 980–1013. https://doi.org/10.1111/JOMS.12354

Khan, H., Rehmat, M., Butt, T. H., Farooqi, S., & Asim, J. (2020). Impact of transformational leadership on work performance, burnout and social loafing: A mediation model. *Future Business Journal*, 6(1), 1–13. https://doi.org/10.1186/S43093-020-00043-8

King, D. R., Wang, G., Samimi, M., & Cortes, A. F. (2021). A meta-analytic integration of acquisition performance prediction. *Journal of Management Studies*, 58(5), 1198–1236. https://doi.org/10.1111/JOMS.12636

Konara, P., & Wei, Y. (2019). The complementarity of human capital and language capital in foreign direct investment. *International Business Review*, 28(2), 391–404. https://doi.org/10.1016/J.IBUSREV.2018.10.009

Konara, P., & Wei, Y. (2021). Does language matter to foreign subsidiary performance? *International Marketing Review,* 38(2), 276–299. https://doi.org/10.1108/IMR-05-2019-0129

Kostova, T., Beugelsdijk, S., Scott, W. R., Kunst, V. E., Chua, C. H., & van Essen, M. (2020). The construct of institutional distance through the lens of different institutional perspectives: Review, analysis, and recommendations. *Journal of International Business Studies*, 51(4), 467–497. https://doi.org/10.1057/S41267-019-00294-W

Latif, K. F., Nazeer, A., Shahzad, F., Ullah, M., Imranullah, M., & Sahibzada, U. F. (2020). Impact of entrepreneurial leadership on project success: Mediating role of knowledge management processes. *Leadership and Organization Development Journal*, 41(2), 237–256. https://doi.org/10.1108/LODJ-07-2019-0323

Lind, S. C., & Lattuch, F. (2021). Trust, leadership and conflict: The impact of HR practices during family firm M&As. *Strategic HR Review*, 20(6), 221–225. https://doi.org/10.1108/SHR-09-2021-0044

Liu, Y., Sarala, R. M., Xing, Y., & Cooper, C. L. (2017). Human side of collaborative partnerships. *Group & Organization Management*, 42(2), 151–162. https://doi.org/10.1177/1059601117695138

Liu, Y., Vrontis, D., Visser, M., Stokes, P., Smith, S., Moore, N., Thrassou, A., & Ashta, A. (2021). Talent management and the HR function in cross-cultural mergers and acquisitions: The role and impact of bi-cultural identity. *Human Resource Management Review*, 31(3). https://doi.org/10.1016/J.HRMR.2020.100744

Ly, A., Esperança, J., & Davcik, N. S. (2018). What drives foreign direct investment: The role of language, geographical distance, information flows and technological similarity. *Journal of Business Research*, 88, 111–122. https://doi.org/10.1016/J.JBUSRES.2018.03.007

Malhotra, S., Reus, T. H., Zhu, P., & Roelofsen, E. M. (2018). The acquisitive nature of extraverted CEOs. *Administrative Science Quarterly,* 63(2), 370–408. https://doi.org/10.1177/0001839217712240

Mount, M. P., & Baer, M. (2021). CEOs' regulatory focus and risk-taking when firms perform below and above the bar. Journal of Management, 48(7), 1980–2008. https://doi.org/10.1177/01492063211016029 (Original work published 2022)

Panibratov, A. (2017). Cultural and organizational integration in cross-border M&A deals: The comparative study of acquisitions made by EMNEs from China and Russia. *Journal of Organizational Change Management*, 30(7), 1109–1135. https://doi.org/10.1108/JOCM-01-2016-0011

Paudel, S. (2019). Entrepreneurial leadership and business performance: Effect of organizational innovation and environmental dynamism. *South Asian Journal of Business Studies*, 8(3), 348–369. https://doi.org/10.1108/SAJBS-11-2018-0136

Renneboog, L., & Vansteenkiste, C. (2019). Failure and success in mergers and acquisitions. *Journal of Corporate Finance*, 58, 650–699. https://doi.org/10.1016/J.JCORPFIN.2019.07.010

Rouine, I. (2018). Target country's leadership style and bidders' takeover decisions. *International Review of Financial Analysis*, 60, 17–29. https://doi.org/10.1016/J.IRFA.2018.08.004

Sarala, R. M., Vaara, E., & Junni, P. (2019). Beyond merger syndrome and cultural differences: New avenues for research on the "human side" of global mergers and acquisitions (M&As). *Journal of World Business*, 54(4), 307–321. https://doi.org/10.1016/J.JWB.2017.10.001

Savović, S., & Babić, V. (2021). Impact of behaviour factors on acquisition performance: Mediating role of speed of post-acquisition change. *Journal of Organizational Change Management*, 34(5), 929–956. https://doi.org/10.1108/JOCM-02-2020-0056

Shi, W., Zhang, Y., & Hoskisson, R. E. (2019). Examination of CEO–CFO social interaction through language style matching: Outcomes for the CFO and the organization. *Academy of Management Journal*, 62(2), 383–414. https://doi.org/10.5465/AMJ.2016.1062

Srivastava, R. K. (2018). Merger, acquisition–right move or emotional move for growth–case study of sun pharmaceutical. *International Journal of Pharmaceutical and Healthcare Marketing*, 12(3), 270–287.

Su, F., Cheng, D., & Wen, S. (2019). Multilevel impacts of transformational leadership on service quality: Evidence from China. *Frontiers in Psychology*, 10(MAY), 463522. https://doi.org/10.3389/FPSYG.2019.01252

Tao, F., Liu, X., Gao, L., & Xia, E. (2017). Do cross-border mergers and acquisitions increase short-term market performance? The case of Chinese firms. *International Business Review*, 26(1), 189–202. https://doi.org/10.1016/J.IBUSREV.2016.06.006

Tenzer, H., Terjesen, S., & Harzing, A.-W. (2017). Language in international business: A review and agenda for future research. *MIR: Management International Review*, 57(6), 815–854. https://www.jstor.org/stable/44985725

Wang, Q., Lau, R. Y. K., & Yang, K. (2020). Does the interplay between the personality traits of CEOs and CFOs influence corporate mergers and acquisitions intensity? An econometric analysis with machine learning-based constructs. *Decision Support Systems*, 139, 113424. https://doi.org/10.1016/j.dss.2020.113424

Yung, K., & Chen, C. (2017). Managerial ability and firm risk-taking behavior. *Review of Quantitative Finance and Accounting*, 51(4), 1005–1032. https://doi.org/10.1007/S11156-017-0695-0

Zahoor, N., Khan, Z., & Sinkovics, R. R. (2022). The role of emotions in cross-border mergers & acquisitions: A systematic review of the inter-disciplinary literature and future research agenda. *Journal of International Management*, 28(4), 100958. https://doi.org/10.1016/j.intman.2022.100958

Zhang, C. M. (2020). Mergers and acquisitions as strategic decisions for a decision-making theory. In S. Finkelstein & C. L. Cooper (eds.), *Advances in Mergers and Acquisitions* (Vol. 19). Emerald Publishing Limited, pp. 49–64. https://doi.org/10.1108/S1479-361X20200000019006

# 3 Assessing the Strategic Orientation towards Ergonomic Hazards in Health and Safety Reporting Systems in Selected Oil and Gas Drilling Sites

*Henry Jonathan and Yahya Fuad Al Abdulsalam*

## 3.1 INTRODUCTION

The oil and gas industry ranks among the most critical and perilous sectors globally, serving as a key driving force of the world's economy. According to recent business research reports, the global oil and gas market is expected to grow from the current $7330 billion in 2023 to $8670 billion by 2027. Much of this share comes from the Asia Pacific region which has the largest oil and gas market in the year 2022 followed by North America. The Middle East stands globally as the leading oil-producing nation substantially contributing towards the global oil revenue expected to reach $180.5 billion by the year 2025. Among the Middle East region, the major oil-producing countries are the Gulf Cooperation Council, principally the revenue coming from fossil fuel resources. The industry involves a myriad of complex processes, including drilling, extraction, transportation, and refining, all of which are labor-intensive tasks, posing considerable health and safety challenges to the workforce. Workers are routinely exposed to a wide range of hazards, such as chemical exposure, heavy machinery accidents, fires, explosions, and physically demanding tasks at every stage of work. Consequently, the potential for accidents and injuries is heightened mostly among workers between 25 and 44 years of age, which necessitates the implementation of robust health and safety management practices.

This multifaceted industry involves a myriad of complex processes, including drilling, extraction, transportation, and refining, all of which pose considerable health and safety challenges to the workforce. Workers are routinely exposed to a wide range of hazards, such as chemical exposure, heavy machinery accidents, fires,


DOI: 10.1201/9781003532293-4

explosions, and physically demanding tasks. Consequently, the potential for accidents and injuries is heightened, necessitating the implementation of robust health and safety management practices.

By adopting comprehensive health and safety protocols, companies operating in Oman's oil and gas sector can effectively safeguard their employees, reduce the likelihood of accidents, and uphold overall operational efficiency. Furthermore, such practices contribute to a culture of safety awareness and promote adherence to both national and international regulations. In turn, this fosters a sustainable and responsible approach to oil and gas extraction, transportation, and processing, thereby enhancing Oman's reputation as a reliable and conscientious energy producer.

The oil and gas industry is an indispensable component of the global economy, with Oman playing a significant role in the Middle East's energy production landscape. Given the inherent risks associated with the industry's various operations, companies must prioritize health and safety management to protect their workforce, optimize efficiency, and maintain a sustainable and responsible approach to energy production.

Ergonomics, a crucial component of occupational health and safety management, is centered on tailoring workspaces, equipment, and processes to suit employees' physiological, psychological, and cognitive abilities. This concept is particularly relevant for industries with intricate and high-risk work environments, such as the oil and gas sector, which stands to gain significantly from the effective implementation of ergonomic practices.

By emphasizing ergonomics, companies in the oil and gas industry can reduce the likelihood of injuries, boost employee productivity, and promote overall job satisfaction. Over the years, there has been a marked improvement in safety standards within the sector, demonstrating the industry's commitment to creating safer work environments. Nonetheless, the importance of ergonomics in health and safety management is often undervalued in the oil and gas industry. This oversight can lead to missed opportunities for enhancing workplace safety and efficiency. By focusing on the human factors involved in work-related tasks, ergonomics enables the development of safer and more efficient workspaces that cater to the diverse needs and capabilities of employees.

## 3.2 LITERATURE REVIEW

The oil and gas industry, being a vital component of the global economy, is characterized by complex processes whether they are related to drilling, processing, or even exploration and that pose inherent risks. Oman plays a significant role in the Middle East's energy production landscape, with ergonomics being a crucial and essential aspect of occupational health and safety management, which focuses on adapting workspaces, equipment, and processes to suit the unique physiological, psychological, and cognitive abilities of employees (Dul et al., 2012).

### 3.2.1 Significance of Ergonomics in Human Factors

In General, the concept of ergonomics has been widely acknowledged as an essential and crucial component of any occupational health and safety management system. As defined by the International Ergonomics Association (IEA, 2000), ergonomics is the discipline that is concerned with understanding the interactions between humans and other elements of a system in the workplace, to optimize human well-being and overall system performance. This can further be asserted by Carayon and Smith (2000) who provide a thorough understanding of work organization and ergonomics, emphasizing the importance of considering both in designing efficient and safe work environments. The authors argue that ergonomics should be integrated with work organization strategies to optimize employee performance and well-being, thus reducing work-related injuries and accidents. Karwowski (2006) posits that ergonomics' primary goal is to ensure optimal human well-being and system performance by designing tasks, systems, products, and environments based on human abilities and limitations. This study outlines the importance of considering various human factors, including physical, cognitive, and organizational aspects, to create effective ergonomic solutions. This notion is further echoed by both Helander (2006) and Dul et al. (2012), who emphasize the benefits of ergonomics, by again, highlighting its positive impact on productivity, safety, and worker satisfaction. We can then see in these studies that by incorporating ergonomic principles into the design of workstations, equipment, and tasks, companies can minimize the risk of work-related injuries and enhance employee well-being, leading to improved overall performance, which highlights the importance of considering human factors in the design and management of work systems to optimize performance and well-being.

### 3.2.2 Ergonomics in the Oil and Gas Industry

While the importance of ergonomics in the oil and gas industry has been recognized, there remains a need for further research and implementation. In a review of 40 years of research on human factors and ergonomics in the oil and gas industry, O'Hara et al. (2016) identified a limited number of studies addressing ergonomics in the sector, suggesting a lack of focus on the topic. Nonetheless, some studies have investigated ergonomic interventions in specific oilfield contexts, demonstrating their potential benefits. For instance, a study by Saurin et al. (2008) emphasizes the role of ergonomics in reducing the likelihood of accidents, especially in offshore oil platforms and rigs, this can be further emphasized by a study by Hignett and Wilson (2004) which further examined the implementation of ergonomic principles in the design of an offshore oil platform, which can imperatively lead to improved working conditions and reduced risk of injuries. This study highlights the need for tailored ergonomic solutions to address the unique requirements of offshore environments, such as space constraints, extreme weather conditions, and extended work hours, which further extends to analyze the importance of ergonomics in the oil and gas industry.

Given the hazardous nature of the oil and gas sector, ergonomics is particularly relevant. Industry often exposes workers to physically demanding tasks, extreme temperatures, and high-pressure environments. Al-Hemoud et al. (2015) also highlighted in their study the importance of ergonomics in maintaining the safety and well-being of workers in the Kuwaiti oil and gas sector to show emphasis on ergonomics. Furthermore, a study by Hämäläinen et al. (2006) underscores the significance of ergonomic practices in the prevention of occupational accidents and diseases in the global oil and gas industry this can also be backed up by a study by Al-Balushi and Dutta (2016) which further emphasize this importance of integrating ergonomics into process safety management in the oil and gas industry. The authors discuss the role of ergonomics in preventing major accidents and improving operational performance, suggesting that companies should consider ergonomics as a crucial component of their safety management systems.

Similarly, a study by Jain et al. (2014) investigated the ergonomic risk factors associated with manual handling tasks in the Indian oil and gas industry, emphasizing the need for ergonomic interventions to reduce musculoskeletal disorders (MSDs). This can be further analyzed in a study by Ghasemi et al. (2017) which further examined the impact of ergonomic interventions on reducing MSDs among Iranian oil refinery workers. The results showed a significant reduction in the prevalence of MSDs following the implementation of ergonomic measures, emphasizing the importance of ergonomics in the oil and gas industry.

### 3.2.3 Ergonomics in Oman's Oil and Gas Industry

In the context of Oman, a few studies have addressed the role and function of ergonomics in the country's oil and gas sector. All these studies showed coextensive results, although most companies are pivotal frontrunners in health and safety management, Ergonomics remains an area requiring improvement. A study by Al-Maashani et al. (2020) echoed this by assessing the effectiveness of health and safety management systems in Oman's oil and gas industry, Al-Maashani et al. (2020) found that while companies generally performed well in implementing health and safety measures, ergonomic practices were often overlooked. The authors recommended incorporating ergonomics as a separate and standalone entity into company policies and training programs to address this gap.

Al-Abri (2017) also conducted a study on the ergonomic practices of Petroleum Development Oman (PDO), the study highlighted the crucial need for improvements in workstation design, work processes, and employee training. The study also revealed a lack of awareness about ergonomics among employees, emphasizing the need for increased education and communication about the benefits of ergonomic principles. Furthermore, Al-Balushi et al. (2019) examined the impact of ergonomic training on workers' knowledge and practices in the Omani oil and gas sector, finding a significant improvement in workers' awareness of ergonomic principles. This is prevalent in many companies and statistics have shown that the lack of ergonomics in the oil and gas sector is positively related to the increase in musculoskeletal injuries, this can be echoed in a study by Al-Harrasi et al. (2018) which investigated the prevalence of MSDs among oil and gas workers in Oman. This study reported that over 40% of the workforce experienced discomfort in the lower back, neck, and shoulders. The study concluded that ergonomic interventions could significantly reduce these issues.

All this existing research collectively indicates and shows a persistent ergonomic deficiency within Oman's oil and gas sector, despite the growing awareness of ergonomics in the oil and gas industry in Oman. Furthermore, there is limited literature conducted on ergonomics in the oil and gas sector. According to Petroleum

Development Oman, the company's health and safety practices have been recognized as exemplary in the Omani oil and gas sector (PDO, 2018). However, the company's approach to ergonomics remains underexplored.

However, recent studies have already begun to shed light on this topic. For example, previous research conducted by Hovden et al. (2008) provided some insights into the health and safety practices in oil and gas companies with the authors reporting that the company had a strong safety culture but lacked a systematic approach to ergonomics. There is a huge gap for improvement in terms of ergonomics both in terms of awareness and implementation and this study could benefit more from comprehensive ergonomic assessments and targeted interventions instead of blending the ergonomic portion within their management systems.

The literature suggests that incorporating ergonomic principles in the design and evaluation of workstations, tools, equipment, and processes can significantly benefit workers in the oil and gas industry. Furthermore, the integration of ergonomics in the health and safety reporting system safety is crucial for a comprehensive approach to employee safety.

### 3.2.4 Background and Study Objectives

The primary aim of the study is to conduct a comprehensive analysis of the current state of ergonomics within the oil and gas industry, specifically focusing on a few large drilling locations in Oman to contribute to the understanding of ergonomics and its potential impact on the sector. In conjunction, the study proceeds to critically clarify the following objectives.

- Probe the role of ergonomics in enhancing employee safety, health, and productivity in the oil and gas industry.
- Examine the critical role ergonomics plays in preventing work-related injuries and promoting worker well-being to explore how ergonomic improvements can lead to increased productivity and overall efficiency.
- Assess the physical work environment, workstations, tools, equipment, and work processes, to identify gaps and shortcomings in current ergonomic practices.
- Examine the present extent and enhance the integration of ergonomic principles within the health and safety monitoring system that is largely in practice across the oil and gas contractors and clients.

## 3.3 METHODOLOGY

The study methodology includes three main components: a survey on ergonomics awareness and its implementation in the oilfield, an analysis of the Stop Card system used by some oil and gas contractors, and an internal ergonomic survey conducted among on-site employees, both in the field site and main corporate office. The simultaneous integration of these three methodological components provides a comprehensive understanding of the current state of ergonomics within the company and offers practical recommendations for continuous improvement in the matter.

The study includes a primary oil and gas exploration company and 12 contractors under the client's management. The contributors to this study are academicians, students, oilfield workers, and workers in the energy sector in Oman.

Site visits and observations are conducted to various oilfield locations to observe the working conditions for both contractors and subcontractors, safety measures that are present, and potential ergonomic risks faced by employees. These visits provided initial and first-hand information on the practical challenges encountered by workers in the oilfield. An internal site survey was conducted by administering a questionnaire to employees, both in the field site and in the corporate office to gather information on their experiences related to ergonomics in the workplace.

### 3.3.1 Statistical Analysis of Survey Data

The data collected from questionnaires and internal surveys are analyzed using descriptive and inferential statistics to identify trends, relationships, and patterns related to ergonomics awareness and implementation. This analysis will help determine the factors contributing to ergonomic success or failure.

## 3.4 RESULTS AND DISCUSSION

### 3.4.1 Demographic Information

From the study conducted, we have received a total of 1,135 responses which include academicians, students, oil and gas field workers, and the energy sector.

### 3.4.2 Survey on Ergonomics Awareness and Its Implementation in the Oil and Gas Industry

Most respondents who participated in the survey are in the age group 21–30 years followed by 31–40 years and 41–50 respectively (Table 3.1).

Of the total respondents, males constitute 54.9% and females 45.0%, of which most of the participation was from university students (44.4%), academicians (5.5%), and 23.6% from oil and gas sector. The awareness of people on ergonomic hazards shows that most of them have hardly any knowledge about ergonomics in the workplace but 45.2% were able to provide a meaningful definition of the term.

**TABLE 3.1**
**Demographic Data of the Survey Respondents**

| | N | | Percent (%) | | Age Range | |
|---|---|---|---|---|---|---|
| | Male | Female | Male | Female | Male | Female |
| Academicians | 213 | 110 | 65.9 | 34.1 | 29–60 | 30–55 |
| Students | 208 | 320 | 39.4 | 60.6 | 21–30 | 20–28 |
| Oil field workers | 227 | – | 100 | – | 28–55 | – |
| Energy sector | 42 | 15 | 73.6 | 26.4 | 26–60 | 31–58 |

Though the definitions given are different, the common description given is ergonomics is an interdisciplinary science that focuses on the design and arrangement of furniture, equipment, tools, and products to optimize human efficiency, comfort, and safety. It involves the study of human physiology, psychology, and the unique needs of the individual, as well as the interactions between the worker and their physical tasks.

The importance of ergonomics takes an important position in the oil and gas sector. The study shows that 62.5% of the respondents regard it as a crucial aspect to consider in the industry. On the contrary, in most of the contractors' health and safety reporting systems, ergonomic interventions are not adequately recognized; in some cases, there are hardly any instances of ergonomic hazard indicators. The poor recognition given to ergonomic aspects in the target contractors is mainly due to a lack of awareness and understanding, availability of limited resources for ergonomic hazard assessment, poor management support, change management and the nature of the complex work environment. The study also assessed the percentage of workers that experienced any pain, discomfort, or MSDs due to lack of ergonomics, 70.8% of the respondents expressed themselves to have experienced lower back pain, neck pain, and eye strain while working with computers and DSE, while the rest have felt the mild effect of ergonomic hazards.

The last part of the survey aims to get recommendations for continual improvements 63.89% of the respondents recommended an increase of awareness and training on ergonomics, 55.56% recommended providing more resources including funding and equipment, 45.83% of the respondents recommended management support and involvement, 36.11% of the respondents recommended regular assessments and evaluations of ergonomic practices and lastly 31.94% recommended developing tailored ergonomic solutions for the unique requirements of the industry (Figure 3.1).

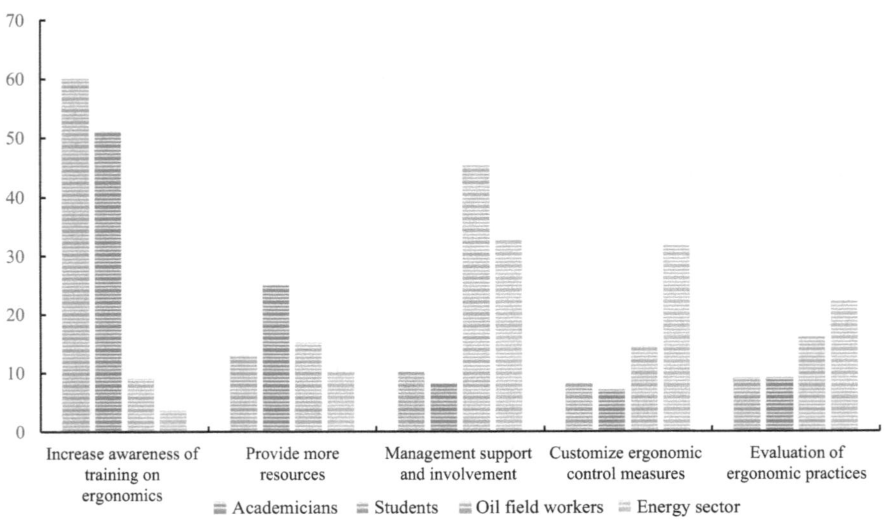


**FIGURE 3.1** Suggested improvements by participants.

### 3.4.3 Analysis of Organizations' Health and Safety Reporting System

The organization's stop card system followed by the client and associated contractors is part of their management system, to analyze the effectiveness to understand the extent and degree of integration of ergonomics in the company's health and safety management system. The client and associated contractor's online reporting system consists of 6 modules, the first is to report Personal Protective Equipment, the second is tools and equipment, the third is the position of people/working environment, the fourth is behaviors, the fifth is driving and the last one is lifesaving rules.

The first module allows users to select all the observations that are based on the PPE, which includes things like hard hats, safety boots and other PPE. The second module observes tools and equipment, which include using the wrong tools for work, using tools in bad condition, and so forth. The third reports everything that has to do with people and their working environment, this includes unsafe acts like pinch points, slips, trips and falls, flying objects, and so forth. In this module, we see only overreaching and unsafe lifting as a small part of ergonomics incorporated into the module. The fourth module is about behavior, and it consists of reporting of workers' behavior like lack of ownership and negligence. This module also displays one aspect of ergonomics which is posture. The fifth module looks at road safety and all the observations and reporting from road safety, it reports things like driving without a license, vehicle condition, journey plan, and even night journeys. The last module is a way to report violations of lifesaving rules, and it consists of all the rules shown in the graphics below (Figure 3.2).

From observing this STOP Card system of the client and associated contractors, out of the 58 actions that employees can observe, only three of them are related to ergonomics, this shows a lack of emphasis on the integration of ergonomics in the

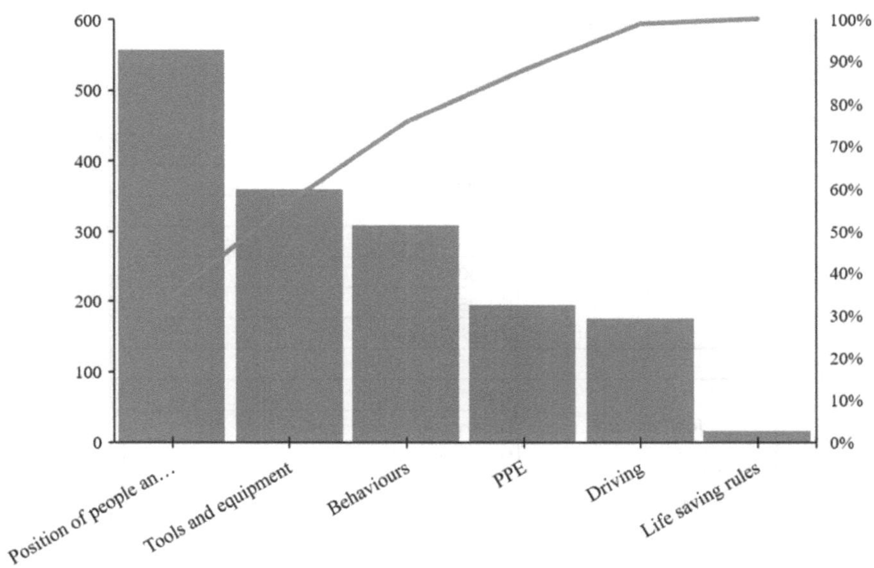


**FIGURE 3.2** Observations of stop card system by category.

management system. Furthermore, based on a review in the year 2022, there were around 1,627 observations. And among those, only 26 in the whole year were attributed to ergonomics.

### 3.4.4 Internal Employee Survey

A survey done for all employees of the client, in the field site operations and corporate office, shows most employees were male, over 35 years and working as staff. While in field site operations, all the employees are male between the age group 26–30 and 31–35 (Figures 3.3 and 3.4).

Moving to the injuries related to MSD or injury type, in both locations, corporate and field site, the data shows the same trend, employees experienced significant lower back and neck pain (Figures 3.5 and 3.6).

| | | |
|---|---|---|
| **Gender :** | Male | **32** |
| | Female | **8** |
| | | |
| **Age:** | (18-25) Years | **1** |
| | (26-30) Years | **9** |
| | (31-35) Years | **10** |
| | (Over 35) Years | **20** |
| | | |
| **Job Position:** | Staff | **22** |
| | Supervisor/ Team leader | **10** |
| | Manager | **8** |
| | | |
| **Work Area:** | Head Office- Muscat | **40** |

FIGURE 3.3 Corporate branch demographics.

| | | |
|---|---|---|
| **Gender :** | Male | **15** |
| | Female | **0** |
| | | |
| **Age:** | (18-25) Years | **1** |
| | (26-30) Years | **6** |
| | (31-35) Years | **2** |
| | (Over 35) Years | **6** |
| | | |
| **Job Position:** | Staff | **9** |
| | Supervisor/ Team leader | **6** |
| | Manager | **0** |
| | | |
| **Work Area:** | Rima Field | **15** |

FIGURE 3.4 Client field demographics.

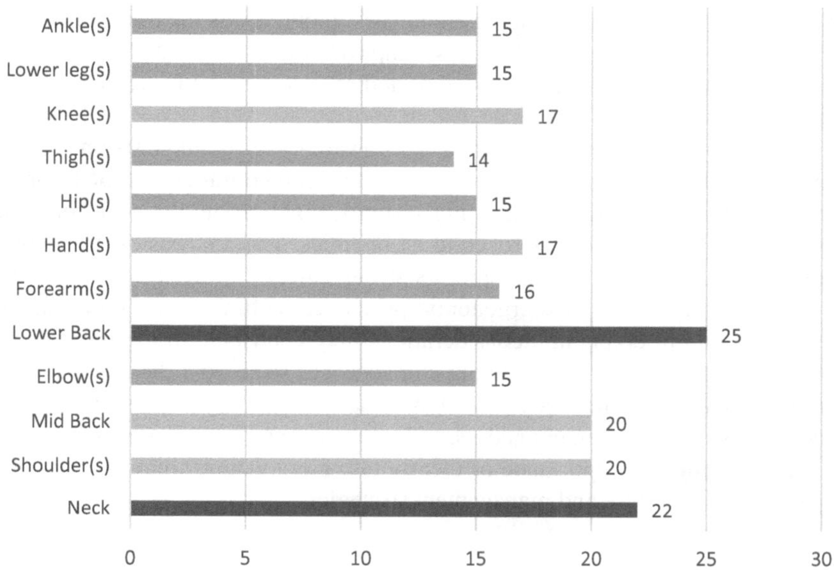


**FIGURE 3.5** Injury/strain types in Muscat branch.

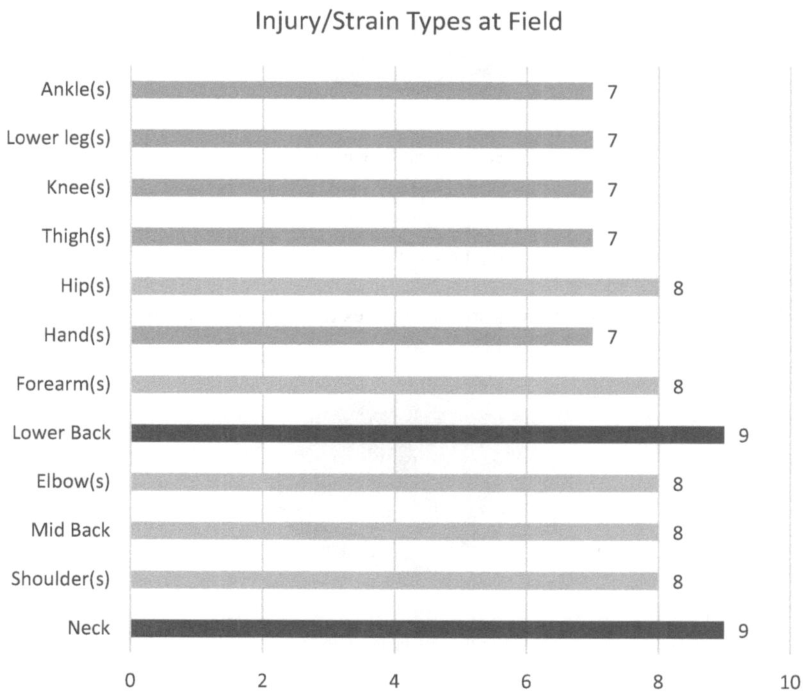


**FIGURE 3.6** Injury/strain types in field office.

The survey also revealed what the employees' thought was the cause of the ergonomic and the results showed that most employees believed that the injuries and strains are caused by working for a long period of time without movement and poor workstation design.

The study analysis reveals a severe lack of awareness and emphasis on ergonomics in the oilfield. This deficiency is evident in the limited integration of ergonomic principles in health and safety reporting systems, highlighting the need for a more comprehensive and targeted approach to ergonomics. The analysis of the reporting system also demonstrates that, despite its strong safety culture, the primary company has yet to fully embrace ergonomic principles within its safety management system. This gap in ergonomic considerations may contribute to the prevalence of work-related injuries and issues among employees as seen from the internal survey (Figures 3.7 and 3.8). The study also shows that the primary causes of ergonomic injuries and issues are often not addressed in the company's safety management system, underscoring the importance of incorporating ergonomics into the core of the company's safety culture and management strategies.

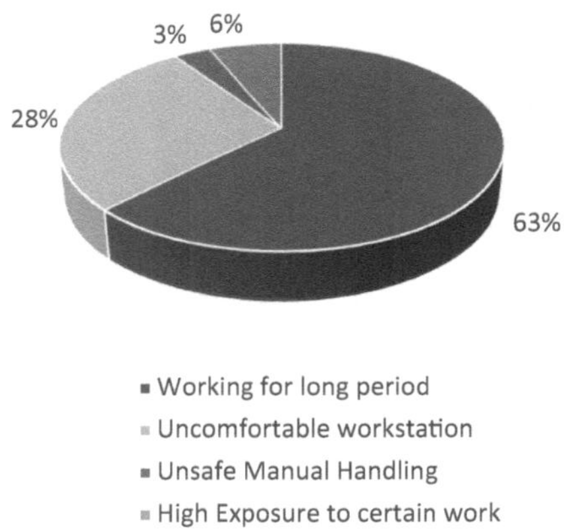


**FIGURE 3.7** Causes of ergonomics strains/injuries in corporate branch.

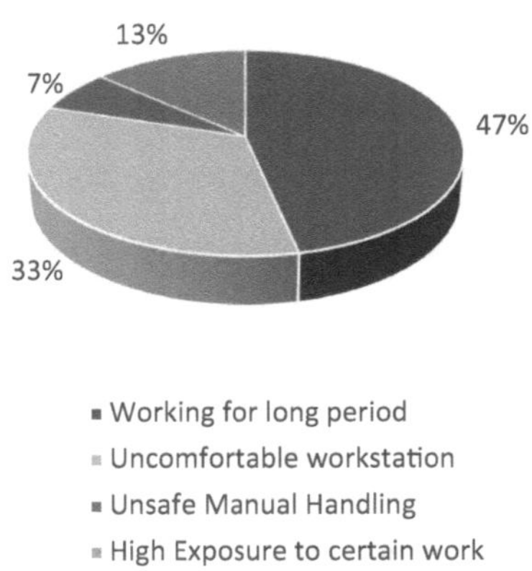


**FIGURE 3.8** Causes of ergonomics strains/injuries in field office.

Considering these findings, it is essential to prioritize ergonomics and integrate its principles into their safety management systems. By doing so, companies can reduce work-related injuries, enhance employee well-being, and improve overall productivity and performance. Moreover, this serves as a starting point for further research and exploration of ergonomics in the oilfield, with the goal of fostering a culture of safety and responsibility that benefits both employees and the industry.

## 3.5 CONCLUSION

This study has delved into the critical aspects of ergonomics in the oilfield, with a particular focus on a target oil and gas exploring company in Oman. The primary aim and objectives were to assess the current awareness and implementation of ergonomics in the oilfield and provide practical recommendations for continual improvement. The findings revealed a gap and a significant lack of awareness and understanding of ergonomics within the oilfield sector.

The analysis of the observations highlights recommendations that are imperative to the oil and gas industry in Oman. The importance of proper posture while working should be emphasized, as well as encouraging the employees to take regular breaks to avoid stress and fatigue. The client and the contractors should engage in regular ergonomic assessments as an initiative-taking monitoring measure to evaluate the health and safety performance. Furthermore, the integration of ergonomics as a separate and essential entity within the reporting system can significantly enhance the focus on ergonomic practices within the company. The findings and recommendations from this study will enable the oil and target company and other oil and gas companies to enhance their commitment to employee well-being and foster a culture of safety and health for the benefit of all stakeholders. In addition, the workstation comfort levels must be enhanced, incorporating ergonomics as an essential element in the health and safety reporting system to include reporting of ergonomic issues, periodically monitoring the effectiveness of ergonomic intervention in work activities, etc. Implementing these actions in the oil and gas companies can enhance their employees' well-being, reduce the risk of work-related injuries, and improve overall productivity.

## REFERENCES

Al-Abri, S. S. (2017). *Ergonomics Practices in PDO: A Case Study*. Muscat: Sultan Qaboos University.

Al-Balushi, S., & Dutta, P. K. (2016). Integrating ergonomics into process safety management: A case study in the Oman oil and gas industry. *Safety Science*, *88*, 170–183. https://doi.org/10.1016/j.ssci.2016.05.013

Al-Balushi, Y. M., Al-Jardani, K., & Al-Abri, S. S. (2019). The impact of ergonomic training on workers' knowledge and practices in the oil and gas sector in Oman. *Applied Ergonomics*, *75*, 87–95. https://doi.org/10.1016/j.apergo.2018.09.010

Al-Harrasi, A. S., Al-Abri, S. S., & Al-Maashani, A. (2018). Musculoskeletal disorders among oil and gas workers in Oman: A cross-sectional study. *Oman Medical Journal*, *33*(2), 104–110. https://doi.org/10.5001/omj.2018.20

Al-Maashani, A. S., Al-Jardani, K. S., & Al-Abri, S. S. (2020). Health and safety management systems in the oil and gas industry in Oman: A case study. *Journal of Loss Prevention in the Process Industries*, *63*, 103975. https://doi.org/10.1016/j.jlp.2019.103975

Al-Hemoud, A., Al-Asousi, M., & Al-Qashan, H. (2015). Ergonomics in the Kuwaiti oil and gas sector: An assessment of work-related musculoskeletal disorders. *Safety Science*, *73*, 28–34. https://doi.org/10.1016/j.ssci.2014.11.010

Carayon, P., & Smith, M. J. (2000). Work organization and ergonomics. *Applied Ergonomics*, *31*(6), 649–662. https://doi.org/10.1016/S0003-6870(00)00040-5

Dul, J., Bruder, R., Buckle, P., Carayon, P., Falzon, P., Marras, W. S., ... & van der Doelen, B. (2012). A strategy for human factors/ergonomics: Developing discipline and profession. *Ergonomics*, *55*(4), 377–395. https://doi.org/10.1080/00140139.2012.661087

Ghasemi, F., Kalatpour, O., Rezaei, M., & Azghani, M. R. (2017). Ergonomics intervention in an Iranian oil refinery. *Work*, *56*(2), 323–328. https://doi.org/10.3233/WOR-172503

Hämäläinen, P., Takala, J., & Saarela, K. L. (2006). Global estimates of occupational accidents. *Safety Science*, *44*, 137–156. https://doi.org/10.1016/j.ssci.2005.08.017.

Helander, M. G. (2006). *A Guide to Human Factors and Ergonomics* (2nd ed.). CRC Press.

Hignett, S., & Wilson, J. R. (2004). The role for qualitative methodology in ergonomics: A case study to explore theoretical issues. *Theoretical Issues in Ergonomics Science*, *5*(6), 473–493. https://doi.org/10.1080/14639220412331303382

Hovden, J., Lie, T., Karlsen, J. E., & Alteren, B. (2008). The safety representative under pressure. A study of occupational health and safety management in the Norwegian oil and gas industry. *Safety Science*, *46*(3), 493–509. https://doi.org/10.1016/j.ssci.2007.06.018

International Ergonomics Association (IEA). (2000). *What Is Ergonomics*? Retrieved from https://iea.cc/whats/

Jain, R., Sain, M. K., Meena, M. L., Dangayach, G. S., & Bhardwaj, A. K. (2014). Ergonomic risk factors and their association with lower back and neck pain among pharmaceutical workers in India. *Work*, *49*(4), 725–732. https://doi.org/10.3233/WOR-131803

Karwowski, W. (2006). The discipline of ergonomics and human factors. In G. Salvendy (Ed.), *Handbook of Human Factors and Ergonomics* (3rd ed., pp. 3–33). John Wiley & Sons.

O'Hara, J. M., Higgins, J., Persensky, J. J., & Tiller, M. (2016). Forty years of research on human factors and ergonomics in the nuclear industry. In P. Arezes (Ed.), *Advances in Safety Management and Human Factors* (pp. 3–15). Springer. https://doi.org/10.1007/978-3-319-41929-9_1

Petroleum Development Oman. (2018). *Health, Safety and Environment*. Retrieved from https://www.pdo.co.om/en/about-pdo/Pages/HSE.aspx

Saurin, T. A., Guimarães, L. B. M., Costella, M. F., & Ballardin, L. (2008). An algorithm for classifying error types of front-line workers based on the error management theory: A study in the construction industry. *Safety Science*, *46*(7), 1003–1018. https://doi.org/10.1016/j.ssci.2007.12.006

# 4 Effect of Personality Traits as Human Factors on Academic Stress of Students

## *An Exploratory Study Using PLS-SEM*

*Bhanupriya Khatri, Charu Saxena, Megha Sharma, Shad Ahmad Khan, and Anand Nayyar*

## 4.1 INTRODUCTION

In the current volatile times of ever-changing needs of the industry, anticipations of the students, aspirations of parents as well as the outcome expectations of the regulatory bodies, the education system all around the world are focusing towards outcome-based education (Davis, 2003). The continuous shift towards more involvement in practical learning and inclusion of the number and variety of assessments in the academic curriculum is rapidly taking place (Spady & Marshall, 1991). Due to these multiple assessments with tight-roped timelines, and higher pressure of scoring better than others in every examination, the learners have become more prone to experiencing stress (Khan & Khan, 2019). This stress faced by students at the academic level can be categorized into five categories viz. frustration, conflict, burden, change, or self-imposed, and they give physiological, emotional, behavioral, or intellectual responses to these stressors (Misra & Castillo, 2004). Moreover, all the students do not have a homogeneous response, and it varies due to several sets of variables including their demographic profile, personality characteristics, and different attitudes and emotional responses to the environment (Hashmi & Naz, 2020). As the students possess distinctive personality traits, helping them learn to adapt to the academic environment, they may perceive different levels of academic stress (Soliemanifar & Shaabani, 2012). The present study explores the relationship between students' personality traits and academic stress.

Stress can be considered an inevitable part of one's life, but when it comes to students it may be multifaceted and can arise from both academic and non-academic activities, which can include social, environmental, cultural, and psychological factors (Klein, 2009; Zareen & Khan, 2023; Magd et al., 2023; Khan & Magd, 2023;

DOI: 10.1201/9781003532293-5 

Khatri et al., 2024). In this highly dynamic and competitive world, where there is cutthroat competition, students encounter various academic problems such as fear of examination, lack of attentiveness in attending classes, and incapability in understanding a subject. Academic stress includes a sense of powerlessness which creates mental distress about expected academic challenges or failures. Moreover, the sense of fear of possible academic failure and any hurdle between them also create distress among the students. The level of Stress may be in significant proportions of some students; however, not all of them may experience the same levels of anxiety with similar severity. Earlier research shows that only 10%–35% of students may practice intolerable stages of examination nervousness (Chapell et al., 2005).

Personality is the outlook or the preference of a person and the way a person thinks, feels, and behaves. Personality is a cluster of traits that refers to any characteristics with individual differences that are relatively permanent and consistent. Personality is considered the strongest predictor of academic success. Some of the personality characteristics are connected with well-being, happiness, discontent, grief, and distress. Big-5 Personality traits are similar patterns of behaviors demonstrated by people of a specific type of personality (Ezeakabekwe & Nwankwo, 2020).

The linkage between personality and academic stress is closely associated with each other. Students who are suffering from academic stress may show emotive and physical difficulties such as hopelessness, unease, nervousness, and stress-related conditions which show a dominating association with certain personality characteristics. The Big-5 Personality Model comes with personality characters (neuroticism, extraversion, openness to experience, agreeability, and conscientiousness) which have been defined as predictors for various aspects of academic development. Academic stress and its relationship with personality have been intensely studied in the past, as they're considered to be a strong agreement between these five traits of the Big-5 Personality Model characteristics and academic stress.

Studies have shown that personality predicts the coping mechanisms of a person with stressful situations (Connor-Smith & Flachsbart, 2007). Recent developments and disruptions in the academic world due to the epidemic of COVID-19 have contributed to a reinvigorated interest in the effects of personality traits on academic stress. Thus, this piece of research is an attempt to explore the impact of academic stress on the Big-5 Personality traits among students.

This study thus has five objectives addressing each factor of the big five personality traits

RO1: To explore the influence of Agreeableness on Academic Stress
RO2: To explore the influence of Conscientiousness on Academic Stress
RO3: To explore the influence of Extraversion on Academic Stress
RO4: To explore the influence of Openness on Academic Stress
RO5: To explore the influence of Neuroticism on Academic Stress

Understanding the interplay between personality traits, strategic orientation, and ergonomic design can lead to more effective interventions aimed at reducing academic stress, promoting student well-being, and enhancing academic performance. Tailoring approaches based on these factors allows for a more holistic educational experience, addressing both individual needs and broader environmental influences. Principles from human factors and ergonomics can optimize learning environments to accommodate diverse personalities, enhancing comfort and productivity (Khatri et al., 2023; Ligori et al., 2022; Magd & Khan, 2022). Understanding the interplay between these traits, strategic orientation, and ergonomic design allows for more effective interventions aimed at reducing academic stress and promoting student well-being (Ansari et al., 2024; Khan & Ansari, 2024; Zheng et al., 2020). By tailoring educational strategies to individual traits and fostering supportive environments, we can create a more holistic educational experience that addresses both personal needs and broader environmental influences, ultimately enhancing academic performance (Bocar et al., 2022; Khan et al., 2023; Khatri et al., 2024)

## 4.2 THEORETICAL BACKGROUND AND HYPOTHESES

The capability of persons to successfully handle life encounters has been a subject of great interest to psychologists and researchers. Stress is perceived as a negative emotional, intellectual, behavioral, and physiological development that ensues when coping up with stressors is attempted (Bernstein, 2013). One of the pioneering works James-Lange: theory of emotion (Lang, 1994) established that emotions follow physical responses to an event. The emotional experience of people is formed only based on what they interpret from that physical reaction. On the contrary, Cannon-Bard's theory of emotion (1927) ascertained those emotional responses and physiological responses occur at the same time. Later, Schachter-Singer's theory of emotion (two-factor theory) stated that emotions are the result of both physiological and cognitive processes. Physiological arousal occurs first, and then the individual must interpret the reason behind this from the past experience or from assessing the surrounding environment, this eventually gives rise to emotional experience.

The transactional model to cope up with stress (Lazarus & Folkman, 1987) is recognized to be more admissible to describe stress in humans. The human experience of stress is a dynamic process that is the cognitive appraisal, evaluation, and interpretation of the interactions among the individual and the environment. The individual, in constant transactions with the environment, experiences stress in the form of opinions, feelings, sentiments, and behaviors based upon the judgment that if the expectations exceed the personal resources, and his ability to hold these surpassed expectations. Moreover, these judgments are bound to change as time passes because of adjustments in the requirements, coping effectiveness, or personal abilities. A person undergoes two stages of cognitive evaluation.

Stage 1 of primary judgment is based on the significance of a particular encounter/situation to that person. The situation may be found to be as irrelevant, where the result is assumed to be favorable with pleasurable sentiments, or distress. Irrelevant and 'relevant, but positive' situations are of the least concern for the individual. Relevant but stressful conditions may further be assessed as challenging, threatening, or harmful/loss-making. Challenging situations are perceived as inciting and stimulating opportunities to prove oneself and the person is optimistic, enthusiastic, and poised for anticipating the achievement of benefits, controls, or personal progress through these. In case of a threat, the person perceives to be in danger, assesses insufficient capabilities to deal with stress, and anticipates future harm or loss i.e., damage to his physical body, valuable thing, or self-esteem. He strives to achieve a favorable outcome of the situation to attain or to restore his or her happiness. In harm/loss situations, some loss has already happened and due to his helplessness, he surrenders which may further lead to emotions of nervousness and despair.

In stage 2 of secondary appraisals, the person evaluates the available competence, support, options, and resources to cope with the situation, and looks forward to adapting to the new conditions and to reestablishing equilibrium between the person and environment. The intensity of stress and emotional response is subject to how much control an individual feels to have, actions likely to improve the circumstances, and the involvement of the stakeholders. In circumstances of inadequate control or high outcome stake, the stress response would be high. In these evaluations, the role played by situational stressors (Terry, 1991) i.e., formal properties, such as uniqueness, event indecision, vagueness, and temporal facets of the stressing conditions, is also predominant.

### 4.2.1 Academic Stress

Stress surfaces when any burden goes beyond the assets in the hands of a person. Academic pressure denotes the stress practiced by students to compete in the examinations. The severe and continuing nature of this may affect the academic performance of a student and in turn his ability to participate in productive curricular and extra-curricular activities and events. This may further give rise to other negative behaviors, including substance abuse (Richlin-Klonsky & Hoe, 2003). Person-Environment Model of stress puts forward, two distinct effects of stressful events on the students. Positively, the challenging aspect of academic goals exerts some stress, which subsequently leads to enhanced individual competence and a higher level of learning capability. Negatively, academic targets are taken as a threat which creates hopelessness and thereon reduces academic performance (Misra & McKean, 2000). Ghazanfari et al. (2020) examine the consequences of mental stress, which creates problems of nervousness, unhappiness, and illness. Distress is the situations that disrupt, or threaten to disrupt, individuals "daily functioning and cause people to make adjustments" (Auerbach & Gramling, 1998). Distress is the procedure of altering to or handling situations that interrupt or portend to disturb an individual's physical or mental running (Selye, 1976). Pressure amongst medical students is reasonable to high. Initial identification of the aspects instigating stress factors may be accommodated in averting upcoming ailment among students Krithika and Rajam (2020). The education may offer improved perceptions to the academic administrators for dropping the concentration of abstract pressure.

### 4.2.2 Personality Traits in the Big-5 Personality Model

Personality characteristics are articulated in learning styles, which are incorporated in learning tactics, which results in a learning outcome in the time being (De Raad & Schouwenburg, 1996). A Personality characteristic is not a single or meticulous behavior (a method of connecting an individual's solitary presentation) but just a

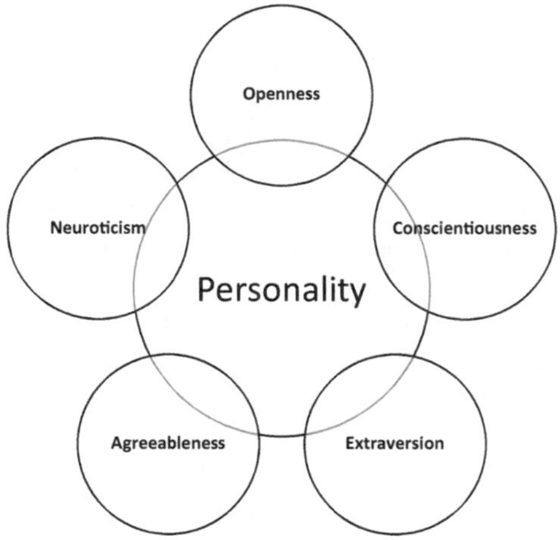


**FIGURE 4.1** Big five personality traits.

summary of activities which are linked and acknowledged by the person who is observing the dependability of such plan from conditions to conditions (Funder & Colvin, 1991). The Big Five model is known to be a cohesive framework as presented in Figure 4.1. for contributing a strong and comprehensive basis for human personality (Vollrath, 2001).

As visible from Figure 4.1, the different traits of this model are explained below:

**Openness**: The first trait of the big-5 model is 'openness to experience'; people with this trait are imaginative and open to exploring new things with divergent and flexible thinking and are engaged in self-learning and thoughtfulness (McCrae, 1987). High openness personality trait people demonstrate creative inclinations, intellectual inquisitiveness, insights, and intuition (Brown, 2009). They are generally unconventional, keen to challenge authority, and always geared up for questioning new philosophies. They also possess an elevated level of imagination, creativity, resourcefulness, and emotional responsiveness to others.

**Conscientiousness:** The second trait is 'conscientiousness' which refers to a person who is hard-working, careful, reliable, self-disciplined, self-directed, self-oriented, and self-managed. People high in this trait are positive fellow social rules, norms with moral obligations, and ethical commitments. This trait is closely associated with predicting academic outcomes in the educational system (Duckworth et al., 2007). Conscientiousness comprises individual talents such as thoroughness, self-control, and individual capability, and people who are conscientious finish their work devotedly, work meticulously, support others, and receive additional accountability (Abraham, 2004). Moreover, highly conscientious personalities are rational, dependable, and risk handlers.

**Extraversion:** The third big five trait is 'extraversion' which is considered an eminent personality predictor traits such as affable, talkative, lively, assertive, adventure lover, etc. The opposite end of this trait is featured as introversion with noiseless, reserved, independent, and even-paced characters (Costa & McCrae, 1992).

Extravert personalities are associated with a more optimistic, energetic, and outgoing nature. The students exhibiting this trait are highly upbeat, motivated, cheerful, and positive. They possess a component of interpersonal dominance, a characteristic that helps them to change and control others (Bono & Judge, 2004).

**Agreeableness:** The fourth trait is 'agreeableness', people with this trait are easily going along with people who are closely associated with positivity, social perception, and behavior such as altruism, teamwork, conviction, and modesty. Agreeableness may also be explained as an attribute of personality, encompassing aspects like confidence, affection, kindness, collaboration, humility, etc. (McCrae & Terracciano, 2005). Derue et al. (2011) theorized agreeableness as being amicable and supportive of others' strong points. An Agreeable person is compassionate and sensitive towards others and their needs.

**Neuroticism:** The last trait in the big-5 model is 'neuroticism **or** emotional stability,' which is the competence to be calm and stable under stress and respond with resilience and cheerfulness when faced with hurdles, encounters, transformations, and ambiguities. People with low levels of this trait are neurotic, evenly tempered, aggressive, and low on emotional resources, tend to perceive events as threatening or undesirable, and are often implicated in occupational stress and burnout (Eschleman et al., 2010). This implies a lower tendency of vulnerability to experience undesirable emotional states that are fear, anger, and irritation. Rather, neurotic personalities exhibit more emotive volatility and self-consciousness (Żurko et al., 2022). Neurotics hold traits, including annoyed, worried, unsociable, anxious, nervous, uncertain, disbelieving, insecure, terrible, and disappointed (Judge & Bono, 2000). Neuroticism escorts an exposure to emotional suffering and specifies a tendency to grasp impracticable philosophies, incompetent conduct of conduct with pressure, and an incapability to control desires (Borja et al., 2009).

## 4.3 PERSONALITY TRAITS, BIG FIVE MODEL, AND ACADEMIC STRESS

The traits in Big-5 Personality Model are eminent in all aspects of distress among people (Vollrath, 2001). The personality traits of students suggest in the direction of feeling a high level of stress resulting in depressed academic performance. Usually, people's reaction to their academic position is in arrangement with the personality, and these personality behaviors are fragments of their theoretic lifetime, which cannot be transformed easily. When the achievement criteria of a learner and behavior feature intersection, and therefore a scholar cannot answer consequently, then it generates a lot of stress on students that can develop an additional educational pressure, anxiety, and stress. Klimstra et al. (2010) inspected the association among the Big-5 personality factors with depression. Whereas depression is not the identical degree of pressure, stress plays a part in the knowledge of despair. This research also creates negative links between traits of personality and depression. Though, no linkage is found on openness to experience and despair.

Dobewall et al. (2014) have conducted one research to discover whether one can find out another person's principle exactly, or whether ideals are too subjective for assessment. They associated self-other promise in individual values with an agreement in the five personality qualities. The consequences endorsed that people could evaluate the morals of others whom they recognize well with extraordinary precision. Consequently, other ratings of individual values can be pragmatic to authenticate and accompaniment self-report value actions. Gadzella et al. (2012) assessed the connection between college students and stress that facilitated them to determine a relationship between the two. A noteworthy negative correlation between stress and academic performance was discovered. Students who notched greater on stress had poorer educational accomplishment than those pupils who gained a decrease on points of stress. With these conclusions in the way of reasoning, it is also important to recognize the cause of stress that is distinctive to students for college.

Khodaei and Shokri (2009) have found that better psychological well-being was being expected via elevated notches in extraversion aspect and bad scores in neuroticism aspect along with the sentiment-founded managing style. Heine and Buchtel (2009) report that the Big-5 Personality Model Factor Theory is a commonly applied instrument utilized as a method for recognizing individuality by analyzing the diverse aspects of personality. Zaidi et al. (2013) conducted one research study on faculty members and concluded that all the big five traits are linked with work engagement. This means that engaged academicians are expressive, considerate, entrusting, established, satisfied, productive, and exceptional. A great academician desires to acquire all the afore-stated attributes. Another landmark research explored the linkages among personality traits, academic stress, and academic performance in scholars, and a noteworthy negative correlation was observed among extraversion and conscientiousness and academic stress. Again, a substantial correlation was confirmed between neuroticism and academic stress (Shokri et al., 2007).

Duckworth and Yeager (2015) denoted personality characteristics as instruments, that are employed to assess the leader's attributes, ethical principles, and capabilities, which he manages to improve his accomplishment and endeavors in the direction of the success of the organizational objective. Personality traits assist managers to succeed the greater accomplishment, prominent to the attainment of the objective of the association. The accomplishment of the company can be measured as when a company accomplishes its ambitions within a specified time duration. Zhang (2013) stated that scholars who favored organized knowledge situations and instinctive dispensation are inclined to nervousness and fear, whereas those wishing a futuristic and realist style are further extrovertive. Strati et al. (2017) also discovered the link of student engagement and accomplishment, concentrating on the linkage of students' apparent encounters, educator care, educator obstacles, and student rendezvous. Eryilmaz (2014) found that being an operative instructor necessitates definite behavior qualities, but he also resolved that limited pieces of training have been attended on the association between definite personality behaviors of instructors to the academic accomplishment of scholars.

The personality traits such as conscientiousness and agreeableness have the likelihood to meaningfully narrate to academic accomplishment. The outcome indicates that students possessing a good level of conscientiousness and agreeableness achieve better than those less (Chowdhury & Amin, 2006). So, the researchers have formed the assumption of a relationship between agreeableness and academic Stress which leads to setting this research hypothesis, as given below:

H1: There is a significant influence of Agreeableness on Academic Stress.

Colquitt et al. (2000) stated conscientiousness as the solitary feature that influences the person's inspiration to acquire knowledge and learning. Trapmann et al. (2007) observed that Conscientiousness was the greatest influence on performance for college students. Thus, the following hypothesis is framed and evaluated in the research.

H2: There is a significant influence of Conscientiousness on Academic Stress.

It cannot be specified with the assurance that extraversion conveys an advantage for stress; somewhat, it can be decided that they can occur together. Also, as long as these cross-sectional designs are unsuccessful to regulate the state optimistic effect, an issue that is frequently linked with extraversion and stress responses, the process of the co-occurrence is further complicated (Costa & McCrae, 1980). Thus, the researchers proposed the below-stated hypothesis:

H3: There is a significant influence of Extraversion on Academic Stress.

Openness is the mark of academic interest, imagination, and an orientation for innovation and variability. Numerous research studies have originated the optimistic linkage between openness and academic accomplishment (Lounsbury et al., 2003). So, researchers established the following hypothesis

H4: There is a significant influence of Openness on Academic Stress.

People having a good level of neuroticism may be further probable to appraise stressors as pressures in its place as encounters, growing the likelihood of sensing adverse effects as a reply (Suls, 2012). Individuals with high neuroticism incline to emphasize added to the undesirable structures of demanding actions than individuals little in neuroticism (Hemenover, 2001). Thus, the proposed hypothesis is put forward as given below:

H5: There is a significant influence of Neuroticism on Academic Stress.

## 4.4 METHODS

The study is conducted among the students of business management from the universities of India. Both public and private universities were targeted for data collection. The simple random sampling technique has been utilized for collecting the responses of the students using a questionnaire with seven-point Likert scale. Two standardized questionnaires were employed for gauging personality and measuring the level of academic stress. The questionnaire was circulated among the undergraduate and post-graduate students of Business School of Management, accompanied by a cover letter. Out of 370 questionnaires administered of each type to students of different Indian universities, a total of 335 students from the business schools completed both of the questionnaires. A conceptual model framework is proposed for understanding the linkages among Personality Traits and Perception of Academic Stress. Based on the planned research framework the latent constructs and the indicators are adopted from past studies, as shown in Figure 4.1. Structural equation modeling is applied to analyze the results.

## 4.5 RESULTS AND DISCUSSION

To compute the partial regression in the measurement and structural model PLS-SEM, (Hair et al., 2011; Mateos-Aparicio, 2011) has been used.

### 4.5.1 Demographic Profile of Respondents

Based on the participants' information in Table 4.1, maximum number of students (71.6%) lie in the range of 21–25 years of age, out of the total 335 respondents. Among the total participants, 63.2% participants are male and the rest of the 36.7 students are females. 66.8% students belong to India whereas 33.1% are international students of the university whose responses are collected. 64.7% of students are facing enough academic stress and 29.2% are facing not much academic stress. The international students may be facing more stress due to additional strain of learning different languages, diverse cultural values in addition to the adoption of the new academic curriculum, so they are also taken as respondents in the sample under study (Mori, 2000). Thus, it has become crucial to analyze the role of personality traits in leading academic stress among students.

**TABLE 4.1**
**List of Indicators with Their Factor Loadings**

| Construct | Indicators | Description | Factor Loadings |
|---|---|---|---|
| **Extraversion** | EXT 1 | Talkative | 0.902 |
| | EXT 2 | Reserved | 0.915 |
| | EXT 3 | Full of energy | 0.923 |
| | EXT 4 | Generates a lot of enthusiasm | 0.916 |
| **Conscientiousness** | CONS 1 | Thorough completion of job | 0.909 |
| | CONS 2 | Somewhat careless | 0.873 |
| | CONS 3 | Reliable and hard worker | 0.906 |
| | CONS 4 | Disorganized | 0.759 |
| **Agreeableness** | AGG1 | Nitpicker | 0.737 |
| | AGG 2 | Helpful and unselfish with others | 0.877 |
| | AGG 3 | Quarrelsome | 0.889 |
| | AGG 4 | Forgiving nature | 0.903 |
| **Openness** | OPP 1 | Innovative | 0.897 |
| | OPP 2 | Inquisitive | 0.901 |
| | OPP 3 | Deep thinker | 0.886 |
| | OPP 4 | Inventive | 0.725 |
| **Neuroticism** | NEU 1 | Emotionally stable, | 0.899 |
| | NEU 2 | Relaxed, handles stress well | 0.909 |
| | NEU 3 | Tensed | 0.874 |
| **Perception of academic stress** | STRESS 1 | Competition is intense | 0.757 |
| | STRESS 2 | Teachers are concerned | 0.857 |
| | STRESS 3 | Unrealistic expectations of parents | 0.896 |
| | STRESS 4 | Unrealistic expectations of teachers | 0.901 |
| | STRESS 5 | Time allocated to classes is not enough | 0.872 |
| | STRESS 6 | More work assignments | 0.846 |
| | STRESS 7 | Examination times are stressful | 0.711 |
| | STRESS 8 | Fear of failure | 0.831 |

### 4.5.2 Measurement Model Assessments

The scale is settled afterward blending the literature linked to the personality traits and academic stress of the students, shown in Table 4.2.

To check the convergent validity of data, the Average variance extracted (AVE) are found to be greater than the required cut-off of 0.5 (Hair et al., 2019; as shown in Table 4.3. The composite reliability of each construct is greater than 0.8, indicating the overall scale reliability (Henseler et al., 2016). As depicted in R-square value of the structural model which is 60.3 for 'Academic Stress,' indicates that the proposed conceptual model has higher explanatory significance and the personality factors have an adequate impact on stress among the students.

The values of CR for each construct (shown in Table 4.4) is more than the suggested limit point 0.7 suggested by Malhotra and Dash (2010). To calculate the CV of the scale that deals with the amount to which the observed variables (that constitute a resolute scale) congregate on the alike construct (Casaló et al., 2008;

**TABLE 4.2**
**Demographic Profile of Learners**

| Demographic Variable | Category | Frequency | Percentage |
|---|---|---|---|
| **Age** | 18–21 | 90 | 26.8 |
| | 21–25 | 240 | 71.6 |
| | Over 25 | 5 | 1.4 |
| **Gender** | Male | 213 | 63.6 |
| | Female | 122 | 36.4 |
| **Nationality** | Indian | 224 | 66.8 |
| | Other | 111 | 33.1 |
| **Level of education** | UG | 90 | 26.8 |
| | PG | 245 | 73.1 |
| **Academic stress** | None | 20 | 5.9 |
| | Not much | 98 | 29.2 |
| | Enough | 217 | 64.7 |

**TABLE 4.3**
**Content Validity**

| Constructs | Adapted From |
|---|---|
| Agreeableness | McCrae and Terracciano (2005) |
| Conscientiousness | Deary et al. (1996), Vollrath (1999), Mark and Ganzach (2014) |
| Extraversion | Deary et al. (1996), Vollrath (1999), and Goldberg (2013) |
| Neuroticism | Matthews et al. (2003) and McCrae and Costa (1985) |
| Openness | Email and Wohlfarth (1991) and Floros and Siomos (2014) |
| Academic stress | Pritchard and Wilson (2003) and Saklofske et al. (2007) |

**TABLE 4.4**
**Construct Reliability, Validity and Cronbach's Alpha**

| Constructs | Cronbach's Alpha | CR | AVE |
|---|---|---|---|
| Agreeableness | 0.87 | 0.91 | 0.729 |
| Conscientiousness | 0.88 | 0.92 | 0.746 |
| Extraversion | 0.93 | 0.95 | 0.835 |
| Neuroticism | 0.87 | 0.92 | 0.799 |
| Openness | 0.87 | 0.91 | 0.731 |
| Stress | 0.88 | 0.92 | 0.613 |

**TABLE 4.5**
**Fornell–Larcker Criterion Results**

| | AGG | CONS | EXT | NEU | OPP | STRESS |
|---|---|---|---|---|---|---|
| **AGG** | 0.854 | | | | | |
| **CONS** | 0.786 | 0.864 | | | | |
| **EXT** | 0.735 | 0.621 | 0.914 | | | |
| **NEU** | 0.729 | 0.723 | 0.747 | 0.894 | | |
| **OPP** | 0.581 | 0.599 | 0.535 | 0.605 | 0.855 | |
| **STRESS** | 0.622 | 0.506 | 0.627 | 0.654 | 0.681 | 0.783 |

Malhotra & Dash, 2010). Also in Table 4.4, all AVE values are informed to be beyond 0.8. Also, the values CR for each construct are greater than its AVE, which is satisfactory (Casalo et al., 2008) as shown in Table 4.4. In addition to check the reliability, Cronbach Alpha was used which is greater than 0.80 for all the latent constructs.

To examine the reliability of each item, the cross-loadings are assessed. The values of factor loading are high on their respective constructs, which proves the reliability of every item as good and offers strengthening to the distribution for every item on the enumerated latent construct. Table 4.5 shows that the items load higher correlation with their own constructs. They also have correlation with other constructs but are less than 0.50. Thus, it is clear that no item is loaded to two or more constructs simultaneously. The VIF measures for each construct are found to be less than five, which proves no multi-collinearity problem in the data (Hair et al., 2011). The coefficient of determination for stress is found to be 0.603, which explains the 60% variance in academic stress due to personality traits in the big five personality model.

PLS-SEM count on a non-parametric bootstrap technique (Efron & Tibshirani, 1986; Davison & Hinkley, 1997) to check the significance of path coefficients, Cronbach's alpha, and $R^2$ values. In this technique, sub-samples are generated annotations randomly drawn (with replacement) from the collected data. To confirm the constancy of outcomes 5,000 sub-samples are used in bootstraping, as suggested by

**TABLE 4.6**
**Path Coefficients**

| Hypothesis | | Std. Beta (β) | Mean Values (*M*) | Standard Deviation (SD) | *T-value* | *p-value* |
|---|---|---|---|---|---|---|
| **H1** | **Agreeableness → STRESS** | 0.240 | 0.241 | 0.073 | 3.307 | **0.001** |
| **H2** | **Conscientiousness → STRESS** | −0.237 | −0.238 | 0.061 | 3.920 | **0.000** |
| **H3** | **Extraversion → STRESS** | 0.172 | 0.175 | 0.066 | 2.592 | **0.010** |
| **H4** | **Neuroticism → STRESS** | 0.260 | 0.257 | 0.071 | 3.640 | **0.000** |
| **H5** | **Openness → STRESS** | 0.435 | 0.436 | 0.058 | 7.501 | **0.000** |

Hair et al. (2017), Davison and Hinkley (1997). In order to test the research hypothesis, the standardized beta values (**β**) of path coefficients are calculated by using the bootstrapping method, to generate standardized loadings and *t*-values (Aibinu & Al-Lawati, 2010), shown in Table 4.6.

The standardized beta values are the path coefficients of the structural model. The *t*-values for the three hypotheses H1, H2, H3, H4, and H5 are greater than 2.52, thus the null hypothesis is rejected. The results show that the personality characteristics, viz agreeableness, extraversion, neuroticism, and openness lead to academic stress ($p$ values less than 0.05), whereas Conscientiousness has a significant negative impact on academic stress ($p$ values less than 0.05, $\beta= -0.237$), which proves that High conscientiousness personalities are rational, dependable, and risk handler, as shown in Table 4.6. (Goldberg, 1990). Thus, it is concluded that the personality trait – Conscientiousness –refers to a person who is hard-working, careful, reliable, self-disciplined, self-directed, self-oriented, and self-managed; it does not lead to academic stress among the students.

However, the other personality traits viz Agreeableness, Extraversion, Neuroticism, and Openness have a significant impact on academic stress among university students. Stone-Geisser's $Q^2$ value has been calculated using the Blindfolding technique to gauge the predictive relevance. The $Q^2$ of latent variables in the PLS path model is attained by means of the blindfolding method, (Geisser, 1974), (Hair et al., 2017), (Stone, 1974). The value of $Q^2$ is 0.362 which is greater than 0.35 (Henseler et al., 2009) showing the large predictive relevance of the results.

## 4.6 COMPLIANCE WITH ETHICAL STANDARDS

The manuscript is prepared keeping in view of the scope of your esteemed journal. The study is concerned with examining the relationship between the personality traits and the academic stress of the students. The study is conducted in India after proposing a conceptual model. The results depict that the four personality traits, agreeableness, extraversion, neuroticism, and openness have a significant impact on academic stress, whereas the conscientiousness personality trait is not associated with academic stress among the students. The findings of the study depict that Personality Traits can help the students to pursue their career objectives.

Students have to work on the keen performance of knowledge, spirits and to present them efficiently. The consequences indicate that the four personality traits viz agreeableness, extraversion, neuroticism, and openness influence the academic stress; however, conscientiousness has no significant impact on academic stress. Informed consent was obtained from all individual participants included in the study. The authors declare that they have no conflict of interest.

## 4.7 MANAGERIAL IMPLICATIONS AND CONCLUSION

Academic Stress is a crucial element on which most of the students are tested. It is very significant to sharpen the personality traits and make the personality, effective so that the stress of the students can be balanced and controlled. Regular exercise, maintaining a timetable, and relaxation techniques may help the students to cope up with stress and make their personality more credible. The academicians can frame strategies for the personality development of the students.

Personality development comprises arrangements that progress responsiveness and uniqueness, development of aptitudes and possibility, framing human investment and simplifying employability, develop the quality of life, and contribute to thoughts, opinions, and determinations.

Personality development is a crucial component that students must obtain to overcome academic stress and maintain psychological well-being. Personality Traits can help the students to pursue their career objectives. Students have to work on the keen performance of knowledge, spirits and to present them efficiently. The consequences indicate that the four personality traits viz., Agreeableness, Extraversion, Neuroticism, and Openness are prime to academic stress, however, Conscientiousness has no significant impact on academic stress.

Fostering supportive settings and putting measures that improve well-being and academic achievement into practice are the managerial implications for tackling the impact of human-related factors on academic stress in students. By establishing mentorship programs, peer counseling, and promoting family participation to lessen isolation and offer emotional support, educational institutions can improve social support networks. Students' emotional regulation and ability to handle academic problems can be improved by incorporating emotional intelligence training into the curriculum and offering workshops on the subject.

In addition to providing mental health assistance through mindfulness and relaxation techniques, it is crucial to promote physical well-being through health programs, fitness activities, and education on nutrition and sleep. Institutions can also help students become more adept at time management by giving them the instruments, workshops, and materials they need to efficiently arrange their work and lessen the anxiety that comes with approaching deadlines. Schools should promote a collaborative culture that prioritizes individual academic improvement over competitiveness in order to lessen peer pressure and the anxiety that comes with comparing oneself to peers. Customizing academic support to address certain personality traits, such as procrastination or perfectionism, can assist students in setting more reasonable objectives and effectively managing their stress.

## 4.8 LIMITATIONS AND FUTURE SCOPE

The data were gathered from college students in the management field so the consequences cannot be generalized for students belonging to all fields. Thus, a future study can be done regarding the impact of personality-based characteristics of students on their academic stress, covering the students of all fields. Further, an additional set of variables such as work commitment, work satisfaction, work-life balance, and work-related stress can be taken into consideration while doing future research on HR policies during the outbreak of COVID-19.

## REFERENCES


Abraham, R. (2004). Emotional competence as antecedent to performance: A contingency framework. *Genetic, Social, and General Psychology Monographs*, *130*(2), 117–145.

Aibinu, A. A., & Al-Lawati, A. M. (2010). Using PLS-SEM technique to model construction organizations' willingness to participate in e-bidding. Automation in construction, 19(6), 714-724.

Ansari, M. S. A., Khan, S. A., & Bhuyan, U. (2024). Exploring the strategic orientation factors influencing the organizational performance through bibliometric analysis. In S. Umair, U. Waqas, B. Mrugalska, & I. R. Al Shamsi (Eds.), *Human Perspectives of Industry 4.0 Organizations* (pp. 68–88). CRC Press.

Auerbach, S. M., & Gramling, S. E. (1998). Stress management: Psychological foundations. *Scholarly Journal of Education*, Prentice-Hall, Inc.

Bernstein, C. (2013). *The relationship between patterns of sex role identity, work stress, social support and wellbeing in South African female managers* (Doctoral dissertation).

Bocar, A. C., Khan, S. A., & Epoc, F. (2022). COVID-19 work from home stressors and the degree of its impact: Employers and employees actions. *International Journal of Technology Transfer and Commercialisation, 19*(2), 270–291. https://dx.doi.org/10.1504/ijttc.2022.124349

Bono, J. E., & Judge, T. A. (2004). Personality and transformational and transactional leadership: A meta-analysis. *Journal of Applied Psychology*, *89*(5), 901.

Borja, S. E., Callahan, J. L., & Rambo, P. L. (2009). Understanding negative outcomes following traumatic exposure: The roles of neuroticism and social support. *Psychological Trauma: Theory, Research, Practice, and Policy*, *1*(2), 118.

Brand, H. S., & Schoonheim-Klein, M. (2009). Is the OSCE more stressful? Examination anxiety and its consequences in different assessment methods in dental education. *European Journal of Dental Education*, *13*(3), 147–153.

Brown, S. A. (2009). Personality and non-suicidal deliberate self-harm: Trait differences among a non-clinical population. *Psychiatry Research*, *169*(1), 28–32.

Casaló, L., Flavián, C., & Guinalíu, M. (2008). The role of perceived usability, reputation, satisfaction and consumer familiarity on the website loyalty formation process. *Computers in Human Behavior*, *24*(2), 325–345.

Chapell, M. S., Blanding, Z. B., Silverstein, M. E., Takahashi, M., Newman, B., Gubi, A., & McCann, N. (2005). Test anxiety and academic performance in undergraduate and graduate students. *Journal of educational Psychology*, *97*(2), 268.

Chowdhury, M. S., & Amin, M. N. (2006). Personality and students academic achievement: Interactive effects of conscientiousness and agreeableness on students performance in principles of economics. *Social Behavior and Personality: An International Journal*, *34*(4), 381–388.

Colquitt, J. A., LePine, J. A., & Noe, R. A. (2000). Toward an integrative theory of training motivation: a meta-analytic path analysis of 20 years of research. *Journal of Applied Psychology*, *85*(5), 678.

Connor-Smith, J. K., & Flachsbart, C. (2007). Relations between personality and coping: A meta-analysis. *Journal of Personality and Social Psychology*, *93*(6), 1080.

Costa Jr, P. T., & McCrae, R. R. (1992). The five-factor model of personality and its relevance to personality disorders. *Journal of Personality Disorders*, *6*(4), 343–359.

Costa, P. T., & McCrae, R. R. (1980). Influence of extraversion and neuroticism on subjective well-being: Happy and unhappy people. *Journal of Personality and Social Psychology*, *38*(4), 668.

Davis, M. H. (2003). Outcome-based education. *Journal of Veterinary Medical Education*, *30*(3), 258–263.

Davison, A. C., & Hinkley, D. V. (1997). *Bootstrap Methods and Their Application* (No. 1). Cambridge University Press.

De Raad, B., & Schouwenburg, H. C. (1996). Personality in learning and education: A review. *European Journal of Personality*, *10*(5), 303–336.

Deary, I. J., Blenkin, H., Agius, R. M., Endler, N. S., Zealley, H., & Wood, R. (1996). Models of job-related stress and personal achievement among consultant doctors. *British Journal of Psychology*, *87*(1), 3–29.

Derue, D. S., Nahrgang, J. D., Wellman, N. E., & Humphrey, S. E. (2011). Trait and behavioral theories of leadership: An integration and meta-analytic test of their relative validity. *Personnel Psychology*, *64*(1), 7–52.

Dobewall, H., Aavik, T., Konstabel, K., Schwartz, S. H., & Realo, A. (2014). A comparison of self-other agreement in personal values versus the big five personality traits. *Journal of Research in Personality*, *50*, 1–10.

Duckworth, A. L., Peterson, C., Matthews, M. D., & Kelly, D. R. (2007). Grit: Perseverance and passion for long-term goals. *Journal of Personality and Social Psychology*, *92*(6), 1087.

Duckworth, A. L., & Yeager, D. S. (2015). Measurement matters: Assessing personal qualities other than cognitive ability for educational purposes. *Educational Researcher*, *44*(4), 237–251.

Efron, B., & Tibshirani, R. (1986). Bootstrap methods for standard errors, confidence intervals, and other measures of statistical accuracy. Statistical science, 54-75.

Eryilmaz, A. (2014). Perceived personality traits and types of teachers and their relationship to the subjective well-being and academic achievements of adolescents. *Kuram Ve Uygulamada Egitim Bilimleri*, *14*(6).

Eschleman, K. J., Bowling, N. A., & Alarcon, G. M. (2010). A meta-analytic examination of hardiness. *International Journal of Stress Management*, *17*(4), 277.

Ezeakabekwe, S. U., & Nwankwo, E. A. (2020). Correlate of personality traits, self-efficacy and work copping among police personnel: the need for psychotherapy services for police personnel in Awka Metropolis. *International Journal for Psychotherapy in Africa*, *4*(1).

Floros, G., & Siomos, K. (2014). Excessive Internet use and personality traits. *Current Behavioral Neuroscience Reports*, *1*, 19–26.

Funder, D. C., & Colvin, C. R. (1991). Explorations in behavioral consistency: Properties of persons, situations, and behaviors. *Journal of Personality and Social Psychology*, *60*(5), 773.

Gadzella, B. M., Baloglu, M., Masten, W. G., & Wang, Q. (2012). Evaluation of the student life-stress inventory-revised. *Journal of Instructional Psychology*, *39*(2).

Geisser, S. (1974). A predictive approach to the random effect model. *Biometrika*, *61*(1), 101–107.

Ghazanfari, E., Kazemnejad, A., Feizi, A., Fesharaki, M. G., Dinu, I., Keshteli, A. H., & Adibi, P. (2020). The relationship between personality traits and psychosomatic complaints in a sample of Iranian adults. Journal of Affective Disorders, 261, 253-258.

Goldberg, L. R. (2013). An alternative "description of personality": The Big-Five factor structure. In Personality and personality disorders (pp. 34-47). Routledge.

Hair, J. F., Ringle, C. M., & Sarstedt, M. (2011). PLS-SEM: Indeed a silver bullet. *Journal of Marketing Theory and Practice*, *19*(2), 139–152.

Hair, J. F., Hult, G. T. M., Ringle, C. M., & Sarstedt, M. (2017). A primer on partial least squares structural equation modeling (PLS-SEM). Thousand Oaks.

Hashmi, A., & Naz, Q. (2020). Impact of Big five personality traits on Academic Achievement of Prospective Teachers. Journal of Arts & Social Sciences, 40–52.

Heine, S. J., & Buchtel, E. E. (2009). Personality: The universal and the culturally specific. Annual review of psychology, 60(1), 369-394.

Henseler, J., Ringle, C. M., & Sarstedt, M. (2016). Testing measurement invariance of composites using partial least squares. International marketing review, 33(3), 405-431.

Hemenover, S. H. (2001). Self-reported processing bias and naturally occurring mood: Mediators between personality and stress appraisals. *Personality and Social Psychology Bulletin*, *27*(4), 387–394.

Henseler, J., Ringle, C. M., & Sinkovics, R. R. (2009). The use of partial least squares path modeling in international marketing. In *New Challenges to International Marketing* (pp. 277–319). Emerald Group Publishing Limited.

Judge, T. A., & Bono, J. E. (2000). Five-factor model of personality and transformational leadership. *Journal of Applied Psychology*, *85*(5), 751.

Khan, S., & Khan, R. A. (2019). Online assessments: Exploring perspectives of university students. Education and Information Technologies, 24(1), 661-677.

Khan, S. A., & Ansari, M. S. A. (2024). Critical factors for relationship between strategic orientation, corporate success, and innovation. In S. Umair, U. Waqas, B. Mrugalska, & I. R. Al Shamsi, (Eds.), *Human Perspectives of Industry 4.0 Organizations* (pp. 89–110). CRC Press.

Khan, S., & Khan, R. A. (2019). Online assessments: Exploring perspectives of university students. *Education and Information Technologies*, *24*(1), 661–677.

Khan, S. A., & Magd, H. (2023). New technology anxiety and acceptance of technology: An appraisal of MS teams. In *Advances in Distance Learning in Times of Pandemic* (pp. 105–134). CRC Press, Taylor & Francis. https://doi.org/10.1201/9781003322252-5.

Khan, S. A., Magd, H., & Batra, M. (2023). An empirical investigation of higher education students' intention towards green products usage/purchase: What can educators do? *International Journal of Pluralism and Economics Education*, *14*(3/4), 297–314. https://dx.doi.org/10.1504/IJPEE.2023.10061177.

Khatri, B., Nandini, G., & Khan, S. A. (2024). Transition from outcome-based learning to global accreditation of top universities of India: Elevating the bar of educational excellence. In A. Naim, A. Saklani, S. Khan, & P. Malik (Eds.), *Evaluating Global Accreditation Standards for Higher Education* (pp. 278–292). IGI Global. https://doi.org/10.4018/979-8-3693-1698-6.ch019

Khatri, B., Saxena, C., Nandini, G., Khan, S. A., Alaskar, K. (2024). Inside The Gamer's Mind: A Comprehensive Exploration of Online Game Addiction on Psychological Wellbeing and The Moderating Effect Of Physical Activity. *Journal of Content, Community and Communication*, *21*, 49–62. https://www.amity.edu/gwalior/jccc/pdf/paper-4-game-addiction.pdf

Khatri, B., Shrimali, H., Khan, S. A., & Naim, A. (2023). Role of HR analytics in ensuring psychological wellbeing and job security: Learnings from COVID-19. In R. Yadav, M. Sinha, & J. Kureethara (Eds.), *HR Analytics in an Era of Rapid Automation* (pp. 36–53). IGI Global.

Khodaei A., & Shokri, O. (2009). Establishing a model for structural relationship among personality traits and copying styles with stress and mental health in male and female students. *New Psychological Research*, *4*(16), 39–46.

Klimstra, T. A., Akse, J., Hale III, W. W., Raaijmakers, Q. A., & Meeus, W. H. (2010). Longitudinal associations between personality traits and problem behavior symptoms in adolescence. *Journal of Research in Personality*, *44*(2), 273–284.

Krithika, A., & Rajam, K. (2020). IMPACT OF JOB SATISFACTION AND JOB STRESS AMONG THE WOMEN COLLEGIATE TEACHERS: AN EMPIRICAL STUDY USING SEM. *International Journal of Management (IJM)*, *11*(12), 4193–4201.

Lang, P. J. (1994). The varieties of emotional experience: A meditation on James-Lange theory. *Psychological Review*, *101*(2), 211.

Lazarus, R. S., & Folkman, S. (1987). Transactional theory and research on emotions and coping. *European Journal of Personality*, 1(3), 141–169.

Ligori, T. A. A., Suresh, N., Khan, S. A., Rabgay, T., & Yezer, K. (2022). The mediating effect of university image on the relationship between curriculum and student satisfaction: An empirical study of the Royal University of Bhutan. *International Journal of Pluralism and Economics Education. Inderscience, 13*(2), 192–208. https://dx.doi.org/10.1504/ijpee.2022.127218

Lounsbury, J. W., Sundstrom, E., Loveland, J. L., & Gibson, L. W. (2003). Broad versus narrow personality traits in predicting academic performance of adolescents. *Learning and Individual Differences*, *14*(1), 65–75.

Magd, H., Jonathan, H., & Khan, S. A. (2023). Education situation in online education before the pandemic and in the time of pandemic. In *Advances in Distance Learning in Times of Pandemic* (pp. 53–72). CRC Press, Taylor & Francis. https://doi.org/10.1201/9781003322252-3

Magd, H., & Khan, S. A. (2022). Effectiveness of using online teaching platforms as communication tools in higher education institutions in Oman: Stakeholders perspectives. *Journal of Content, Community and Communication, 16*, 148–160.

Malhotra, N. K., & Dash, S. J. M. R. (2010). An applied orientation. *Marketing Research*, *2*(4), 109–122.

Mark, G., & Ganzach, Y. (2014). Personality and Internet usage: A large-scale representative study of young adults. *Computers in Human Behavior*, *36*, 274–281.

Mateos-Aparicio, G. (2011). Partial least squares (PLS) methods: Origins, evolution, and application to social sciences. *Communications in Statistics-Theory and Methods*, *40*(13), 2305–2317.

Matthews, G., Deary, I. J., & Whiteman, M. C. (2003). *Personality Traits*. Cambridge University Press.

McCrae, R. R. (2003). *Personality in Adulthood: A Five-Factor Theory Perspective*. Guilford Press.

McCrae , R. R., & Costa, P. T. (1985). Updating Norman's" adequacy taxonomy": Intelligence and personality dimensions in natural language and in questionnaires. *Journal of Personality and Social Psychology*, *49*(3), 710.

McCrae, R. R., & Terracciano, A. (2005). Universal features of personality traits from the observer's perspective: Data from 50 cultures. *Journal of Personality and Social Psychology*, *88*(3), 547.

Misra, R., & Castillo, L. G. (2004). Academic stress among college students: Comparison of American and international students. *International Journal of Stress Management*, *11*(2), 132.

Misra, R., & McKean, M. (2000). College students' academic stress and its relation to their anxiety, time management, and leisure satisfaction. *American Journal of Health Studies*, *16*(1), 41.

Mori, S. C. (2000). Addressing the mental health concerns of international students. *Journal of Counseling & Development*, *78*(2), 137–144.

Pritchard, M. E., & Wilson, G. S. (2003). Using emotional and social factors to predict student success. *Journal of College Student Development*, *44*(1), 18–28.

Richlin-Klonsky, J., & Hoe, R. (2003). Sources and levels of stress among UCLA students. *Student Affairs Briefing*, *2*, 31–39.

Saklofske, D. H., Austin, E. J., Galloway, J., & Davidson, K. (2007). Individual difference correlates of health-related behaviours: Preliminary evidence for links between emotional intelligence and coping. *Personality and Individual Differences*, *42*(3), 491–502.

Selye, H. (1976). Further thoughts on "stress without distress". *Medical Times*, *104*(11), 124–144.

Shokri, O., Kadivar, P., Naghsh, Z., Ghanai, Z., Daneshvarpour, Z., & Molaei, M. (2007). Personality traits, academic stress, and academic performance. *Journal of Educational Psychology Studies*, *3*(3), 25–48.

Soliemanifar, O., & Shaabani, F. (2012). The relationship between of personality traits and academic burnout in postgraduate students. *Journal of Life Science and Biomedicine*, *3*(1), 60–63.

Spady, W. G., & Marshall, K. J. (1991). Beyond traditional outcome-based education. *Educational Leadership*, *49*(2), 67–72.

Stone, M. (1974). Cross-validatory choice and assessment of statistical predictions. *Journal of the Royal Statistical Society: Series B (Methodological)*, *36*(2), 111–133.

Strati, A. D., Schmidt, J. A., & Maier, K. S. (2017). Perceived challenge, teacher support, and teacher obstruction as predictors of student engagement. *Journal of Educational Psychology*, *109*(1), 131.

Suls, J. (2012). Affect, stress, and personality. In *Handbook of Affect and Social Cognition* (pp. 397–414). Psychology Press.

Terry, D. J. (1991). Coping resources and situational appraisals as predictors of coping behavior. *Personality and Individual Differences*, *12*(10), 1031–1047.

Trapmann, S., Hell, B., Hirn, J. O. W., & Schuler, H. (2007). Meta-analysis of the relationship between the big five and academic success at university. *Zeitschrift für Psychologie/Journal of Psychology*, *215*(2), 132–151.

Vollrath, M. (2001). Personality and stress. *Scandinavian Journal of Psychology*, *42*(4), 335–347.

Vollrath, M., Knoch, D., & Cassano, L. (1999). Personality, risky health behaviour, and perceived susceptibility to health risks. *European Journal of Personality*, *13*(1), 39–50.

Zaidi, N. R. et al. (2013). The big five personality traits and their relationship with work engagement among public sector university teachers of Lahore. *African Journal of Business Management, 7*(15), 1344–1353.

Zareen, S., & Khan, S. A. (2023). Exploring dependence of human resource management (HRM) on internet of things (IoT) and digital marketing in the digital era. In A. Naim & V. Devi (Eds.), *Global Applications of the Internet of Things in Digital Marketing* (pp. 51–66). IGI Global. https://doi.org/10.4018/978-1-6684-8166-0.ch004

Zhang, L. (2013). Does the big five predict learning approaches? *Personality and Individual Differences*, *34*, 1431–1446.

Zheng, J., Jiang, N., & Dou, J. (2020). Autonomy support and academic stress: A relationship mediated by self-regulated learning and mastery goal orientation. *New Waves-Educational Research and Development Journal*, *23*, 43–63.

Żurko, M., Słowińska, A., Senejko, A., Madeja-Bień, K., & Łoś, Z. (2022). Pandemic-activated psychological growth: significance of extraversion, self-consciousness and COVID-19 related anxiety. *Current Issues in Personality Psychology*, *11*(3), 182.

# 5 Synergizing GHRM and Human Factors

## *A Paradigmatic Approach in Contemporary Business*

*Priyanka Indoria, Mohd Danish Chishti, Abdullah Bin Junaid, and Ruchi Agarwal*

## 5.1 INTRODUCTION

The combination of Green Human Resource Management (GHRM) and Human Factors givesasupremeprospect to excelin routine business paradigms, indicating a transformative period of organizational synergy and sustainability. This approach embodies a meticulous commitment to both environmental supervision and the well-being of human resources, thereby developing apleasant ecosystem where ecological imperatives impeccablyentwined with human-centric methodologies (Douglas& Renwick, 2012). By aligning the strategies of GHRM with ergonomic considerations, the businesses can redefine organizational dynamics, where the intricate relation between environmental sustainability and the performance of human resource forms the foundation of progressive and resilient enterprises (Arulrajah et al., 2015; Hosain & Rahman, 2016).

The integration of GHRM and Human Factors engenders a regeneration of operational brilliance, underpinned by anintense understanding of the intricate nexus betweenenvironmental imperatives and human dynamics (Patrick Neumann & Dul, 2010). Through a careful incorporation of refined methodologies and theoretical frameworks, the businesses are hovering to unchainthe insights and good actionable strategies, propelling them towarda sustainable management practice. This interdisciplinary junction not only increases organizational effectiveness and employee well-being but also acts as a catalysttoward a broader societal shift to ethical and environmentally cognizant corporate citizenship. Thus enlightening a path toward anopportunity where prosperity is harmonized with environmental stewardship (Al-Minhas et al., 2020).

The purpose of this study is to illuminate the symbiotic potential inherent in the integration of GHRM and Human Factors within the milieu of contemporary business paradigms (Ren et al., 2018). The study aims to unveil a fresh insight into how the merger of environmental sustainability principles with ergonomic considerations can stimulate a re-constitutional change in organizational dynamics (Somaieh Alavi, 2023).

DOI: 10.1201/9781003532293-6

The study endeavors to propose actionable recommendations and best practices personalized to authorize businesses in navigating the multifaceted terrain of sustainable management (Douglas & Renwick, 2012).

## 5.2 LITERATURE REVIEW

### 5.2.1 Green Human Resource Management (GHRM)

The foundation of "Green Human Resource Management" can be accredited to Wehrmeyer's coinage in 1996. However, its recognition remains obscure among the common people, with limited comprehension even among those familiar with its existence. Regardless of sporadic advancements, mainly outside the Indian Territory, the understanding of GHRM's minutiae remains modest, necessitating acrucial call for heightened scholarly inquisition (Douglas& Renwick, 2012). Positioned as an evolvingbranch of learning within the extensive realm of management, particularly within HRM literature, there exists a pressing need for deep insight and knowledge regarding the nuances of GHRM (Chapol Ali et al., 2020).

Environmental preservation has been an escalating distress since the 1990s. This prompts the organizations and businesses to set off endeavors aimed at extenuating waste and pollution through the establishment of robust Environmental Management Systems (EMS). Strategies encompass a diverse range of measures, including the utilization of less-contaminated materials, innovative product designs, streamlined packaging, and the adoption of energy-efficient practices (Wong et al., 2013; Bibi, 2022). Scholars in the HRM domain have proffered diverse definitions of GHRM, conceptualizing it as an assimilation of HRM practices and strategies aimed at realizing social, financial, and ecological objectives, as articulated by Wikhamn (2019).

GHRM is a transformative process geared toward cultivating environmentally-conscious employees through the adoption of eco-friendly HRM practices and policies. Thus, it fosters collective benefits for society, businesses, individuals, and the environment (Ahmad, 2023). This paradigm shift toward GHRM entails the acceptance of numerous environmentally sustainable initiatives, ranging from electric filing and teleconferencing to recycling, ridesharing, and telecommuting (Mampra, 2013). GHRM practices havebeen acknowledgedas the critical drivers for boosting employee morale and job satisfaction. It plays a central role in fostering increased levels of employee engagement (Jacob, 2012). This progressive paradigm not only underscores organizational commitment to sustainable practices but also serves as a beacon of ethical leadership in an era marked by heightened environmental awareness (Jacob, 2012).

GHRM has challenges like change resistance and resource constraints but it promises substantial benefits that includeimproved corporate reputation, improved employee satisfaction, and fortified competitive edge (Kodua et al., 2022). Through the planned incorporation of GHRM principles, organizations find the way of a quicklybudding eco-conscious landscape, transcending standard HRM paradigms to createa synergistic relation between human capital management and environmental sustainability. On the other hand, the understanding of GHRM's transformative potential necessitates a combined effort to prevailover obstacles and embrace innovation in organizational practices (Hussain, 2020). By promoting a culture of continuous learning and leveraging specificproficiency, organizations can plot a courseof intricacies of GHRM implementation (Hussain, 2020). Moreover, the development of robust assessment frameworks and metrics enables organizations to determine the substantial impact of GHRM practices on both organizational performance and environmental outcomes. It facilitates informed decision-making and strategic alignment (Imran Tanveer et al., 2024). In this questfor sustainable fineness, GHRM emerges not purely as a strategic necessity but as a moral imperative, signaling a commitment to responsible administration of both human resources and the planet (Tsymbaliuk et al., 2021).

### 5.2.2 Human Factors: Psychological and Ergonomic Aspects

Human factors, synonymous with ergonomics, constitute a foundationincurrent business frameworks, embodying the scrupulouslearning of human communication with tasks, systems, and environments. Its soul lies in the synthesis of art and science. It aimed at optimizing workplace design, equipment, and processes to promote human performance, safety, and on the whole well-being (Karwowski & Zhang, 2021). This requisite field not only mitigates risks and improve sefficiency but also improves a culture of organizational efficacy through its holistic methodology. By astute the interaction between cognitive, physical, and environmental factors determining human behavior and performance, organizations can engineer innate and well-organized workspaces and processes that prioritize the satisfaction of employee and employee engagement. From manufacturing protocols to office layouts, human factors considerations seep intoeach organizational component, ensuring systems are customized with the end-user in mind, hence minimizing errorsand accidents and maximizing efficiency (Nayak, 2019).

### 5.2.3 Integration of GHRM and Human Factors with the Help of CaseStudies

Illustrating the principles from fields such as organizational behavior, environmental psychology, and industrial engineering, researchers have found how GHRM practices can be integrated with ergonomic interventions to createworkplaces that encourage sustainability asimprovingworkers' well-being and performance (Yong & Yusliza, 2020).

An ideal case study symbolic of the flawless integration of GHRM and Human Factors principles can be gleaned from the sustainability enterprise undertaken by **Interface Inc.,** a paramountworldwide entity in business flooring solutions. Embarking on their imaginativepursuit christened "Mission Zero," Interface set forth to eliminate its environmental footprint by the year 2020. Central to this futurist odyssey was the incisivesynthesis of GHRM practices with Human Factors considerations. Interface incisively discerned that cultivating a culture of sustainability necessitated not only environmental administration but also the comprehensive well-being and fanatical engagement of its human resources (Interface, 2022).

In order to accomplish this noble objective, the conglomeratecunningly engineered ergonomic workplace designs imbued with natural light, biophilic elements, and flexible workstations, thus improving an environment favorable to improved employee wellbeing and enhanced productivity, while concomitantly curbing the consumption of energy. Furthermore, Interface masterfully entwinessustainability benchmarks into its HRM functions, meticulously integrating them into recruitment, training, and performance management protocols to magnetize and retain talent fervently devoted to the company's environmental ethos. By immaculately aligning GHRM strategies with the veneratedbelief of Human Factors, Interface witnessed anauthenticphenomenon unfurl as it attained outstandingachievement in appeasing its environmental footprint, uplifting employee satisfaction and welfare to extraordinary heights, and promoting anenvironment pulsating with innovation and sustainability (Arulrajah et al., 2015).

Another case study highlighting the integration of GHRM and Human Factors can be brought into being in the renowned legacy of **Toyota Motor Corporation** and its pioneering strides in sustainable manufacturing practices.

Implementing the essence of environmental sustainability and employee well-being with steadydetermination, Toyota embarked on a holistic journey designed at infusing GHRM principles and Human Factors considerations into the fabric of its manufacturing processes. Pioneering aperiodic paradigm shift, Toyota dignified ergonomic workplace designs and lean production systems as the linchpin of its transformative ethos, meticulously refining workspaces, equipment, and tasks to minimize physical strain and accentuate productivity, all while curtailing waste and energy consumption. Additionally, Toyota skillfullyentangled sustainability metrics into its HRM operations, catalyzing employee contribution in eco-conscious accomplishments and lauding hand-outs to sustainable practices through prudently crafted performance evaluations and incentives. Through this combination of GHRM strategies with Human Factors imperatives, Toyota witnessed a suddenrise, registering immense enhancements in operational efficacy, product distinction, and employee morale, all asdeveloping a diminished ecologicalfootstep and cementing its repute as a model of corporate social responsibility (IvyPanda, Human Resource Management Structure of Toyota, 2023).

The abovementioned case studies focus on the organizations at the forefront of innovation, dexterouslybalancing environmental sustainability imperatives with human-centric place of work design and strategies of management. Through the judiciousacceptance of eco-friendly policies and practices, joined with the detailed optimization of workplaces, equipment, and processes to encourage employee health, safety, and efficiency, these imaginative entities illustrate the concrete benefits of integrating GHRM with Human Factors principles. By studying such case studies scholars bring togethervaluable insights and lessons, furnishing a treasure trove of managing for enterprises embarking on alikeintegration activities. Nevertheless, regardless of strides in conception, prominent gaps in research persevere (Wong et al., 2013).

A prominent shortfall lies in the insufficiency of empirical studies assessing the effect of integrated GHRM and Human Factors interventions and theiroutcomes. Additionally, an absence of research delving into the nuanced roles of organizational culture, leadership, and stakeholder engagement poses a considerablebarrier. Opportunities abound for interdisciplinary collaboration among environmental science, luminaries in HRM, ergonomics, and organizational psychology to create holistic frameworks for integrating GHRM and Human Factors.It thus propels organizations toward realizing the infinite potential of mixing sustainability and human-centric design paradigms in the place of work (Imran Tanveer et al., 2024).

### 5.2.4 Theoretical Framework: Development of a Conceptual Model for Synergizing GHRM and Human Factors

The flawless synthesis of GHRM and ergonomics finds a strong theoretical support in numerous significant perspectives that emphasize the resemblance between the objectives of environmental sustainability and human-centric design principles. The Ecological Systems Theory proposes that by improving eco-friendly work environments and ergonomic layouts, organizations can exercise a constructiveimpact on employee behavior and well-being, thusincreasing both environmental sustainability and output.

Additionally, the Resource-Based View (RBV) proposes philosophical insights into the strategic allotment of resources for attaining viable advantage. In the region of integrating GHRM and ergonomics, this theory explains that organizations own the ability to influence their human capital and environmental resources to carve sustainable business paradigms.

Organizations can promote employee contentment, restrict turnover rates, and eventuallyshape out a discrete competitive niche in the market landscape just by investing in GHRM initiatives and ergonomic infrastructure (Collins, 2021).

The Social Exchange Theory underlines the implication of give-and-take dynamics between individuals and the surroundings. This theory explains that organizations championing employee well-being and ecological administration will develop a nurturing organizational culture, magnetizing capable individuals who can resonate with these principles. By presenting eco-conscious workspaces and ergonomic configurations, organizations cultivateanenvironment of trust and adherence among their workers, precipitating high motivation, effectiveness, and overarching organizational success (Cross, 2019).

Within this intricate tapestry, the variable of workplace design and ergonomics emerges as a cornerstone, delineating the physical manifestation of human-centric principles within organizational settings. Through ergonomic equipment, lighting arrangements, and workstation layouts carefullycustomized to human physiology, organizations desire to develop environments that not only improveoutput but also nurture worker health and vitality. This emphasis on ergonomic excellence dovetails seamlessly with GHRM's overarching goal of sustainability, as eco-friendly workplace designs not only promote employee well-being but also contribute to environmental conservation efforts. Consequently, the nexus between ergonomic excellence and environmental sustainability forms a symbiotic relationship, each reinforcing and complementing the other in a virtuous cycle of organizational enhancement (IvyPanda, Human Resource Management Structure of Toyota, 2023).

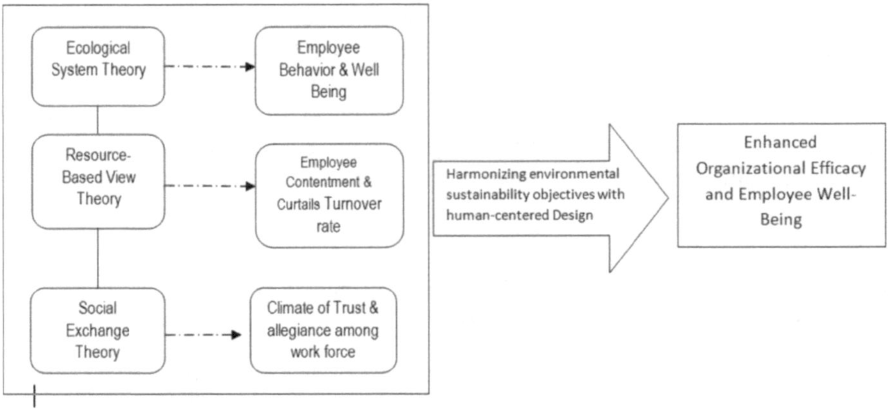


**FIGURE 5.1** GHRM and human capital leads to enhanced organizational efficacy and employee well-being. (Source: Authors Contribution.)

Moreover, at the heart of this integration lie the organized variables of employee health and well-being, organizational culture, and performance outcomes, each strongly influenced by the relationship between GHRM and ergonomics. A culture steeped in sustainability principles and ergonomic mindfulness fosters an atmosphere where employee well-being flourishes, developing a sense of belonging, contentment, and engagement between workforce constituents. In turn, high employee well-being catalyzes organizational performance, motivating productivity, innovation, and effectiveness to unmatched heights.

On the contrary, organizations that prioritize sustainability and ergonomic brillianceharvest the fruit of improved performance outcomes, additionally reinforcing the collegial relationship between GHRM and ergonomics. Thus, as these variables interweave and develop within the organizational ecosystem, the incorporation of GHRM and ergonomics emerges not simply as a planned imperative but as a transformative expedition toward holistic organizational brilliance and sustainability (Arulrajah et al., 2015).

## 5.3 RESEARCH METHODOLOGY

The present study has a comprehensive approach integrating both qualitative and descriptive research methodologies to propose a completeinvestigation of GHRM. Qualitative methods provide a keystone, allowing for anintense investigation into the complex aspects of organizational culture and the complexities entailed in implementing sustainable HR practices.Concurrently, the study employs a descriptive research methodology to scientificallyexamine existing frameworks and case studies applicable to GHRM. The literature review gives a panoramic view of the subject matter, encapsulating the extent of information within the academic and professional discourse on GHRM.

## 5.4 DISCUSSION

The study showsa positive correlation between the adoption of GHRM practices and environmental sustainability outcomes in organizational frameworks (Mousa & Othman, 2020). The research underscores that the organization accepting green initiatives in HRM domains for example recruitment & selection, training & development and performance management, exhibit high levels of environmental awareness and commitment (Rafiq et al., 2024). Organizations that prioritize green initiatives within HRM functions exhibit a high level of environmental awareness and commitment, nurturinga tradition of sustainability among the workers. This highlights the significance of integrating GHRM practices to cultivate environmental stewardship among employees (Nayak, 2019).

The study exhibits that ergonomic interventions, spanning ergonomic furniture, lighting modifications, and task adaptations, yield palpable reductions in physical strain, enlarged comfort levels, and improved productivity among employees (Mehmood, 2024). Such revelations underscore the key of nurturing ergonomically optimized work environments, serving as a pillar to strengthen employee health and performance, thusstimulating organizational resilience. Ergonomic interventions lead to better employee well-being and organizational performance, emphasizing the significance of creating work environments that maintain human health and productivity (Cai Li, 2023). This underscores the importance of integrating Human Factors considerations to improve employee well-being and organizational effectiveness.

Through the above case studies of enterprises adept at balancing GHRM practices with ergonomic design belief, the research explains significant enhancements in employee satisfaction, engagement, and overall organizational effectiveness. These findings corroborate the paradigmatic proposal of mixing GHRM and Human Factors, heralding a transformative path toward sustainable business practices and augmented organizational performance within the modern business setting.Additionally, the synergistic effects observed between GHRM and Human Factors accentuate the transformative potential of their integration. Organizations that successfully unite GHRM practices with ergonomic propose principles experience major improvements in employee satisfaction, engagement, and generally organizational efficiency. This highlights the holistic approach needed to achieve sustainable business practices and enhance organizational performance (Moktadir et al., 2020).

## 5.5 MANAGERIAL IMPLICATION

The integration of GHRMwith Human Factors proposes a transformative prospect for the organizations and shows potential noteworthy managerial implications that can drive long-term success. Managers should understand the need to prioritize aligning HRM strategies with environmental sustainability goals. This will nurture a culture of ecological balance across the workforce. This design can be achieved by the completion of green initiatives like sustainable recruitment and selection practices, eco-friendly training and development programs, and promoting environmentally accountable behaviors within the organization (Rafiq et al., 2024).

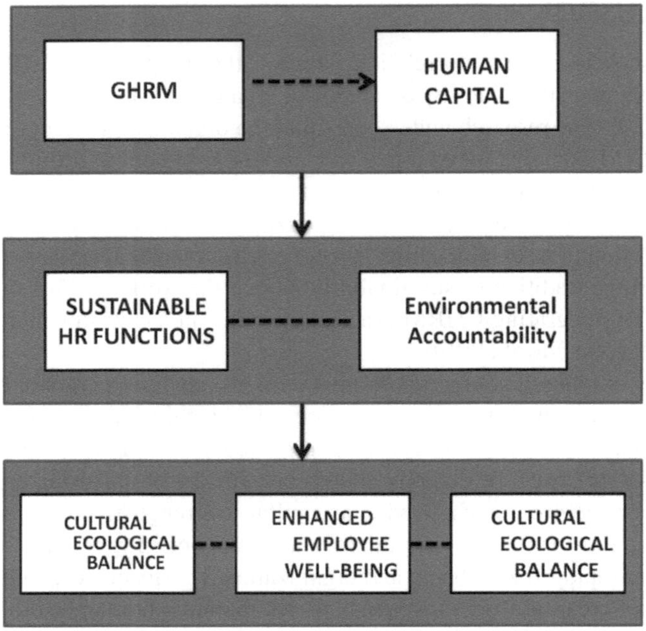


**FIGURE 5.2** Proposed model.

This includes developing ergonomic workspaces, optimizing job designs to cut physical stress, and ensuringthat the environment of the workplace enhances employee well-being. Such interventions not only develop employee engagement and satisfaction but also ensure higher levels of organizational efficiency and productivity. By focusing more on human-centric design, the managers can concentrate on both the physical and psychological needs of employees, which may bring in a more motivated and resilient workforce (Rafiq et al., 2024).

There is a need to develop comprehensive strategies that merge environmental consciousness with human-centered design principles. This approach can develop organizational resilience, giving companies a competitive edge in a vibrant marketplace. For instance, managers can develop policies and practices that promote energy preservation, waste reduction, and use of sustainable resources that improve employee health and safety (Collins, 2021).

In order to utilize the advantagesof this integration, managers need to be proactive in addressing potential challenges, like resistance to change and conflicting organizational priorities. Effective change management strategies, communication, and continuous training and development can help in mitigating these challenges. By identifying and overcoming these challenges, managers can facilitate the successful implementation of integrated GHRM and Human Factors practices (Rafiq et al., 2024).

## 5.6 LIMITATIONS AND FUTURE RESEARCH DIRECTIONS

The study exhibits the positive outcomes of integrating GHRM and Human Factors but highlights the requirement for future research to explore the long-term effects of this integration. It suggests that understanding these lasting impacts is essential for strategic decision-making and policy development. The study's dependence on quantitative methods may have ignored important qualitative insights, demonstrating that a mixed-methods tactic could offer a more nuanced understanding of the integration process (Collins, 2021). Additionally, the study lacks a thorough examination of potential barriers, such as resistance to change or conflicting organizational priorities, which are essential for a comprehensive understanding of the integration's complexities. Future research should address these barriers and identify strategies to overcome them, thereby enhancing the integration of sustainability principles and human-centered design in organizational practices (Kodua et al., 2022).

## 5.7 CONCLUSION

The empirical investigation presented a comprehensive exploration of the integration between GHRM and Human Factors, employing robust research methodologies to unveil its practical implications. The findings underscored a symbiotic relationship between the adoption of GHRM practices and environmental sustainability outcomes, revealing heightened environmental consciousness and commitment within organizations prioritizing green initiatives in HRM domains (Douglas & Renwick, 2012). Furthermore, the study illuminated the salutary impact of integrating Human Factors principles into workplace design and management, elucidating tangible benefits such as improved employee well-being and organizational efficacy through ergonomic interventions. These revelations underscored the transformative potential of harmonizing GHRM and Human Factors, as evidenced by substantial enhancements in employee satisfaction, engagement, and overall organizational efficiency observed in adeptly integrated enterprises (Somaieh Alavi, 2023).

In discussing the findings within the study's context, a clear imperative emerged for organizations to align HRM strategies with environmental sustainability goals, fostering a culture of ecological stewardship among their workforce. Moreover, the empirical evidence accentuated the significance of integrating Human Factors considerations to bolster employee welfare and organizational effectiveness, epitomizing the holistic approach essential for achieving sustainable business practices. Ultimately, the interpretation of the findings underscored the profound importance of integrating GHRM with Human Factors to drive organizational success, enhance employee well-being, and cultivate a culture of sustainability in contemporary business landscapes (Chapol Ali et al., 2020).

The integration of GHRM and Human Factors heralds a new dawn in organizational strategy, offering a paradigm shift that transcends traditional boundaries and fosters a holistic approach to sustainable business practices. By marrying environmental consciousness with human-centric design principles, organizations can pave the way for a brighter future characterized by enhanced employee well-being, operational efficiency, and ecological stewardship. This innovative fusion not only elevates organizational resilience but also propels them toward a competitive edge in today's dynamic marketplace. The study's revelations unlock a treasure trove of insights, empowering managers to navigate the complexities of sustainability with finesse and foresight (Nayak, 2019).

In practical terms, the integration of GHRM and Human Factors equips organizations with a potent arsenal of tools to cultivate a workplace culture that champions both environmental responsibility and employee welfare. From implementing ergonomic workspaces to fostering green initiatives in HRM functions, the possibilities are boundless. This transformative synergy lays the foundation for a thriving ecosystem where employees flourish, productivity soars, and the planet thrives. By embracing this innovative approach, managers can steer their organizations toward a future where sustainability is not merely a buzzword but a guiding principle deeply ingrained in the fabric of their operations.

## REFERENCES

Ahmad, J. (2023). Modeling the workplace pro-environmental behavior through. *Heliyon, 9*(9), e19134.

Al-Minhas, U., Ndubisi, N. O., & Barrane, F. Z. (2020). Corporate environmental management: A review and integration of green human resource management and green logistics. *Management of Environmental Quality*, *31*(2), 431–450.

Arulrajah, A. A., Opatha, H., &Nawaratne, N. (2015). Green human resource management practices: A review. *USJP - Academic Journals*, *5*(1), 1–16.

Bibi, K. (2022). Exploring the nexus of human resource management and legal studies: An in-depth analysis. *The Traditional Journal of Humanities, Management, and Linguistics (TJHML)*, *1*(02), 01–11.

Cai Li, F. A. (2023). Employee green behavior: A study on the impact of corporate social responsibility (CSR) on employee green behavior, green culture: The moderating role of green innovation. *Environmental Science and Pollution Research,* 105489–105503.

Chapol Ali, Md., et al. (2020). A study of green human resources management (GHRM) and green creativity for human resources professionals. *International Journal of Business and Management Future*, *4*(2), 57–67.

Collins, C. J. (2021). Expanding the resource based view model of strategic human resource management.*The International Journal of Human Resource Management,* 32(2), 331–358.

Cross, C., & Dundon, T. (2019). Social exchange theory, employment relations and human resource management. In K. Townsend, K. Cafferkey, A. McDermott, & T. Dundon (Eds.), *Elgar Introduction to Theories of Human Resources and Employment Relations* (pp. 264–279).Edward Elgar Publishing.

Douglas, W.S. & Renwick, T.R. (2012). Green human resource management: A review and research agenda. *International Journal of Management Reviews (*IJMR*), 15*, 1–14.

Hosain, S., & Rahman, Md. S. (2016). Green human resource management: A theoretical overview. *IOSR Journal of Business and Management (IOSR-JBM)*,18(6)Ver. III, 54–59.

Hussain,F. S. (2020). Striving to implement green human resource management (GHRM) policies and practices: A study from HR managers perspective (FMCG Sector). *International Journal of Management Excellence, 14*, 2037–2051.

Imran Tanveer, M., et al. (2024). Green HRM and hospitality industry: Challenges and barriers in adopting environmentally friendly practices. *Journal of Hospitality and Tourism Insights*, *7*(10), 121–141.

Interface. (2022, March 3). *Interface Publishes 2022 Environmental, Social, and Governance (ESG) Report*. Retrieved March 3, 2024, from https://www.interface.com/GB/en-GB/more-from-interface/press-room/2022-esg-report-press-release.html

IvyPanda. (2023). *Human Resource Management Structureof Toyota*. Retrieved March 3, 2024, from https://ivypanda.com/essays/hrm-in-toyota-motor-corporation-2/

Jacob, J. C. (2012). A study of green HR practices and its effective implementation in. *International Journal of Business and Management*,7(21),25–33.

Karwowski, W. & Zhang, W. (2021). The discipline of human factors and ergonomics. In G. Salvendy (Ed.), *Handbook of Human Factors and Ergonomics* (pp. 1–37). Wiley online.

Kodua, L. T., Xiao, Y., Adjei, N. O., Asante, D., Ofosu, B. O., & Amankona, D. (2022). Barriers to green human resources management (GHRM) implementation in developing countries. Evidence from Ghana. *Journal of Cleaner Production, 340*, 130671.

Mampra, M. (2013). Green HRM: Does it help to build a competitive service sector? A study. In *Proceedings of Tenth AIMS International Conference on Management* (Vol. 3, No. 8, pp. 1273–1281). The Association of Indian Management Scholars International and Indian Institute of Management, Bangalore.

Mehmood, S. (2024). Government regulations and top management influence on manufacturing firms environmental performance: Nexus of green supply chain, culture and technology adoption. In *Enviornmental Science and Pollution Research*.

Moktadir, Md. A., Dwivedi, A., Ali, S.M., Paul, S.K., Kabir, G., & Madaan, J. (2020). Antecedents for greening the workforce: Implications for green human resource management. *International Journal of Manpower*, *41*, 1135–1153.

Mousa, S. K. & Othman, M. (2020). The impact of green human resource management practices on sustainable performance in healthcare organisations: A conceptual framework. *Journal of Cleaner Production*, *243*(37), 118595.

Nayak, A. T. (2019). Human factors and ergonomics systems-based tools for understanding and addressing global problems of the twenty-first century. *Ergonomics,* 63(3),2020.

Patrick Neumann, W. &Dul, J. (2010). Human factors: Spanning the gap between OM and HRM. *International Journal of Operations & Production Management*, *30*, 923–950.

Rafiq, M., Naz, S., & Maqbool, S. (2024). Sowing sustainability through innovative hr practices to cultivate green workplaces: Converting linear to circular activities. In S. Iqbal, K. Khalid, & A. Nur (Eds.), *Innovative Human Resource Management for SMEs* (pp. 165–184). IGI Global Scientific Publishing. https://doi.org/10.4018/979-8-3693-0972-8.ch010

Ren, S., Tang, G., & Jackson, S. (2018). Green human resource management research in emergence: A review and future directions. *Asia Pacific Journal of Management*, 769–803.

Somaieh Alavi, H. A. (2023). Identifying the effect of green human resource management practices on lean-agile (LEAGILE) and prioritizing its practices. *International Journal of Productivity and Performance Management*, *35*(3), 599–624.

Tsymbaliuk, S.,et al. (2021). Green human resource management: How to implement environmental issues into HR practices. In *International Conference on Sustainable, Circular Management and Environmental Engineering*, Odesa, Ukraine.

Wikhamn,W. (2019). Innovation, sustainable HRM and customer satisfaction. *The International Journal of Hospitality Management,* 102–110.

Wong, K.-L. Tan, P. S.-H., Ng, Y.-K., & Fong, C.-Y.(2013). The role of HRM in enhancing organizational. *Human Resource Management Research*, *3*(1), 11–15.

Yong, J. Y. &Yusliza,M.-Y. (2020). Green human resource management: A systematic literature review from 2007 to 2019. *Benchmarking: An International Journal*, *27*(1), 2005–2027.

# 6 Potential Strategies for Corporate Success in Dealing with WMSDs at Workplace

## *A Study of Ergonomic Practices in Hospitals in India*

*Obaidur Rahman, Manik Batra, Farah Ahmad, and Anshu*

## 6.1 INTRODUCTION

Ergonomics may be understood as the science of modeling the job for creating better conditions for workers. In simpler terms, it is termed as "the scientific study of human work". The field of Ergonomics has attained priority in recent decades owing to increasing awareness about overall employee well-being. This is the age of machines and technology and human interaction with them has only increased. Thus, the study of Ergonomics deals with how employment conditions, machineries, and equipment are arranged in a way that employees can work more efficiently. Long working periods in an ergonomically challenged workplace may lead to several psychological and physical problems like stress, anxiety, musculo-skeletal disorders (MSDs), etc. Not paying attention to the principles of ergonomics may lead to inefficiency and pain at the workplace. The adoption of ergonomic principles helps improve efficiency, comfort, and well-being without jeopardizing the health and safety of the employees.

### 6.1.1 Background

World Health Organization (WHO) has found that roughly three-fourths of adults in industrialized countries suffer from lower back pain. Institute of Health Metrics and Evaluation IHME (2019) found that Musculo-skeletal ailments (ailments affecting the muscles, bones, or joints) are the chief source of disabilities in the last three decades, and also the sixth biggest reason for disability-adjusted life years (www.wellfitconsultancy.com, 2023). Global Burden of Disease Study (2017) found that lower back pain is the primary reason for disability since 1990 (www.wellfitconsultancy.com, 2023). These numbers are only the tip of the iceberg as the problems


DOI: 10.1201/9781003532293-7

associated with non-compliance of ergonomic principles go mostly unreported and this phenomenon is scarcely studied in the developing nations.

In India, the study of Ergonomics is a comparatively novel notion and many stakeholders are still indifferent from considering it a crucial dimension of employee well-being. Nursing centers and hospitals comprise nearly 49 percent of the Indian healthcare sector (Singh & Rangnekar, 2020). The healthcare employees working in these setups are under severe pressure to be more competent even though the resources are less and management skills are poor (Govindaras et al., 2023). Doctors and nurses are the primary workers in the healthcare system and are more susceptible to Musculo-skeletal problems due to their work stress (Talapatra et al., 2023). Doctors perform several tasks at the hospitals including going from ward to ward for patient visits, regularly leaning over the bed to examine the patients, and extended sitting hours in outpatient departments apart from report maintenance and record keeping (Das et al., 2022). Nurses, on the other hand, are referred to as the backbone of the hospitals because they perform a wide range of activities including patient care and implementation of doctor's plan for the patient (Carayon et al., 2021). They sometimes work in back-to-back shifts to keep the hospitals functional in the absence of active workforce (Das et al., 2022). This results in high rates of work-related physical and psychological issues reported by healthcare workers who are involved in direct patient care (Das et al., 2022). Though it is a relatively less explored area of research, a better understanding of the critical ergonomics factors will be helpful in designing better work systems to improve safety and employee well-being.

This study intends to explore the latent sentiments of healthcare employees and their expectations about ergonomic practices at the hospitals. The researchers first intend to crystallize the body of knowledge in the area and develop an understanding of ergonomics practices through an exhaustive review of literature.

#### 6.1.1.1 Objectives

The review of extant literature brought forth the following objectives for the current study:

1. Developing insights into Ergonomics practices among hospital employees.
2. Identification of Latent Factors which influence Ergonomic practices among hospital employees
3. Developing and validating a scale of Ergonomics practices in hospitals.

## 6.2 LITERATURE REVIEW

The current study explored several research papers in the area of Ergonomics and its application areas for developing a better understanding of the area. Research studies in the area of ergonomics, though scant and scattered, are primarily focused on a variety of settings in human-work interaction such as stonecutters (Gangopadhyay et al., 2010), IT professionals (Sirajudeen et al., 2013), potters and sculptors (Sahu et al., 2013), dental professionals (Kalghatgi et al., 2014), students (Sirajudeen & Siddik, 2017), footwear artisans (Jadhav et al., 2020), dental students (Kumar et al., 2020), periodontists (Ketkar & Malaiappan, 2020), leather industry workers

**TABLE 6.1**
**Summary of Major Studies Reviewed[a]**

| Author(s) | Year | Country | Remarks |
|---|---|---|---|
| Zeverdegani et al. | 2022 | Iran | MSDs and ergonomic designing |
| Ketkar and Malaiappan | 2020 | India | Knowledge of ergonomics |
| Kumar et al. | 2020 | India | Practices related to ergonomics |
| Kujerdi et al. | 2022 | Iran | Identification of ergonomics problems |
| Kalghatgi et al. | 2014 | India | Awareness towards ergonomics |
| Gangopadhyay et al. | 2010 | India | Ergonomics assessment |
| Jadhav et al. | 2020 | India | Postural analysis |
| Heidarimoghadam et al. | 2022 | Iran | Strategic literature review |
| Sahu et al. | 2013 | India | Postural assessment |
| Patel and Mukhopadhyay | 2021 | India | Ergonomic analysis of symbols |
| Sirajudeen and Siddik | 2017 | India | Awareness of ergonomics |

[a] Prepared by the researcher.

(Karuppiah et al., 2020), corporate workers (Zeverdegani et al., 2022). The research works related to ergonomics have gradually shifted from the unorganized sector to the organized sector, as evident from the above-mentioned studies. A summary of the important studies in the area of Ergonomics has been prepared by the researcher (Table 6.1).

The studies related to the application of ergonomics in hospitals are scant, especially in the Indian context. Patel and Mukhopadhyay (2021) in their study titled *Ergonomic analysis and design intervention in symbols used in hospitals in Central India* dealt with highlighting the problems associated with existing symbols in the hospital. The researchers conducted two experiments to achieve the study objectives. The first experiment stressed upon identifying new concepts based on principles of Ergonomics and feedback from the users. The second experiment dealt with testing of these newly developed concepts on the intended users. Based on the results of the study they proposed new ergonomic friendly symbols and signages to be used in the hospitals.

Ketkar and Malaiappan (2020) in their study titled *Knowledge Attitude and Practice of Ergonomics and Musculo-skeletal Disorders as an Occupational Hazard among Periodontists in India – A Questionnaire Based Survey* explored MSDs as a work-related risk in the case of periodontists. They gauged the understanding and implementation of ergonomics among 150 periodontists in India and the data was examined and deciphered. The outcomes of the study revealed that most of the respondents had an adequate level of understanding and attitude towards Ergonomics.

Kalghatgi et al. (2014) in their research paper titled *Insights into Ergonomics Among Dental Professionals of a Dental Institute and Private Practitioners in Hublie Dharwad Twin Cities, India* explored the understanding and implementation of ergonomics practices among dental professionals. Researchers developed a self-administered, closed-ended questionnaire for conducting a survey on a sample of 250 respondents. The results revealed that the respondents had substantial knowledge of ergonomics.

Zeverdegani et al. (2022) published a cross-sectional study titled *Latent Class-derived Patterns of Musculo-skeletal Disorders in Sedentary Workers and Chair Ergonomic Design.* They conducted a survey on office-going employees and identified precise imprints of multisite MSDs. They examined how such imprints contribute to designing an ergonomic chair. They employed Latent Class Analysis to investigate the relationship between patterns of MSDs with ergonomic chair design. The study outcomes revealed that MSDs may be explained with hidden class-derived imprints among office-goers. The results also revealed that ergonomic chair design was significantly related to the type of MSD imprints.

The literature review revealed that there exist several gaps with regard to the study of Ergonomics and its application in India especially in the context of hospital employees. The current study would focus on a few of these research gaps and would like to develop insights into the understanding of this area.

### 6.2.1 Research Gaps

After a thorough review of extant literature regarding Ergonomics and its application in various areas, the following research gaps were identified:

1. Lack of research with respect to Ergonomics practices in hospital employees.
2. Lack of studies on awareness of Ergonomic practices in hospital employees.
3. Lack of research on factors influencing ergonomic practices in hospitals.

## 6.3 METHODOLOGY

### 6.3.1 Text Analysis

It has been observed that most of the previous studies on scale development and validation have mostly used literature reviews and interviews in the initial stages of scale development to identify the dimensions and to determine the constructs. However, interviews may bring sampling bias as respondents are confined to a particular convenience area (Xu et al., 2019). Therefore, this research work uses big data to enhance the data gained from healthcare workers' online reviews to attain a better understanding of their workstation ergonomics. Online reviews are typically ambiguous text data; thus, this study uses a type of text analytics approach to categorize distinct themes of text delineating ergonomics elements (Shah et al., 2020). An in-depth analysis of a significant number of online reviews may yield insightful data regarding the identification of crucial components that can be utilized to understand healthcare workers' expectations from the hospital. According to Ashton and Prybutok (2020), scale domains and scale items of interest are identified using the concepts found in text analytics.

### 6.3.2 Methods

This study uses a sequential multi-stage exploratory process involving qualitative and quantitative methods to create and test a scale for determining the ergonomic practices among healthcare workers. The qualitative part of the study was based on the review of extant literature followed by text-mining analysis and then expert panel feedback. The text-based data employed in the study was extracted from Glassdoor.com, one of the world's major websites focusing on employees' work-life quality. Reviews of employees working in seven private healthcare units located in the National Capital Region (NCR) of Delhi and its neighboring cities were used. Hospitals containing more than 300 beds were used for the study.

Data was collected from January 2023 to August 2023. The hospital reviews and stories of healthcare workers were copied from Glassdoor.com. Libraries including TextBlob, Fitz, and Sklearn in Python were used to crawl the reviews of each of the seven hospitals. A total of 665 reviews were obtained. There were 125 reviews that either had less than 20 words or were not written in English. Thus, a total of 540 reviews were considered for analysis.

The Term Frequency-Inverse Document Frequency (TF-IDF) approach was utilized to extract essential workstation dimensions. TF-IDF helps measure a word's significance in an article (Shi & Li, 2011). The TF-IDF weight is better suited for qualitative analysis than word cloud as it increases in proportion to the number of occurrences of a word in the article but is balanced by the term's frequency (Shi & Li, 2011).

## 6.4 DATA ANALYSIS

### 6.4.1 Item Generation

The TF-IDF analysis of healthcare workers' online reviews and the previous literature was used to determine the initial workstation ergonomics dimensions. Table 6.2 lists the first 50 terms relevant to the healthcare workers' experience together with their accompanying TF-IDF values. The research team evaluated the ergonomic features of the healthcare workstations by applying the *Standards for Planning* and *Designing Safe Hospitals guidelines*.

These words with high TF-IDF values were combined with domains found in previous literature to create the health ergonomics dimensions following the criteria of Standards for Planning and Design in Safe Hospitals (Mokarami et al., 2021). The encoding and retrieving of the domains were done independently after verifying the themes. The reliability of the domains was checked using Holsti's formulae. Reliability is equal to $2m/N1 + N2$, where $m$ is the total number of coding choices made by coders, and $N1$ and $N2$ are the number of choices made by the programmers) (Holsti, 1969). The reliability of the developed domains was satisfactory (>80%).

Once a consensus was formed on the item statements and their division into six main dimensions, operational definitions were developed following accepted worldwide concepts and definitions found in the literature (Table 6.3).

**TABLE 6.2**
**Top 50 Words Extracted from Online Views with Their TF-IDF Score**

| Word | TF-IDF | Word | TF-IDF | Word | TF-IDF | Word | TF-IDF | Word | TF-IDF |
|---|---|---|---|---|---|---|---|---|---|
| Hospital | 0.577 | Location | 0.078 | Staff | 0.030 | Hygiene | 0.068 | Wires | 0.035883 |
| Room | 0.419 | Floorplan | 0.070 | Reception | 0.032 | Fully-equipped | 0.064024 | Safety | 0.033954 |
| Service | 0.384 | Waiting rooms | 0.073 | Environment | 0.025 | Spacious | 0.061082 | Satisfied | 0.031775 |
| Quality | 0.282 | Noise | 0.071 | Cozy | 0.023 | Staff | 0.052041 | Ventilation | 0.029384 |
| Physical space | 0.209 | Lighting | 0.061 | Facilities | 0.021 | Machines | 0.049764 | Windows | 0.027196 |
| Physical safety | 0.204 | Experience | 0.060 | Support | 0.020 | Patient | 0.047332 | Desk | 0.025614 |
| Physical structures | 0.203 | Dedicated | 0.0521 | Signs | 0.019 | Help | 0.043785 | Furniture | 0.024112 |
| Noise | 0.123 | Infrastructure | 0.0474 | Value | 0.01851 | Flowers | 0.041348 | Toilet facilities | 0.023218 |
| Smell | 0.104 | Colors | 0.043 | Bed | 0.07135 | Environment | 0.039986 | Staff | 0.019611 |
| Comfort | 0.080 | Monitor | 0.041 | Helpful | 0.06995 | Service | 0.037865 | Ease of sitting | 0.017621 |

**TABLE 6.3**
**Identified Factors of Hospital Ergonomics**

| Factors | Definition |
|---|---|
| | **Physical Servicescape** |
| Ambience | It refers to the background circumstances and non-visual aspects of the interaction. |
| Spatial layout | It includes the layout of machinery, equipment and furnishings. |
| Location | It refers to the floor plan and the hospital wards. |
| Safety and security | It refers to the physical safety of the employees working in the hospital |
| Chair | It refers to the furniture provided for sitting in the hospital |
| Desk | It refers to the physical workspace for sitting and performing work. |

### 6.4.2 Scale Validation

The data collection for scale validation was done between November 1, 2023, and December 31, 2023 using the survey sample method. We received 418 completed surveys in total. A total of 360 valid replies were kept for additional analysis following data screening. 205 men (57%), and 155 women (43%), made up the 360 participants. We made an effort to collect responses from a range of age groups (Hair et al., 2012).

Prior to conducting CFA, the normality of data is tested to check whether the received data follows the normal probability distribution or not. The significance of the relationships may be affected if the data is far from being normal (Hair et al., 2016). To assess the normality, we measure the skewness and kurtosis of the data collected. Skewness is the measure of the asymmetry of a distribution of data. Similarly, kurtosis measures the peakiness of the normal curve distribution. The distribution of variables is considered approximately normal when the value of skewness is less than $\pm 1$ and the value of kurtosis is below 10 (Leech et al., 2005). In this data, all values of the variables are within the range for skewness and kurtosis (Table 6.4).

For this study, the principal component analysis (PCA) technique along with Varimax rotation method is used for carrying out factor analysis. Only constructs that accounted for a variance of greater than one (eigenvalue >1) along with a cumulative percentage of total variance of more than 40% were extracted (Kim & Muller, 2020) (Table 6.5).

CFA has been carried out using *SEMinR* package in R software. *SEMinR* means Structure Equation Modelling in *R* language. The Chi-square value for the distribution was 136.16 ($p < 0.05$). The proposed items of the scale significantly converged on their expected variables. The loadings for all the items were greater than 0.5 (Table 6.6).

Further, a Multi-factor Congeneric Model was developed using Structural Equation Modeling (SEM). This Model is developed using R language via the *SEMinR* package.

**TABLE 6.4**
**Indicator Analysis**

| Name | Missings | Mean | Standard Deviation | Kurtosis | Skewness |
|---|---|---|---|---|---|
| A1 | 0 | 3.708 | 1.602 | –0.812 | 0.089 |
| A2 | 0 | 3.8 | 1.643 | –0.901 | 0.104 |
| A3 | 0 | 3.775 | 1.635 | –0.909 | 0.09 |
| A4 | 0 | 3.769 | 1.633 | –0.912 | –0.079 |
| A5 | 0 | 3.689 | 1.615 | –0.902 | 0.013 |
| CL1 | 0 | 3.958 | 1.633 | –0.949 | –0.052 |
| CL2 | 0 | 3.856 | 1.615 | –0.802 | –0.082 |
| CL3 | 0 | 3.883 | 1.6 | –0.794 | –0.013 |
| SL1 | 0 | 3.758 | 1.626 | –0.899 | 0.005 |
| SL2 | 0 | 3.806 | 1.65 | –0.942 | –0.041 |
| SL3 | 0 | 3.847 | 1.659 | –0.837 | 0.054 |
| SL4 | 0 | 3.864 | 1.632 | –0.764 | 0.12 |
| SL5 | 0 | 3.758 | 1.683 | –0.907 | 0.146 |
| SS1 | 0 | 3.789 | 1.504 | –0.729 | 0.098 |
| SS2 | 0 | 3.8 | 1.513 | –0.737 | 0.048 |
| F1 | 0 | 3.8 | 1.554 | –0.749 | 0.154 |
| F2 | 0 | 3.703 | 1.516 | –0.604 | 0.181 |
| F3 | 0 | 3.706 | 1.552 | –0.614 | 0.163 |
| SQ1 | 0 | 3.825 | 1.542 | –0.695 | 0.113 |
| SQ2 | 0 | 3.839 | 1.555 | –0.591 | 0.155 |
| SQ3 | 0 | 3.908 | 1.558 | –0.727 | 0.104 |
| SQ4 | 0 | 3.797 | 1.546 | –0.763 | 0.066 |

**TABLE 6.5**
**Eigen Values of All Variables in *R*-Console**

```
> print (z$values, digits=1)
[1] 7.40  4.21  3.84  2.77  2.58  2.23  0.81  0.70  0.88  0.83  0.80  0.78  0.75
[14] 0.69  0.67  0.62  0.59  0.59  0.58  0.52  0.50  0.49
```

**TABLE 6.6**
**SS Loadings and Variance Values of Rotated Components in *R*-Console**

| | RC1 | RC3 | RC2 | RC4 | RC8 | RC6 |
|---|---|---|---|---|---|---|
| SS loadings | 4.21 | 3.84 | 3.72 | 2.36 | 2.15 | 1.52 |
| Proportion var | 0.19 | 0.17 | 0.17 | 0.11 | 0.10 | 0.07 |
| Cumulative var | 0.19 | 0.37 | 0.54 | 0.64 | 0.74 | 0.81 |

**TABLE 6.7**
**Values ($R^2$, AVE, CR, and Cronbach Alpha) of All Variables**

| Research Construct | R2 | AVE | CR | Cronbach's Alpha |
|---|---|---|---|---|
| A | 0.47 | 0.760 | 0.950 | 0.921 |
| L | 0.59 | 0.814 | 0.931 | 0.886 |
| SL | 0.68 | 0.783 | 0.953 | 0.931 |
| SS | 0.46 | 0.873 | 0.932 | 0.855 |
| C | 0.42 | 0.825 | 0.947 | 0.894 |
| D | 0.47 | 0.806 | 0.948 | 0.920 |

This model is tested by adopting the following measures.

- Goodness of Fit ($R^2$)
- Loading of every item should be greater than 0.5 (Fornell & Larcker, 1981).
- The value of Average Variance Extracted (AVE) for every item should be greater than 0.5. Its value is calculated as per equation (i).
- The value of Composite Reliability (CR) for every item should be greater than 0.7.

**Assessment of Uni-dimensionality of the Variables:** Uni-dimensionality is assessed by conducting R Square correlation ($R^2$) test on every proposed variable. $R^2$ Values of every variable are greater than 0.4 and its range is lying from 0.4 to 0.69. So, all values lie in the significant range. Hence variables have good fitness and they are also uni-dimensional. All these values are mentioned in Table 6.7.

**Average Variance Extracted (AVE):** This parameter is the measure of the convergent validity test. The value of AVE greater than 0.5 represents a significant level of convergent validity (Fornell & Larcker, 1981). The value AVE for all elements is calculated and mentioned in Table 6.7.

**Composite reliability (CR):** CR determines the measure of internal reliability. The value of CR is greater than 0.7, which is considered significant. Cronbach's alpha coefficient for each factor is also calculated to determine internal reliability. Cronbach's alpha coefficient for all the factors is greater than 0.7 (Table 6.7).

**Discriminant Validity Test:** Discriminant validity is also tested for the constructs used in our study (Table 6.7). It was observed that the value of Cronbach alpha coefficient is greater than the mean of the correlation of latent constructs and no coefficient has a correlation greater than 0.9, as shown in Table 6.8. This confirms the discriminant validity test.

## 6.5 DISCUSSION

The study explores ergonomic practices at workplace in the context of hospitals in India. The study has explored the latent sentiments of healthcare employees. The researchers aim to crystallize knowledge in the field and enhance understanding of ergonomics practices through a thorough review of literature.

**TABLE 6.8**
**Discriminant Validity test**

| Constructs | Cronbach's Alpha | A | L | SL | SS | C | D |
|---|---|---|---|---|---|---|---|
| A | 0.921 | 0.71 | | | | | |
| L | 0.886 | 0.54 | **0.83** | | | | |
| SL | 0.931 | 0.63 | 0.788 | **0.77** | | | |
| SS | 0.855 | 0.684 | 0.743 | 0.739 | **0.67** | | |
| C | 0.894 | 0.682 | 0.707 | 0.699 | 0.61 | **0.68** | |
| D | 0.920 | 0.659 | 0.698 | 0.668 | 0.28 | 0.39 | **0.898** |

Work-related musculo-skeletal disorders (WMSDs) pose significant challenges in various industries, particularly in healthcare settings where staff often perform physically demanding tasks. In India, where the healthcare sector is rapidly expanding, addressing WMSDs through effective ergonomic practices is essential for enhancing worker well-being and operational efficiency. This discussion explores potential strategies for corporate success in managing WMSDs within hospitals in India, emphasizing the integration of ergonomic principles.

This research work used big data to enhance the data extracted from healthcare workers' online reviews to gather a better understanding of their workstation ergonomics. It applies text analytics to recognize significant text patterns that describe elements of ergonomics. For conducting the study, it has used a sequential multi-stage exploratory process involving qualitative and quantitative methods to develop a scale (Table 6.9) for determining the ergonomic conditions of healthcare workers.

The research team evaluated the ergonomic features of the healthcare workstations by applying the *Standards for Planning and Designing Safe Hospitals guidelines.* The Term Frequency-Inverse Document Frequency (TF-IDF) approach was applied to extract essential workstation dimensions. The terms with high TF-IDF values were combined with domains found in previous literature to create the health ergonomics dimensions following the criteria of Standards for Planning and Design in Safe Hospitals (Mokarami et al., 2021). The encoding and retrieving of the domains were done independently after verifying the themes.

Different tools have been applied in order to assess the data like Assessment of Uni-dimensionality of the Variables. Uni-dimensionality is assessed by conducting *R* Square correlation ($R^2$) test on every proposed variable. $R^2$ Values of every variable are greater than 0.4 and its range is lying from 0.4 to 0.69. Another method was CR—it determines the measure of internal reliability. The value of composite reliability (CR) is greater than 0.7, which is considered significant. Next was AVE—this parameter is the measure of the convergent validity test. The value of AVE greater than 0.5 represents a significant level of convergent validity and Discriminant Validity Test- It was found that the value of Cronbach alpha coefficient is greater than the mean of the correlation of latent constructs and no coefficient has a correlation greater than 0.9.

**TABLE 6.9**
**Developed Scale for Maternity Hospitals**

| | Items | Factor |
|---|---|---|
| A1 | The hospital has sufficient lighting in the corridors. | |
| A2 | The hospital has a satisfying smell. | |
| A3 | The hospital has a satisfying temperature. | |
| A4 | The hospital has sufficient noise barriers between the wards. | AMBIENCE |
| A5 | The noise level in the hospital is comfortable. | |
| SL1 | The nurses should have clear access to the drug store. | |
| SL2 | The workstations should have proper ventilation. | |
| SL3 | The hospital should allow an easy entry and exit to the employees. | SPATIAL LAYOUT |
| SL4 | The hospital staff should have a dedicated space for meetings. | |
| SL5 | There should be sufficient space for the medical equipment and tools | |
| L1 | The workstations should be at the center of the ward | |
| L2 | The entrance and exit of the ward should be visible from the workstation. | LOCATION |
| L3 | The workstation should allow a clear view of all patients. | |
| SS1 | The floor in the hospital should prevent slipping. | |
| SS2 | All wires and cables should be safely placed in a duct. | SAFETY AND SECURITY |
| C1 | The hospital has sufficient number of chairs for the staff | |
| C2 | The chairs should not have sharp edges. | CHAIR |
| C3 | All chairs in the hospital are in proper condition. | |
| D1 | The employees in the hospital should have a charting desk | |
| D2 | The employees in the hospital should have a window at the desk. | DESK |
| D3 | The desk should have drawers for storage. | |
| D4 | The trolley containing the patient files should be near the desk. | |

## 6.6 CONCLUSION

The current study explored several research papers in the area of Ergonomics and its application areas for developing a better understanding of the area. The primary focus of the study was the Indian environment. The review of literature revealed that there exist several gaps with regard to the study of Ergonomics and its application in India especially in the context of hospital employees. Addressing WMSDs in hospitals through ergonomic practices not only enhances employee well-being but also contributes to overall organizational success. By implementing comprehensive strategies that involve assessment, training, equipment integration, open communication, leadership commitment, cultural promotion, and ongoing evaluation, hospitals in

India can create a safer and more productive workplace. Ultimately, these efforts will lead to better patient care, reduced absenteeism, and lower healthcare costs, positioning hospitals for long-term success in a competitive landscape.

## REFERENCES

Ashton, T., & Prybutok, V. R. (2020). Developing and validating e-retailing satisfaction scales with text-mining. *Journal of Modelling in Management*, *15*(4), 1655–1677.

Carayon, P., Wust, K., Hose, B. Z., & Salwei, M. E. (2021). Human factors and ergonomics in health care. In G. Salvendy (Ed.), *Handbook of Human Factors and Ergonomics* (5th ed., pp. 1417–1437). Hoboken, NJ: Wiley & Sons.

Das, S. K., Pathak, V., & Chakraborty, T. (2022). Ergonomic mapping of work-related Musculo-skeletal disorders and stress among Indian healthcare workers: An overview. *Journal of Public Affairs*, *22*(2), e2478.

Fornell, C., & Larcker, D. F. (1981). Evaluating structural equation models with unobservable variables and measurement error. *Journal of Marketing Research*, *18*(1), 39–50.

Gangopadhyay, S., Das, B., Das, T., Ghoshal, G., & Ghosh, T. (2010). An ergonomics study on posture-related discomfort and occupational-related disorders among stonecutters of West Bengal, India. *International Journal of Occupational Safety and Ergonomics*, *16*(1), 69–79.

Govindaras, B., Wern, T. S., Kaur, S., Haslin, I. A., & Ramasamy, R. K. (2023). Sustainable environment to prevent burnout and attrition in project management. *Sustainability*, *15*(3), 2364.

Hair, Jr, J. F., Sarstedt, M., Matthews, L. M., & Ringle, C. M. (2016). Identifying and treating unobserved heterogeneity with FIMIX-PLS: Part I–method. *European Business Review*, *28*(1), 63–76.

Hair, J. F., Sarstedt, M., Ringle, C. M., & Mena, J. A. (2012). An assessment of the use of partial least squares structural equation modeling in marketing research. *Journal of the Academy of Marketing Science*, *40*, 414–433.

Heidarimoghadam, R., Mohammadfam, I., Babamiri, M., Soltanian, A. R., Khotanlou, H., & Sohrabi, M. S. (2022). What do the different ergonomic interventions accomplish in the workplace? A systematic review. *International Journal of Occupational Safety and Ergonomics*, *28*(1), 600–624.

Holsti, O. R. (1969). *Content Analysis for the Social Sciences and Humanities*. Reading: Addison-Wesley (content analysis).

Jadhav, G. S., Arunachalam, M., & Salve, U. R. (2020). Ergonomics and efficient workplace design for hand-sewn footwear artisans in Kolhapur, India. *Work*, *66*(4), 849–860.

Kalghatgi, S., Prasad, K. V. V., Chhabra, K. G., Deolia, S., & Chhabra, C. (2014). Insights into ergonomics among dental professionals of a dental institute and private practitioners in Hubli–Dharwad twin cities, India. *Safety and Health at Work*, *5*(4), 181–185.

Karuppiah, K., Sankaranarayanan, B., Ali, S. M., & Kabir, G. (2020). Role of ergonomic factors affecting production of leather garment-based SMEs of India: Implications for social sustainability. *Symmetry*, *12*(9), 1414.

Ketkar, G. N., & Malaiappan, S. (2020). Knowledge attitude and practice of ergonomics and Musculo-skeletal disorders as an occupational hazard among periodontists in India–a questionnaire based survey. *Journal of Pharmaceutical Research International*, *32*(20), 162–183.

Kim, T. H., & Muller, C. (2020). Inconsistency transmission and variance reduction in two-stage quantile regression. *Communications in Statistics-Simulation and Computation*, *49*(4), 1044–1077.

Kujerdi, M. F., Mokarami, H., Keshtkar, V., Ziaei, M., Petramfar, P., & Choobineh, A. (2022). Improving working conditions in an Iranian hospital: a participatory ergonomics approach. *International Journal of Occupational Safety and Ergonomics*, *28*(3), 1683–1689.

Kumar, P. M., Sahitya, S., Penmetsa, G. S., Supraja, S., Kengadaran, S., & Chaitanya, A. (2020). Assessment of knowledge, attitude, and practice related to ergonomics among the students of three different dental schools in India: An original research. *Journal of Education and Health Promotion*, *9*(1), 266.

Leech, N. L., Barrett, K. C., & Morgan, G. A. (2005). *SPSS for Intermediate Statistics: Use and Interpretation* (2nd ed.). Mahwah, NJ: Lawrence Erlbaum Associates.

Mokarami, H., Eskandari, S., Cousins, R., Salesi, M., Kazemi, R., Razeghi, M., & Choobineh, A. (2021). Development and validation of a Nurse Station Ergonomics Assessment (NSEA) tool. *BMC Nursing*, *20*(1), 83.

Patel, G., & Mukhopadhyay, P. (2021). Ergonomic analysis and design intervention in symbols used in hospitals in central India. *Applied Ergonomics*, *94*, 103410.

Sahu, S., Moitra, S., Maity, S., Pandit, A. K., & Roy, B. (2013). A comparative ergonomics postural assessment of potters and sculptors in the unorganized sector in West Bengal, India. *International Journal of Occupational Safety and Ergonomics*, *19*(3), 455–462.

Shah, K., Patel, H., Sanghvi, D., & Shah, M. (2020). A comparative analysis of logistic regression, random forest and KNN models for the text classification. *Augmented Human Research*, *5*(1), 12.

Shi, Y., & Lei, L. (2021). Lexical use and social class: A study on lexical richness, word length, and word class in spoken English. *Lingua*, 262, 103155.

Shi, H. X., & Li, X. J. (2011). A sentiment analysis model for hotel reviews based on supervised learning. In *2011 International Conference on Machine Learning and Cybernetics* (Vol. 3, pp. 950–954). IEEE.

Singh, A., & Rangnekar, S. (2020). Enhancing proactivity among hospital employees: A serial mediation model. *International Journal of Business Excellence*, *21*(1), 64–80.

Sirajudeen, M. S., Pillai, P. S., & Vali, G. M. (2013). Assessment of knowledge of ergonomics among information technology professionals in India. *Age (Years)*, *20*(29), 135.

Sirajudeen, M. S., & Siddik, S. S. M. (2017). Knowledge of computer ergonomics among computer science engineering and information technology students in Karnataka, India. *Asian Journal of Pharmaceutical Research and Health Care*, *9*(2), 64–70.

Talapatra, S., Parvez, M. S., Saha, P., & Kibria, M. G. (2023). Assessing the impact of critical risk factors on the development of Musculo-skeletal disorders: A structural equation modelling approach. *Theoretical Issues in Ergonomics Science*, *25*(3), 343–368. https://doi.org/10.1080/1463922X.2023.2219297

www.wellfitconsultancy.com (2023). "Ergonomics in Numbers: What does the data say?". https://wellfitconsultancy.com/ergonomics-in-numbers-what-does-the-data-say/. Accessed on 29 March, 2023.

Xu, X., Liu, W., & Gursoy, D. (2019). The impacts of service failure and recovery efforts on airline customers' emotions and satisfaction. *Journal of Travel Research*, *58*(6), 1034–1051.

Zeverdegani, S. K., Yazdi, M., & MollaAghaBabaee, A. H. (2022). Latent class-derived patterns of Musculo-skeletal disorders in sedentary workers and chair ergonomic design. *International Journal of Occupational Safety and Ergonomics*, *28*(3), 1636–1641.

# 7 The Integration of Strategic Orientation with Human Factors and Ergonomics Interventions in Workplace

*Shivani Kapoor, Fehmina Khalique, Ankita Kapoor, and Sanshya Kapoor*

## 7.1 INTRODUCTION

In today's vibrant and modest competitive business environment, strategic course assists as a directorial principle for organizations targeting for competitive advantage and long-term sustainable growth. It consists of aligning the goal of the organization, resources, and the activity with external market demands and internal capabilities in a logical manner (Baker & Sinkula, 2009). At the same time, interventions in Human Factors and Ergonomics (HF&E) play an important role in improving the interactions between people, technology, and their work surroundings to boost productivity, contentment, and security (Salvendy, 2012). By combining strategic direction with HF&E measures, companies can encourage and cultivate a workplace ethos that emphasizes the importance of employee well-being and productivity. This will greatly enhance the overall achievements of the organizations.

Ergonomics takes into account a multidisciplinary approach which is a culmination of information, findings, and principles of sciences like physiology, anatomy, physiology, physics, and engineering. Ergonomics helps in optimizing the human—system work, in order to find healthy, safe and comfortable working environment. 'Human Factor & Ergonomic' (HF&E) is devised for healthy and safe work environment for better productivity of the employees. Therefore, efficient ergonomic design is an integral part of the organizational setup in order to prevent injuries and disorders that can lead to disability among the employees.

Strategic orientation embraces the strategic choices and selections made by an organization to achieve its aims and objectives and maintain its competitive spot in the market (Khan & Ansari, 2024; Ansari et al., 2024). This process involves understanding of the dynamics of the market, identification of the opportunities, and leveraging of the internal capabilities so as to create value for all the stakeholders. Baker and Sinkula (2009) underscore the importance of market orientation and entrepreneurial

DOI: 10.1201/9781003532293-8

orientation in driving organizational profitability and growth. Market orientation involves a deep understanding of customer needs and preferences, while entrepreneurial orientation emphasizes innovation, risk-taking, and proactive decision-making.

On the other hand, HF&E interventions emphasize on enhancing the work environment to boost the performance and the well-being of employees. According to Salvendy (2012), HF&E interventions incorporate a wide range of strategies, comprising the design of ergonomic workstation, psychosocial interventions, task analysis, and workload management. These interventions are committed to create a work environment that restricts physical stress, inspires psychological well-being, and benefits from maximized productivity.

### 7.1.1 Introduction to Workplace Well-Being

Workplace well-being has become a progressively important aspect of organizational studies, with scholars stressing its impact on employee satisfaction and productivity. Karasek and Theorell (1990) were among the first to highlight the link between workplace stress and productivity, finding the need for healthy work environments. Schulte and Vainio (2010) extended this by stating an all-inclusive synopsis of well-being at work, connecting it to positive employee performance and the success of the organization.

### 7.1.2 Organizational Culture and Employee Performance

Organizational culture considerably influences employee performance and total business results. Beheshtifar and Shariatifar (2013) claimed that a positive organizational culture nurtures greater performance heights, bringing into line employees with the goals of the organization. Schneider and Barbera (2014) additionally exemplified how organizational climate contours the experiences of the employees thus persuading productivity. Additionally, Bartsch et al. (2013) emphasized that strategic orientation plays a crucial role in improving organizational performance.

### 7.1.3 Market Orientation and Profitability

The affiliation of market orientation and business profitability is well-documented in the literature. Narver and Slater (1990) verified that companies with robust market positioning attain improved financial outcomes, underlining the significance of supporting organizational practices with market demands. In line with this, Li et al. (2023) provided a literature review presenting the positive correlation between strategic orientation and firm performance.

### 7.1.4 Human Resource Management Practices

Actual 'human resource management' (HRM) practices are crucial for enhancing employee well-being and performance (Zareen & Khan, 2023). Almutairi (2016) emphasized the intervening role of organizational assurance in the connection between HR practices and employee performance, suggesting that well-made HR strategies are important for enhancing productivity. Chen and Peng (2021) reinforced this statement, indicating how employee engagement is positively influenced by robust HR practices.

### 7.1.5 Work-Life Balance

Work-life balance has a deep effect on satisfaction and retention of the employee. Ho and Tzeng (2011) establish that a positive work-life balance considerably contributes to job satisfaction, especially in high-stress professions. This is resonated by Lingard and Turner (2017), who deliberated on the effect of organizational factors on employee well-being, illuminating that helpful workplace policies endorse healthier work-life assimilation

### 7.1.6 Job Stress and Performance

Job stress is an acute factor affecting employee performance. Gupta et al. (2015) delivered evidence of the unfavorable effects of job stress on employee productivity, emphasizing the prerequisite for actual stress management interventions. Morrow et al. (2023) discovered the relationship between job demands and resources, indicating that balancing these factors can improve employee performance and engagement.

### 7.1.7 Leadership and Employee Well-Being

Leadership plays an essential role in stimulating employee well-being and nurturing a conducive and supportive work environment. Parker and Griffin (2011) piloted a meta-analysis enlightening that effective leadership behaviors positively correlate with employee performance and satisfaction of the employee. Zohar (2002) recommended a leadership-based intervention model directed at improving workplace safety, moreover emphasizing the significance of positive leadership in improving employee well-being (Khatri et al., 2024; Khatri et al., 2023).

### 7.1.8 Interventions for Work-Related Stress

Tuckey and Neall (2014) discovered the role of organizational culture in determining employee well-being, signifying that directed interventions can lessen work-related stress. Catapano et al. (2023) studied current practices in healthcare settings, underlining the inevitability for methodical approaches for stress management, while Van der Klink et al. (2001) scrutinized the effects of occupational health interferences on work-related stress and health outcomes.

The role of ergonomics in improving workplace safety and productivity cannot be ignored. Amar (2011) emphasized the importance of ergonomics in avoiding workplace injuries, while Citro and Ling (2012) discoursed the integration of ergonomic practices in healthcare settings to augment employee comfort and efficiency

## 7.2 LITERATURE REVIEW

Workplace ergonomics studies have revealed several factors contributing to musculoskeletal and visual discomfort among employees using computer. These factors include increased job demands and more hours working at a computer (e.g., Bernard et al., 1994; Faucett & Rempel, 1994), sustained awkward head and arm postures (Marcus & Gerr, 1996; Tittiranonda et al., 1999), increased levels of psychological stress and poor psychosocial work environment, (e.g., Bongers et al., 1993; Carayon & Smith, 2000; Faucett & Rempel, 1994; Marcus & Gerr, 1996), work organizational factors (e.g., Lassen et al., 2009; Punnett & Bergqvist, 1997), a lack of specific ergonomic features in the workstations and office buildings, and poor lighting (e.g. Nelson & Silverstein, 1998; Sauter et al., 1990). Typically, these studies are cross-sectional in design and only describe the work, safety, and health experiences of computer and office workers at one time period (Demure et al., 2000).

Few longitudinal field or lab studies have examined the effects of office ergonomics interventions on workers' health, safety, and performance (Brewer et al., 2006; Buckle et al., 1997; Karsh et al., 2014). Although there is a growing interest among employers to improve office workplaces, studies that investigated the effects of workstation, eyewear, and behavioral interventions on upper body musculoskeletal and visual symptoms are of mixed quality (Brewer et al., 2006; Karsh et al., 2014).

There is some evidence, however, that ergonomics training (Brisson et al., 2000) in workstation and building design (e.g., Hagberg et al., 2007; Nelson & Silverstein, 1998; Sauter et al., 1990) can prevent or reduce musculoskeletal and visual discomforts and symptoms in office environments. A suggestive way for reducing the musculoskeletal and visual symptoms among the employees is to formulate special training programs for workstation changes. These training programs in ergonomics help employees to understand proper workstation set-up and postures (Bohr, 2000; Ketola et al., 2004). However, a significant decrease in work-related musculoskeletal disorders (WMSDs) and visual discomfort has been observed when workers were given an adjustable/flexible work environment coupled with ergonomics training (Amick et al., 2003; Robertson et al., 2013). Further, the provision of control over the work environment through adjustability and knowledge may enhance worker effectiveness as well as their health and safety (Hedge & Ray, 2004; McLaney & Hurrell, 1988; O'Neill et al., 1994; Robertson et al., 2013; Smith & Bayeh, 2003).

According to a study, the trained group who used workstations sitting/standing showed lower musculoskeletal and visual uneasiness as compared to an untrained reference group also working in the same configuration. Moreover, the trained group had significantly higher performance and effectiveness than the reference group as exhibited by higher accuracy and quality control scores (Robertson et al., 2013). These findings, which indicate the ability to mitigate symptoms, change behaviors, and enhance performance through training combined with a sit/stand workstation, have implications for preventing discomforts in office and computer workers (Robertson et al., 2013). These results are also supported by the conclusions of a systematic review conducted by Brewer et al. (2006) with their critical and strict 11-point inclusion criteria of 31 studies. A positive effect of alternative pointing devices was found. They further concluded that for all other workplace interventions, a mixed or insufficient evidence of effect was observed (Brewer et al., 2006). Other related areas of workplace interventions, such as participatory ergonomics interventions, have also shown mixed results (Rivilis et al., 2008).

The literature indicates a composite relationship between organizational culture, leadership, HR practices, and employee well-being. Organizations highlighting these factors can promote an environment that encourages employee satisfaction, lessens stress, and eventually boosts performance.

## 7.3 STRATEGIC ORIENTATION AND HF&E INTERVENTIONS

Strategic orientation is the guiding torch that escorts organizations toward attaining competitive advantage and sustainable growth by aligning their objectives, resources, and activities. As defined by Narver and Slater (1990), it encompasses a practical approach to acclimating to market changes and influencing internal capabilities to meet organizational objectives effectively. This section will investigate deeper into the components and importance of strategic orientation, stressing its role in directing the complexities of the business environment and driving the success of the organization.

At its base strategic orientation exemplifies a progressive mind-set that forestalls and reacts to changes in the market landscape. Strategic orientation empowers organizations to probe the external environment, recognize evolving trends, and benefit from opportunities while modifying risks. By staying accustomed to market dynamics, organizations can position themselves strategically to advance a competitive edge.

Furthermore, strategic orientation requires the orientation of organizational goals with internal competences and resources. It requires an assessment of SWOT (Strengths, Weaknesses, Opportunities & Threats) Analysis to develop a clear strategic direction. This orientation confirms that organizational resources are positioned proficiently to achieve anticipated outcomes. For instance, if an organization recognizes innovation as a significant strategic objective, it should allocate resources headed for research and development initiatives to nurture a culture of innovation.

In today's unpredictable business environment, organizations cannot afford to remain stationary. They must adapt to change and innovation to stay ahead. Strategic orientation inspires organizations to finance research and development, discover new markets, and expand product offerings to withstand long-term growth. By nurturing a culture of innovation, organizations can set apart themselves from rivals and generate value for customers.

The additional grave aspect of strategic orientation is the hands-on exploitation of internal capabilities. Organizations retain unique resources, comprising human capital, technology, and intellectual property, which can be controlled to gain a competitive advantage. Strategic orientation comprises identifying and connecting these capabilities to produce value for stakeholders. For example, if an organization outshines in customer service, it can use this forte to discriminate itself from contestants and increase customer loyalty.

Besides, strategic orientation nurtures a customer-centric method of decision-making. Organizations should understand the needs and likings of their target customers to deliver value efficiently. Strategic orientation encourages organizations to section their target market, tailor products and services to meet customer needs and build strong relationships with customers. By prioritizing customer satisfaction and loyalty, organizations can enhance brand reputation and drive revenue growth.

In short, strategic orientation is an essential driver of organizational success, empowering organizations to attain competitive advantage and sustainable development in today's vibrant business environment. By orienting the organizational goals towards internal capabilities and market opportunities, organizations can steer complexities commendable and exploit emerging trends. Strategic orientation nurtures a culture of innovation, unceasing improvement, and customer-centricity, placing organizations for long-term success and resilience.

### 7.3.1 HF&E Interventions

HF&E interventions play a crucial role in optimizing the interactions between individuals, their work environment, and technology to enhance efficiency, safety, and satisfaction. As highlighted by Carayon (2006), these interventions encompass a range of strategies aimed at improving various aspects of the work environment to promote employee well-being and productivity. This section will explore the key components and significance of HF&E interventions, emphasizing their impact on organizational performance and employee satisfaction.

At the heart of HF&E interventions lies the principle of optimizing the fit between individuals and their work environment. Task analysis is a foundational element of HF&E interventions, involving the systematic examination of job tasks to identify potential sources of strain or inefficiency. By breaking down complex tasks into their constituent elements, organizations can identify opportunities for improvement and design workflows that enhance productivity and minimize errors.

Moreover, ergonomic workstation design is a cornerstone of HF&E interventions, focusing on creating work environments that support employee health and comfort. Ergonomic principles dictate the design of workstations, seating arrangements, and equipment to minimize physical strain and reduce the risk of musculoskeletal disorders. For example, flexible desks and chairs allow employees to improvise their workspace to suit their individual needs, promoting comfort and productivity.

Psychosocial interventions are another essential component of HF&E interventions, addressing the psychological aspects of the work environment. These interventions aim to promote employee well-being by addressing factors such as job stress, workload, and interpersonal relationships. Strategies may include implementing flexible work arrangements, providing stress management training, and fostering a supportive organizational culture. By addressing psychosocial risk factors, organizations can foster satisfied employees, help decrease absenteeism, and improve overall organizational performance.

Furthermore, workload management is critical for ensuring employee well-being and productivity. HF&E interventions focus on optimizing workloads to prevent burnout and fatigue while maintaining high levels of performance. This may involve workload balancing, task prioritization, and the implementation of time management strategies. By ensuring that workloads are manageable and sustainable, organizations can prevent employee burnout and maintain a motivated and engaged workforce.

One of the key benefits of HF&E interventions is their potential to improve organizational performance and competitiveness. Research suggests that organizations that prioritize employee well-being and safety experience higher levels of productivity, lower absenteeism rates, and reduced turnover (Bakker et al., 2008). By investing in HF&E interventions, organizations can create a conducive working environment.

In short, HF&E interventions play a critical role in optimizing the interactions between individuals, their work environment, and technology to enhance efficiency, safety, and satisfaction. By implementing strategies such as task analysis, ergonomic workstation design, psychosocial interventions, and workload management, organizations can create a supportive work environment that promotes employee well-being and productivity. Investing in HF&E interventions not only benefits employees but also contributes to organizational performance and competitiveness, making it a strategic imperative for organizations in today's competitive business environment.

### 7.3.2 Linking Human Factors to an Organizations' Strategy

Human factors (HF) considerations can be accepted internalized in organizations if they regarded by managers as an important factor contributing directly to the companies' strategies (Dul & Neumann, 2009). HF groups are actively involved in design activities in order to prevent problems associated with ergonomics as these contribute to the organizational success and long-term objectives. However, the challenge is that in their often limited "support" role, HF groups are frequently disconnected from management strategies and others affecting business such as engineering groups (Perrow, 1983; Dixon et al., 2009; Jensen, 2002). HF engineers (used synonymously here with Ergonomists) may be perceived as defenders of operators (Perrow, 1983), rather than being essential to achieving the organizations objectives, or integrated throughout the production systems design.

While there is evidence that engineering changes are more difficult and costly when HF considerations are incorporated late in the design lifecycle (Seim and &, 2010), HF engineers have difficulty accessing design groups. HF engineers are often consulted too late in the design process; they are seen as critical of the design (Broberg, 2007; Kirwan, 2000), thereby increasing costs and delays (Perrow, 1983). HF engineers can reduce the conflict described if they know the strategic goals of the design process and align HF with these goals.

Waterson and Kolose (2010), in a large-scale military defense organization, discussed some of the challenges experienced by the HF team with management perceptions of their function and purpose. Although the team had been in existence and had supported many functions for several years, the authors reported that it still lacked visibility and prominence in the overall organizational structure. It was referred to as a "bubble" on a flow diagram of an organization chart. It appeared that others in the organization were unaware of the value of involving the HF team. Integrating HF into an organization, like other business reengineering and quality efforts, is really an "organizational change" initiative because it affects multiple levels in an organization and requires the interaction of different agents.

The research on organizational change strongly suggests that successful initiatives are the ones that are integrated into the "psychological" dimension, or mindset, of the organization (Zink et al., 2008). Similarly, to facilitate the integration of HF among groups and functions throughout an organization, and especially to gain management commitment, HF must be aligned with the strategic objectives and business function strategies (Dul & Neumann, 2009; Drury & Reicher, 2000; Genaidy et al., 2009; Neumann et al., 1999).

## 7.4 ERGONOMICS IN WORKPLACE

Jobs designed ergonomically help in achieving higher productivity among employees as well as result in quality work and improved safety at workplace. Therefore, the organizations must manage health and safety process that involves better designing of tasks, workstations, tools, and equipment in order to minimize and reduce the risk associated with work that can lead to injury and disability. The following guidelines are adapted from the 2002 final draft of ASC Z365 Management of Work-Related Musculoskeletal Disorders and describe the elements of an ergonomics program and process for managing WMSDs to reduce frequency and disability.

### 7.4.1 Employer Responsibilities

The implementation of effective ergonomics process requires the intervention of employers in order to prevent and control activities that can lead to ergonomics issues in the workplace. This can be addressed by highlighting the workplace problems already prevailing. Some organizations focus on the management of identified WMSD cases and thereby evaluating and intervening. Others, in workplaces with few or no WMSD cases or high attrition rate, may involve proactive job surveys like employee interviews, checklists method etc. so that potential problems associated with particular jobs can be identified and addressed before new WMSD cases appear.

### 7.4.2 Training

Training programs are necessary for employees and managers that can help in surveillance, job analysis, job design, and medical management. Training may include recognition and reporting of the signs and symptoms of WMSDs that may be work related or for record-keeping processes for identifying WMSDs. Such training programs will ensure that employees know whom to contact for further assistance in case of WMSD. They help in the recognition and management of WMSD risk factors and for further job analysis and design processes. These can help train the employees on proper use, adjustment, and maintenance of tools, work equipment, and work stations along with job interventions and best-work procedures and practices for minimizing the risk of WMSDs

### 7.4.3 Employee Involvement

The employees should be given the opportunity to actively participate in the management of WMSDs. The employees can be involved by asking for suggestions or concerns and also by their participation in discussions related to their workplace and work methods. Employee surveys and formal meetings can also be arranged to discuss the cases of WMSDs. The employees can also get involved in the design of work, equipment, and procedures and identify the risk associated with them in a timely fashion so that the employers can take corrective action.

### 7.4.4 Injury and Hazard Surveillance

If timely surveillance is carried in the workplace, it can help determine when and where job analysis is needed and possible ergonomic interventions can be carried. Organizations may carry out periodic job surveys or health surveillance that can indicate the need for a job analysis for setting intervention priorities and assessing the program. Surveillance in the workplace includes the initial review of existing records of work-related illnesses and injuries such as 'Occupational Safety and Health Administration' (OSHA) log and worker's compensation records. This analysis will help determine where WMSDs are occurring and will help prioritize jobs needing further analysis. Moreover, Employee reports on WMSD symptoms or employee concerns about WMSD risk factors can be periodically reviewed. Another method is to carry out job surveys that can be included in the existing programs such as regular safety, health, team problem solving, or quality inspections, and can expand their scope to include identification of WMSD risk factors.

### 7.4.5 Evaluation and Management of WMSD Cases

If the WMSDs are assessed at an early stage it can help in initiating the timely treatment and also limit the severity, help in effective treatment, and timely recovery. Moreover, early diagnosis of WMSDs can help the management to take corrective action like job analysis of that particular job. The management should examine existing policies, practices, and programs to make sure that they help in identifying potential WMSD risk factors without reprisals.

### 7.4.6 Job Analysis & Job Design

Job analysis identifies potential exposures to work-related risk factors and evaluates their characteristic properties. Job analysis should be performed when it is identified that an MSD is work related or when a problem persists after changes have been implemented. Also, it can be conducted during the design or acquisition phase of equipment, processes, or jobs. Another way to implement effective job design is to reduce exposure to risk factors that can lead to the severity of WMSDs. When work-related risk factors are identified from job analysis, jobs can be redesigned or reengineered or administrative changes can be done in order to eliminate or reduce exposure.

## 7.5 CONCLUDING REFLECTIONS

The amalgamation of strategic orientation with HF&E interventions suggests ample benefits for the organizations. First, it aligns the strategies of the organization with the requirements and inclinations of the employee thus generating a work culture that emphasizes well-being and satisfaction of the employees. By taking into consideration the ergonomic design of workstations and application of the psychosocial interventions, organizations show their obligation and reassurance to create a favorable and compassionate work environment.

Secondly, assimilating strategic orientation with HF&E interventions improves organizational performance and its competitiveness. Research proposes that organizations that highlight employee well-being and safety witness reduced turnover rates, shrunk absenteeism, and higher levels of productivity (Bakker et al., 2008). By capitalizing on HF&E interventions, organizations can encourage employees and effectively involve the employees for job satisfaction, leading to enhanced organizational performance and competitive advantage.

Moreover, the incorporation of strategic orientation with HF&E interventions shows an environment of unceasing innovation and improvement. Through employee enablement in the decision-making process and imploring feedback on ergonomic interventions, organizations can recognize chances for optimization and innovation. This strategic policy for employee well-being improves organizational performance along with firming the commitment and loyalty of the employees.

In short, strategic orientation and HF&E interventions are indispensable constituents of the success of business firms and possessing cutting edge over competitors. Organizations can craft a work culture that fosters innovation, collaboration, and continuous improvement by aligning organizational strategies with employee well-being and productivity, investing in HF&E interventions not only improves employee satisfaction and performance but also reinforces organizational competitiveness and sustainability. Therefore, organizations must prioritize strategic orientation and HF&E interventions to achieve long-term success and growth.

## REFERENCES

Almutairi, D. O. (2016). The mediating effects of organizational commitment on the relationship between transformational leadership style and job performance. *International Journal of Business and Management*, *11*(1), 231.

Amar, P. (2011). Turning the gendered politics of the security state inside out? Charging the police with sexual harassment in Egypt. *International Feminist Journal of Politics*, *13*(3), 299–328.

Amick III, B. C., Robertson, M. M., DeRango, K., Bazzani, L., Moore, A., Rooney, T., & Harrist, R. (2003). Effect of office ergonomics intervention on reducing musculoskeletal symptoms. *Spine*, *28*(24), 2706–2711.

Ansari, M. S. A., Khan, S. A., & Bhuyan, U. (2024). Exploring the strategic orientation factors influencing the organizational performance through bibliometric analysis. In S. Umair, U. Waqas, B. Mrugalska, & I. R. Al Shamsi (Eds.), *Human Perspectives of Industry 4.0 Organizations* (pp. 68–88). CRC Press.

Baker, W. E., & Sinkula, J. M. (2009). The complementary effects of market orientation and entrepreneurial orientation on profitability in small businesses. *Journal of Small Business Management*, *47*(4), 443–464. https://doi.org/10.1111/j.1540-627X.2009.00279.x

Bakker, A. B., Demerouti, E., & Euwema, M. C. (2008). Job resources buffer the impact of job demands on burnout. *Journal of Occupational Health Psychology*, *13*(3), 293–311.

Bartsch, V., Ebers, M., & Maurer, I. (2013). Learning in project-based organizations: The role of project teams' social capital for overcoming barriers to learning. *International Journal of Project Management, 31*(2), 239–251.

Beheshtifar, M., & Shariatifar, F. (2013). A study of relationship between organizational structure and culture with corporate entrepreneurship. *Entrepreneurship and Innovation Management Journal, 1*(1), 19–32.

Bernard, G. R., Artigas, A., Brigham, K. L., Carlet, J., Falke, K., Hudson, L., ... & Spragg, R. (1994). The American-European consensus conference on ARDS. Definitions, mechanisms, relevant outcomes, and clinical trial coordination. *American Journal of Respiratory and Critical Care Medicine, 149*(3), 818–824.

Bohr, P. C. (2000). Efficacy of office ergonomics education. *Journal of Occupational Rehabilitation, 10*, 243–255.

Bongers, P. M., de Winter, C. R., Kompier, M. A., & Hildebrandt, V. H. (1993). Psychosocial factors at work and musculoskeletal disease. *Scandinavian Journal of Work, Environment & Health, 19*(5), 297–312.

Brewer, S., Eerd, D. V., Amick Iii, B. C., Irvin, E., Daum, K. M., Gerr, F., ... & Rempel, D. (2006). Workplace interventions to prevent musculoskeletal and visual symptoms and disorders among computer users: A systematic review. *Journal of Occupational Rehabilitation, 16*, 317–350.

Brisson, J., Brisson, B., Coté, G., Maunsell, E., Bérubé, S., & Robert, J. (2000). Tamoxifen and mammographic breast densities. *Cancer Epidemiology Biomarkers & Prevention, 9*(9), 911–915.

Broberg, O. (2007). Integrating ergonomics into engineering: Empirical evidence and implications for the ergonomists. *Human Factors and Ergonomics in Manufacturing & Service Industries, 17*(4), 353–366.

Buckle, A. M., Zahn, R., & Fersht, A. R. (1997). A structural model for GroEL–polypeptide recognition. *Proceedings of the National Academy of Sciences, 94*(8), 3571–3575.

Carayon, P. (2006). *Handbook of Human Factors and Ergonomics in Health Care and Patient Safety*. CRC Press.

Carayon, P., & Smith, M. J. (2000). Work organization and ergonomics. *Applied Ergonomics, 31*(6), 649–662.

Catapano, P., Cipolla, S., Sampogna, G., Perris, F., Luciano, M., Catapano, F., & Fiorillo, A. (2023). Organizational and individual interventions for managing work-related stress in healthcare professionals: a systematic review. *Medicina, 59*(10), 1866.

Chen, S. W., & Peng, J. C. (2021). Determinants of frontline employee engagement and their influence on service performance. *The International Journal of Human Resource Management, 32*(5), 1062–1085.

Citro, C. F., & Ling, R. (2012). Workplace ergonomics: A review of current practices and future directions. *Workplace Health & Safety, 60*(11), 503–511. https://doi.org/10.1177/216507991206001102

Demure, B., Luippold, R. S., Bigelow, C., Ali, D., Mundt, K. A., & Liese, B. (2000). Video display terminal workstation improvement program: I. Baseline associations between musculoskeletal discomfort and ergonomic features of workstations. *Journal of Occupational and Environmental Medicine, 42*(8), 783–791.

Dixon, J., Braun, H. J., Kosina, P., & Crouch, J. H. (Eds.). (2009). *Wheat Facts and Futures 2009*. Cimmyt.

Drury, J., & Reicher, S. (2000). Collective action and psychological change: The emergence of new social identities. *British Journal of Social Psychology, 39*(4), 579–604.

Dul, J., & Neumann, W. P. (2009). Ergonomics contributions to company strategies. *Applied Ergonomics, 40*(4), 745–752.

Faucett, J., & Rempel, D. (1994). VDT-related musculoskeletal symptoms: Interactions between work posture and psychosocial work factors. *American Journal of Industrial Medicine*, *26*(5), 597–612.

Genaidy, A. M., Rinder, M. M., Sequeira, R., & A- Rehim, A. D. (2009). The work compatibility improvement framework: Theory and application of improvement action and intervention strategies. *Ergonomics*, *52*(5), 524–559.

Gupta, R. K., Gangoliya, S. S., & Singh, N. K. (2015). Reduction of phytic acid and enhancement of bioavailable micronutrients in food grains. *Journal of Food Science and Technology*, *52*, 676–684.

Hagberg, M., Vilhemsson, R., Tornqvist, E. W., & Toomingas, A. (2007). Incidence of self-reported reduced productivity owing to musculoskeletal symptoms: Association with workplace and individual factors among computer users. *Ergonomics*, *50*(11), 1820–1834.

Hedge, A., & Ray, E. J. (2004, September). Effects of an electronic height-adjustable worksurface on computer worker musculoskeletal discomfort and productivity. In *Proceedings of the Human Factors and Ergonomics Society Annual Meeting* (Vol. 48, No. 8, pp. 1091–1095). SAGE Publications

Ho, C. H., & Tzeng, Y. M. (2011). The impact of work-life balance on job satisfaction: A case study of health care professionals. *Journal of Business Research*, *64*(7), 693–700. https://doi.org/10.1016/j.jbusres.2010.08.006

Jensen, M. C. (2002). Value maximization, stakeholder theory, and the corporate objective function. *Business Ethics Quarterly*, *14*(3), 235–256.

Karasek, R. A., & Theorell, T. (1990). *Healthy Work: Stress, Productivity, and the Reconstruction of Working Life*. Basic Books.

Karsh, K. L., Trull, T. W., Sigman, D. M., Thompson, P. A., & Granger, J. (2014). The contributions of nitrate uptake and efflux to isotope fractionation during algal nitrate assimilation. *Geochimica et Cosmochimica Acta*, *132*, 391–412.

Ketola, R., Toivonen, R., Luukkonen, R., Takala, E. P., & Viikari-Juntura, E. (2004). Expert assessment of physical ergonomics at video-display unit workstations: Repeatability, validity and responsiveness to changes. *International Archives of Occupational and Environmental Health*, *77*, 437–442.

Khan, S. A., & Ansari, M. S. A. (2024). Critical factors for relationship between strategic orientation, corporate success, and innovation. In S. Umair, U. Waqas, B. Mrugalska, I. R. Al Shamsi (Eds.), *Human Perspectives of Industry 4.0 Organizations* (pp. 89–110). CRC Press.

Khatri, B., Arora, S., & Khan, S. A. (2024). Analysis of servant leadership and organizational citizenship behaviour among social service sector employees. In D. Belias, I. Rossidis, & C. Papademetriou (Eds.), *Organizational Behavior and Human Resource Management for Complex Work Environments* (pp. 443–465). IGI Global. https://doi.org/10.4018/979-8-3693-3466-9.ch021

Khatri, B., Shrimali, H., Khan, S. A., & Naim, A. (2023). Role of HR analytics in ensuring psychological wellbeing and job security: Learnings from COVID-19. In R. Yadav, M. Sinha, & J. Kureethara (Eds.), *HR Analytics in an Era of Rapid Automation* (pp. 36–53). IGI Global. https://doi.org/10.4018/978-1-6684-8942-0.ch003

Kirwan, B. (2000). Soft systems, hard lessons. *Applied Ergonomics*, *31*(6), 663–678.

Lassen, M. R., Raskob, G. E., Gallus, A., Pineo, G., Chen, D., & Portman, R. J. (2009). Apixaban or enoxaparin for thromboprophylaxis after knee replacement. *New England Journal of Medicine*, *361*(6), 594–604.

Li, S., Shi, Y., Wang, L., & Xia, E. (2023). A bibliometric analysis of brand orientation strategy in digital marketing: Determinants, research perspectives and evolutions. *Sustainability*, *15*(2), 1486.

Lingard, H., & Turner, M. (2017). Work and well-being in the construction industry. In R. J. Burke & K. M. Page (Eds.), *Research Handbook on Work and Well-Being* (pp. 189–215). Edward Elgar Publishing.

Marcus, M., & Gerr, F. (1996). Upper extremity musculoskeletal symptoms among female office workers: Associations with video display terminal use and occupational psychosocial stressors. *American Journal of Industrial Medicine*, *29*(2), 161–170.

McLaney, M. A., & Hurrell Jr, J. J. (1988). Control, stress, and job satisfaction in Canadian nurses. *Work & Stress*, *2*(3), 217–224.

Morrow, W. J., Adams, B. L., & Vickovic, S. G. (2023). Exploring the gendered nature of work-family conflict on job stress, job satisfaction, and organizational commitment among correctional officers. *Criminal Justice Studies*, *36*(4), 370–395.

Narver, J. C., & Slater, S. F. (1990). The effect of a market orientation on business profitability. *Journal of Marketing*, *54*(4), 20–35. https://doi.org/10.1177/002224299005400402

Nelson, N. A., & Silverstein, B. A. (1998). Workplace changes associated with a reduction in musculoskeletal symptoms in office workers. *Human Factors*, *40*(2), 337–350.

Neumann, G., Watanabe, T., Ito, H., Watanabe, S., Goto, H., Gao, P., ... & Kawaoka, Y. (1999). Generation of influenza A viruses entirely from cloned cDNAs. *Proceedings of the National Academy of Sciences*, *96*(16), 9345–9350.

O'Neill, E. M., Rebay, I., Tjian, R., & Rubin, G. M. (1994). The activities of two Ets-related transcription factors required for Drosophila eye development are modulated by the Ras/MAPK pathway. *Cell*, *78*(1), 137–147.

Parker, S. K., & Griffin, M. A. (2011). Understanding the role of leadership in promoting employee well-being: A meta-analysis. *Journal of Organizational Behavior*, *32*(2), 139–156. https://doi.org/10.1002/job.642

Perrow, C. (1983). The organizational context of human factors engineering. *Administrative Science Quarterly*, *8*(4), 521–541.

Punnett, L., & Bergqvist, U. (1997). *Visual Display Unit Work and Upper Extremity Musculoskeletal Disorders*. National Institute for Working Life.

Rivilis, I., Van Eerd, D., Cullen, K., Cole, D. C., Irvin, E., Tyson, J., & Mahood, Q. (2008). Effectiveness of participatory ergonomic interventions on health outcomes: A systematic review. *Applied Ergonomics*, *39*(3), 342–358.

Robertson, M. M., Ciriello, V. M., & Garabet, A. M. (2013). Office ergonomics training and a sit-stand workstation: Effects on musculoskeletal and visual symptoms and performance of office workers. *Applied Ergonomics*, *44*(1), 73–85.

Salvendy, G. (2012). *Handbook of Human Factors and Ergonomics*. John Wiley & Sons.

Sauter, N. P., Toni, R., Mclaughlin, C. D., Dyess, E. M., & Lechan, R. M. (1990). Isolated adrenocorticotropin deficiency associated with an autoantibody to a corticotroph antigen that is not adrenocorticotropin or other proopiomelanocortin-derived peptides. *The Journal of Clinical Endocrinology & Metabolism*, *70*(5), 1391–1397.

Schneider, B., & Barbera, K. M. (2014). *The Oxford Handbook of Organizational Climate and Culture*. Oxford University Press.

Schulte, P., & Vainio, H. (2010). Well-being at work–overview and perspective. *Scandinavian Journal of Work, Environment & Health*, *36*(5), 422–429.

Seim, R., & Broberg, O. (2010). Participatory workspace design: A new approach for ergonomists? *International Journal of Industrial Ergonomics*, *40*(1), 25–33.

Smith, M. J., & Bayeh, A. D. (2003). Do ergonomics improvements increase computer workers' productivity? An intervention study in a call centre. *Ergonomics*, *46*(1–3), 3–18.

Tittiranonda, P., Rempel, D., Armstrong, T., & Burastero, S. (1999). Effect of four computer keyboards in computer users with upper extremity musculoskeletal disorders. *American Journal of Industrial Medicine*, *35*(6), 647–661.

Tuckey, M. R., & Neall, A. M. (2014). Workplace bullying erodes job and personal resources: between-and within-person perspectives. *Journal of Occupational Health Psychology*, *19*(4), 413.

Van der Klink, J. J., Blonk, R. W., Schene, A. H., & Van Dijk, F. J. (2001). The benefits of interventions for work-related stress. *American Journal of Public Health*, *91*(2), 270.

Waterson, P., & Kolose, S. L. (2010). Exploring the social and organisational aspects of human factors integration: A framework and case study. *Safety Science*, *48*(4), 482–490.

Zareen, S. & Khan, S. A. (2023). Exploring dependence of human resource management (HRM) on internet of things (IoT) and digital marketing in the digital era. In A. Naim & V. Devi (Eds.), *Global Applications of the Internet of Things in Digital Marketing* (pp. 51–66). IGI Global. https://doi.org/10.4018/978-1-6684-8166-0.ch004

Zink, C. F., Tong, Y., Chen, Q., Bassett, D. S., Stein, J. L., & Meyer-Lindenberg, A. (2008). Know your place: Neural processing of social hierarchy in humans. *Neuron*, *58*(2), 273–283.

Zohar, D. (2002). Modifying supervisory practices to improve safety behavior in the workplace: A leadership-based approach. *Journal of Safety Research*, *33*(2), 139–160. https://doi.org/10.1016/S0022-4375(02)00014-2

# 8 Exploring the Role of Employee Participation in Ergonomic Transformations

## *A Study in Change Management*

*Sumit Narula, Dhruv Sabharwal, and Robin Kabha*

## 8.1 INTRODUCTION

In any company or organization, the overall efficiency depends largely on the level of contribution that employees make to ensure that everything runs smoothly during a transition and adjustment period (Zareen & Khan, 2023; Khatri et al., 2023). With ever-changing markets, technological advancements, and competitive landscapes, transforming businesses in order to remain relevant and agile requires several change initiatives, but these transformation projects can never succeed without the active participation and commitment of people within the business. Therefore, it is important to understand what employee engagement is and how it plays a role in the transformation of an organization. The term "employee engagement," which (Kahn, 1990) defined as the mental and emotional commitment that employees feel towards their organization, was elucidated in use. They are more than happy; they demonstrate how engaged people are across all organizational levels. Management shibboleths like enthusiasm, proactivity, and the extra mile are central to engaged employees (see, for example, Bakker & Demerouti, 2017). Employee engagement is comprised of multiple dimensions and elements that all contribute to the motivation and loyalty of your employees. Key factors, according to Saks (2006), are job satisfaction, organizational commitment, participation in decision-making processes, and opportunities for personal growth and development (Rich et al., 2010) focused on the fact that employee engagement refers to a multifaceted concept influenced by individual, job, and organizational aspects. However, recognizing all these different aspects allows organizations to create strategies that work on the employee engagement and sustainability front.

The idea of employee engagement and its dynamics have been studied by a number of theoretical frameworks and models. According to Bakker and Demerouti (2017),

DOI: 10.1201/9781003532293-9

the model suggested is called Job Demand Resources, which says that job demands, such as workload and time pressure, interact with job resources, like autonomy and social support, to affect employee engagement. The JD-R model states that when job resources are plentiful and well-managed, employee engagement levels rise.

There is a well-known concept called the engagement-disengagement continuum, which divides workers into three groups: actively disengaged, not engaged, and engaged (Gallup, n.d.). This model highlights the urgency of handling disengagement to avoid negative outcomes as well as fostering engagement in order to achieve positive ones (Charteris, 1998). With a strong grasp of these theoretical frameworks and models in place, firms can create interventions and practices that promote employee engagement for any organizational transformation (Berkowitz et al., 1999). By aligning change objectives with engagement criteria, companies can create an environment where employees are inspired, respected, and actively involved in the change process.

### 8.1.1 Background of the Study

For businesses to succeed in today's fast-paced business environment, there are several effective methods of change management that must be used. Such as the active involvement of the people who work for these organizations in such activities. This literature review covers the factors that influence employee engagement during organizational changes, the importance of employee engagement in change management, and the link between employee engagement and the success of change initiatives (Beek et al., 1992). These factors can facilitate better results from their changes by improving organizations' understanding of them and maintaining high levels of employee engagement during the transformation process.

Effective management is necessary with regards to changing ideas since this affects how open or closed one is in regards to new ideas. Employee engagement makes workers more liable towards adopting alterations, which they handle positively, exhibit an exceeding drive for assisting successful conclusions, and are loyal throughout difficult phases (Alfes et al., 2013). In addition, Rich et al. (2010) found that motivated employees had higher adaptability rates as well as being quick learners and behavior adjusters. It follows, then, that businesses must emphasize enabling worker participation if they want to ensure that change management initiatives produce the best possible outcomes.

According to certain studies, motivated employees are more likely to adjust well to changes inside the company. Workers who have a stake in the company's success are more likely to accept the change and be fully committed to it (Eisenbeiss et al., 2008). By offering to demonstrate initiative, they offer suggestions, look for opportunities to help out, and work together to facilitate the change process (Rafferty et al., 2013). The presence of engaged employees enhances positive societal effects and a change-supportive culture (Bakker et al., 2014). Macey and Schneider, on the other hand, add that companies that encourage their employees to participate in the change process see improvements in overall company performance, employee satisfaction, and implementation success.

In order for there to be effective management of change initiatives, it is important to understand the factors that influence employee engagement during organizational change. The degree of leadership participation and support greatly affects employee buy-in to a change program (Herscovitch & Meyer, 2002). Employee confidence and trust are built through leaders who actively communicate about the need for change, provide guidance on how to go about it, and show commitment (Appelbaum et al., 2012). Caldwell et al. (2012) also noted that empowering employees to make decisions helps improve engagement levels and facilitates workers' adjustment to changes. Timely provision of information as well as resolution of concerns are essential for keeping staff engaged during organizational changes through effective communication (Van den Heuvel et al., 2015). Ford et al. (2011) claim that transition training resources, support services, and accessibility are key determinants of employee involvement levels in the workforce. The involvement of employees in their work is more likely if organizations can develop their skills, offer enough support, and deploy resources in line with the objectives they want to achieve.

To successfully manage change, there must be some level of employee buy-in. For a change to be effectively implemented within an organization, it requires the engagement of its entire labor force, as this leads to positive results for all stakeholders. Some aspects that impact the commitment to change during organizational changes include leadership encouragement, staff participation, transparency, and the availability of supportive systems and facilities. By dealing with these issues, companies can build a culture that promotes participation, thus enabling the successful implementation of change plans.

### 8.1.2 Research Objectives

1. To understand how employee involvement affects the durability and efficacy of ergonomic changes in the workplace.
2. To examine the methods of change management that successfully encourage staff involvement in ergonomic projects.

## 8.2 LITERATURE REVIEW

The present chapter contains the literature review that aims at presenting the state of prior researches concerning the notion of employees' involvement in the change management processes at organizational level with a specific emphasis on ergonomical changes. In the present review of literature, consideration is made regarding several theoretical frameworks and models that postulate employee engagement as a critical factor in the success of change initiatives. According to Kahn (1990) and Bakker and Demerouti (2017), employee engagement is a personal and psychological state of being that employees experience toward their work and organization In addition, the review also pinpointed that it enhances motivation, organizational commitment, and the alignment to the organizational objectives (Ansari et al., 2024; Khan & Ansari, 2024).

It also discusses the Job Demands Resources model and the Engagement Disengagement Continuum in giving a better understanding of the aspects that engender employee engagement during change processes. Moreover, the literature reviews the results of empirical studies as to the positive effects of EPKs on such criterion variables as change recipient's acceptance of change, job satisfaction, and organizational success (Anton et al., 2002). While the current review establishes the rationale for the current study by providing a background and theoretical framework to guide its development, it also builds the theoretical framework of the current research using the synthesis and critical analysis of previous studies. The support, development, and help during these changes (Ferguson et al., 1996). This can be regarded as a further indication of the argument that commitment is more likely to be established when firms endeavor to enhance skill, provide adequate support, and link applications to change goals.

## 8.3 RESEARCH METHODOLOGY

This study used a mixed-methods assessment in evaluating the effects of employees' involvement in ergonomic improvement endeavors complemented by web-based surveys and face-to-face interviews.

### 8.3.1 Participants

Self-administered questionnaires are utilized in the quantitative approach of the study to assess the attitudes of employees and their interaction with ergonomics in their workplace. Therefore, data gathering through surveys was used to obtain structured information on the current level of happiness among employees, the extent to which they are engaged in processes, and perceived effectiveness of changes. The questions posed in the survey were carefully formulated based on such theoretical pillars as managerial and organizational psychology, and change management as well as deriving from the current literature. Surveys were administered to 70 employees originating from different divisions of various organizations, going through ergonomic enhancement. As well as structural surveys, quantitative interviews were also conducted to collect information related to the personal experiences, beliefs and perceptions of the workers concerning the ergonomic advancement projects. This shows why the interviews were conducted in the first place in order to collect data that would have otherwise been difficult to measure using simple quantitative variables (Creswell & Clark, 2017). In order to have some structure while not limiting the different trains of thought that we planned to pursue, semi-structured interview guidelines were developed (Kallio et al., 2016). These interviews focused on the possible motivation for the employees to participate in the change process, the challenges and resources brought by the engagement, and how it could alter their organizational experiences.

### 8.3.2 Procedure

The sample for the study includes workers who work for various departments of organizations that are in the process of implementing ergonomics. In the view of Teddlie and Tashakkori (2009), this method ensured that a cross-sectional sample of organizational participants was achieved and as such enriched the data to allow for a deeper inspection of the specified phenomenon.

Descriptive and self-administered questionnaires followed by focus group discussions were used in the analysis of the data collected. In focusing on the quantitative survey, there were significant patterns, correlation, and trends identified by the means of descriptive statistics, correlation analysis, and other inferential statistics (Braun & Clarke, 2006). The qualitative interviews conducted were coded for themes, whereby code is conversely the procedure of identifying recurring themes, patterns, and/or meanings within data (Braun & Clarke, 2006). By employing this tactic, insights into the more pressing elements shaping workers' engagement in ergonomic change processes were obtained. In this study, a mixed-methodologies research design was employed in order the provide a more comprehensive understanding of how employee participation impacts ergonomic change with the strengths of both quantitative and qualitative analysis being incorporated into this research design as suggested by Johnson and Onwuegbuzie (2004). The generalizability of the data was enhanced by combining different forms of data gathering and analysis, which offered useful knowledge for change management theory and praxis organizational development (Denzin, 2014).

## 8.4 RESULTS

### 8.4.1 To Understand How Employee Involvement Affects the Durability and Efficacy of Ergonomic Changes in the Workplace

#### 8.4.1.1 Strategies for Employee Engagement during Change Management

Effective employee engagement techniques are essential for successful change management. Transparent and timely communication is critical; leaders must clearly describe the reasons for the change, the expected consequences, and the organization's future orientation. This openness reduces anxiety and fosters trust (Bakker, 2017). Involving employees in decision-making increases engagement because it offers them a sense of ownership and agency (Caldwell et al., 2012). Leaders should empower their staff by giving them liberty and responsibility, which promotes a proactive and engaged staff (Wong & Cummings, 2009). Recognizing and rewarding efforts during the shift is also important for maintaining strong morale and motivation (Le & Lei, 2018). Continuous learning and development opportunities help employees feel supported and competent in adapting to new responsibilities or procedures (Chinowsky et al., 2018; Arora et al., 2024). Together, these tactics foster a climate in which employees feel valued, knowledgeable, and invested in the organization's success, resulting in smoother transitions and more effective change implementation. Employee buy-in is critical for any change management initiative to succeed. Employees who are invested in their jobs are more likely to embrace and promote change, which benefits the firm as a whole. Organizations can implement a variety of measures to increase employee participation, motivation, and commitment during times of transition. This section examines some of these strategies and their significance in supporting employee engagement during change management, based on pertinent research findings.

#### 8.4.1.2 Transparent and Timely Communication

Transparent and timely communication is essential for keeping employees motivated during transitions. Clear explanations of why a change is occurring, how it will affect employees, and the organization's future direction are critical for gaining employee support (Bakker, 2017). Timely communication helps to dispel rumors and clarify concerns, lowering resistance and anxiety (Beer & Nohria, 2000). Leaders who keep staff informed, encourage input, and create open conversation opportunities foster trust, respect, and a sense of belonging (Herold et al., 2008).

#### 8.4.1.3 Employee Involvement and Participation

Employees who have contributed to the formulation of new rules or procedures are more likely to support them. Employees feel more empowered and engaged when they are involved in decision-making (Caldwell et al., 2012). This interactive method can improve problem-solving, generate more innovative ideas, and increase receptivity to change (Weiner et al., 2012). Focus groups, workshops, and cross-functional teams can improve employee engagement by instilling a sense of shared responsibility and purpose (Wong & Cummings, 2009).

#### 8.4.1.4 Leadership Support and Visible Sponsorship

Leadership support is critical for achieving employee buy-in for change (Ansari & Khan, 2023). Leaders build trust and confidence by publicly endorsing change, sharing the vision, and involving their staff in the process (Eisenbeiss et al., 2008). Leaders act as role models, showing the behaviors they want to see in their teams (Appelbaum et al., 2012). To establish a friendly and collaborative environment, executives should be approachable and sensitive to employee problems and suggestions (Herold et al., 2008).

#### 8.1.4.5 Recognition and Rewards

Recognizing and rewarding employees' efforts and contributions during transitions can sustain engagement and motivation. Employees feel valued when their efforts are acknowledged, they receive feedback, and they are offered incentives or promotions (Le & Lei, 2018). An effective recognition program should consider employees' needs and goals, reinforcing the connection between their contributions and the organization's success (Gentry et al., 2017).

#### 8.1.4.6 Continuous Learning and Development

Investing in staff learning and development during the change management process increases participation and adaptability (Dimov et al., 2000). Training programs, workshops, and tools that assist employees in learning new skills prepare them for a dynamic work environment (Ford et al., 2011). Continuous learning opportunities promote a growth mentality, increase self-esteem, and reflect the organization's commitment to employee development (Eisenbeiss et al., 2008). Coaching and mentoring can boost employee engagement and help them navigate change (Chinowsky et al., 2018).

Organizations that implement these techniques can boost employee buy-in for change management efforts. Open and timely communication, employee involvement, leadership support, acknowledgment, and ongoing learning foster an environment in which employees feel appreciated, driven, and actively engaged in the change process. According to research, involving employees in the change process leads to beneficial results such as increased acceptability, dedication, and performance (Rafferty et al., 2013). Prioritizing personnel during change management ensures a smooth transition and long-term success.

### 8.4.2 To Examine the Methods of Change Management that Successfully Encourage Staff Involvement in Ergonomic Projects

#### 8.4.2.1 Role of Leadership in Employee Engagement during Change

Leadership involves encouraging employees, and their participation during times of change is critical to the success of organizational transformations. Effective leaders are critical in creating a compelling vision for change, articulating the rationale for the transition, and outlining the benefits that will result (Appelbaum et al., 2012). When leaders show excitement, devotion, and active participation in the change process, employees gain confidence and trust (Herold et al., 2008). Furthermore, executives who lead by example and have a favorable attitude toward change foster a supportive environment in which employees are encouraged to embrace new ideas (Beer & Nohria, 2000). By defining clear goals, providing direction, and delivering assistance, leaders create a sense of security and stability, which supports employee involvement during periods of uncertainty (Eisenbeiss et al., 2008). Moreover, leaders who are approachable and responsive to employee concerns foster an environment of open communication, which further enhances engagement levels (Appelbaum et al., 2012).

Overall, good leadership plays an important role in establishing an employee engagement culture in which people feel appreciated, motivated, and empowered to contribute to organizational transformation initiatives. Leadership is critical for encouraging worker participation throughout transitions (Aaras et al., 1997). Leaders who prioritize employee participation throughout organizational transitions can significantly improve change management outcomes. This section addresses the theoretical and empirical data that supports the importance of leadership in increasing employee engagement throughout transitions.

#### 8.4.2.2 Setting a Compelling Vision

Leaders have an important role in developing and communicating a compelling vision for change. They should clearly state the rationale for the change and the desired consequences (Appelbaum et al., 2012). Employees are more inclined to invest in organizational transformation when they have a clear and compelling vision to strive for (Eisenbeiss et al., 2008). Leaders can increase staff buy-in during the transition by stressing the good outcomes of their efforts.

#### 8.4.2.3 Leading by Example

Employees are more inclined to support change if their bosses are passionate and devoted to it. Leaders motivate and encourage their staff by demonstrating dedication, passion, and active participation (Herold et al., 2008). Leaders who engage in change activities, ask employee feedback, and maintain a positive attitude foster a supportive and engaged atmosphere in which people can grow (Beer & Nohria, 2000). Employee trust, credibility, and buy-in are improved when leaders set a good example.

Organizations may effectively manage change and foster an engaged and collaborative culture by focusing on clear communication, employee involvement, leadership support, recognition, and continual learning. These tactics not only make transitions smoother but also contribute to long-term organizational success.

## 8.5 DISCUSSION

### 8.5.1 Effective Strategies for Employee Engagement during Change Management

The steps used for change management in organizations are critical and thus need to be implemented with efficiency and impact when it comes to engaging employees. These solutions rest primarily on assurances and updates(Bakker, 2017). Organization whits its employees when leaders facilitate an understanding of the specific reasons for change, probable effects, and potential organizational outlook. Through, engagements such as consultation, workers get the opportunity to contribute their inputs to the organizational decisions (Caldwell et al., 2012). Second, creating possibilities for ongoing competencies also increases to endorse the employees and prepares to availabilities to changes or new responsibilities or manner (Chinowsky et al., 2018).

Another reason for the need to observe and acknowledge the workers during the shift is equally important is that workers need to feel motivated all through the shift, and this is to recognize their efforts (Le & Lei, 2018). It is worth to note that firms that employ these strategies develop an atmosphere where every worker is appreciated, kept abreast, and considered crucial in the execution of change strategies In order for any change management program to be effective, there must be perceived management support from the employees involved. These are the points that show that motivated and committed employees are ready to embrace as well as support change in the organization for the overall development of the company. Many strategies can be used in organizations to enhance the employees' involvement, engagement and the extent of their commitment during organizational change. This section provides details on some of these strategies and whether they have value in enhancing employee engagement during change management, supported by evidence from studies.

### 8.5.2 Effective Communication

One other essential leadership skill in managing transitions requires one to be able to communicate effectively in order to gain the support of staff. The heads of these organizations need to be open and forthright about the process of change throughout all the stages (Herscovitch & Meyer, 2002). Informing and responding to the concerns of other employees, as well as getting advice from the rank and file are all critical responsibilities (Appelbaum et al., 2012). The employees are more likely to participate and share new ideas because they will feel that they are valued and are an essential part of the whole process that leads to the betterment of the company (Bakker, 2017). It is important for leaders who will be managing the projects to engage with the various teams and make sure that everyone understands and supports the projects.

### 8.5.3 Empowering Employees

Empowering people is an important leadership approach for increasing buy-in during times of transition. Leaders should empower employees to make decisions and participate in the change process, giving them more responsibility and autonomy (Caldwell et al., 2012). When employees are given the authority to make decisions, take initiative, and use their skills to achieve change, they become more invested in the company's success (Wong & Cummings, 2009). Leaders build empowerment to encourage participation, teamwork, and creativity in the face of change.

### 8.5.4 Recognizing and Supporting Employees

Effective leaders publicly acknowledge their employees' efforts and triumphs. Employee motivation increases when they believe their efforts are respected and appreciated (Le & Lei, 2018). Leaders who offer opportunities for personal and professional development throughout transitions show a commitment to employee well-being and productivity (Chinowsky et al., 2018). Recognizing and meeting employees' needs can dramatically increase loyalty, motivation, and engagement.

### 8.5.5 Engaging the Workforce and the Third Fact of Change Management

The following are some of the elements of organization change essential for the success of organizational transformation: By its nature, top-down change is often imposed on employees, but when the employees have the opportunity to participate in change, their self-interest, motivation, and volition tend to increase significantly. Part 9 examines the justification of employee engagement/empowerment for change initiatives, based on literature and prior studies.

### 8.5.6 Change Management: Change Adoption by Employees

Employee engagement can be defined as enabling employees to participate in decision-making, issue identification and resolution, change implementation, and generation. Organization performers are useful in that those employees who participate in the process can then be more capable of making better judgments, being open to change, and yielding better results (Caldwell et al., 2012). Engagement in the organizational transformation compels the worker to be committed, take pride, and feels responsible (Jain & Thakur, 2016). The knowledge-oriented employees know and comprehend the advantages and repercussions of change, thus there is less or no resistance to change and the process is easier to work through (Choi, 2011).

This practice enables organizational change because it promotes a culture of mutual obligation and teamwork, thus allowing people to accept change more easily (Rafferty et al., 2013).

### 8.5.7 Change Initiatives during Empowerment

Examples of employee empowerment on change projects include: allowing the employees to work freely during the change process by offering them certain levels of autonomy to manage the change on their own, giving the employees certain responsibilities with regard to the change and offering the employees certain tools they will use to manage change on their own. Employees with mandated empowerment are more committed towards employment and are more determined to overcome all sorts of stress (Caldwell et al., 2012). They are more risk takers compared to their counterparts in the traditional workplace setting, as they take calculated risks and approach their tasks in different ways (Ford et al., 2011). Employees that role percept the organization as trusting them and making them responsible for decisions are proactive in identifying solution to arising issues as well as proactively searching for ideas to present during transitions (Chinowsky et al., 2018). An effective change process fosters organizational culture that fosters accountability, learning, and revised implementation (Weiner et al., 2012).

### 8.5.8 Employee Involvement and Empowerment during Change: Benefits

Including and engaging employees in transformation programs have numerous advantages. First, it increases employee loyalty, enthusiasm, and involvement, making them feel that they are actively involved in the change process (Choi, 2011). This enhanced involvement boosts performance and productivity during transition periods (Jain & Thakur, 2016). Second, companies with high levels of employee involvement and empowerment have more responsible and proactive staff (Rafferty et al., 2013). These employees take the initiative to discover problems, implement solutions, and accept responsibility for results. Third, allowing and encouraging people to experiment, fail, and try again promotes a culture of continuous learning and innovation (Ford et al., 2011). This adaptability is critical for quickly adjusting to new settings.

### 8.5.9 Implementing Effective Strategies

To effectively involve and empower employees in change initiatives, businesses should provide clear information, training, and support throughout the process. Employees must understand why the change is taking place, its goals, their roles, and the resources and support available to them (Choi, 2011). Employees should be given training and development opportunities to help them acquire the skills and knowledge required to adapt to change (Caldwell et al., 2012). Leaders should foster an environment of trust and open communication in which employees feel free to share their thoughts and concerns while actively engaging in the transformation process (Jain & Thakur, 2016).

Finally, effective transformation efforts necessitate widespread employee involvement and autonomy. When employees have autonomy, they become more invested, dedicated, and cooperative. Organizations may build an engaged and collaborative culture by prioritizing effective communication and empowerment, recognition, and continual learning, resulting in smoother transitions and long-term success.

### 8.5.10 Opportunities for Employee Involvement in Change Management

Employee participation in change management is essential for the successful implementation of organizational transitions. Allowing employees to participate in the change process not only fosters a sense of ownership and commitment but also ensures that their diverse perspectives and ideas are considered, resulting in more effective decision-making and smoother implementation. According to the study, when employees participate actively in the planning, implementation, and evaluation of change projects, they are more likely to support and accept the changes. This involvement can take many forms, including engaging in focus groups, serving on change implementation teams, offering feedback through surveys or suggestion boxes, and participating in collaborative problem-solving sessions.

Organizations can benefit from employees' expertise, creativity, and competence by incorporating them at various stages of the change process, resulting in more innovative and sustainable solutions (Cummings & Worley, 2015). Furthermore, employee involvement promotes a culture of transparency, trust, and empowerment, all of which are required to create a positive organizational climate conducive to change (Herold et al., 2008). As a result, businesses should actively seek opportunities to engage employees in change management in order to maximize their contributions and increase the likelihood of positive change outcomes.

Giving employees the ability to participate in strategic planning, issue-solving, and decision-making boosts their engagement, motivation, and initiative. Employees feel more valued and integrated when they are given more responsibility, freedom, and resources to carry out their jobs.

This empowerment brings several benefits, including increased dedication, responsibility, flexibility, and innovation. Organizations that develop a culture that encourages employee participation and autonomy are more adaptable, have more employee engagement, and are more likely to see their initiatives through (Table 8.1).

## 8.6 CONCLUSION

In conclusion, this study emphasizes the need to engage employees in change initiatives, as noted by Australian organizations. The outcomes suggest that engagement for all workers affects a variety of other factors of change in the organization including attitudes, behaviors, and productivity levels. Where employees have commitment towards their job and the objectives of change, they are more likely to agree to alter mode, embrace change initiatives, and stand to gain from such efforts. It has been noted that ergonomics implementation in organizations requires a positive involvement of employees. These outcomes suggest that learners' active engagement in the course raises employees' acceptance and subsequently job satisfaction

**TABLE 8.1**
**Provides a Structured Overview of the Various Facets of Employee Engagement and Change Management, Highlighting Strategies, Leadership Roles, Effective Practices, Opportunities for Involvement, and Ergonomic Considerations**

| Strategies For Employee Engagement During Change Management | Role Of Leadership in Employee Engagement During Change | Effective Strategies for Employee Engagement During Change Management | Opportunities For Employee Involvement in Change Management | Ergonomics |
|---|---|---|---|---|
| Transparent communication: Clearly explain the reasons for the change, its impact, and the future direction of the company (Bakker, 2017). | Setting a compelling vision: leaders articulate a clear and motivating vision for change (Appelbaum et al., 2012). | Timely communication: Provide accurate and timely information to prevent rumors and address concerns (Beer & Nohria, 2000). | Strategic planning: involve employees in the planning stages of change initiatives to ensure their insights and concerns are considered (Caldwell et al., 2012). | Workplace design: tailor workspaces to fit employees' physical needs, enhancing comfort and productivity. |
| Employee involvement: engage employees in the change process to enhance their sense of ownership and commitment (Caldwell et al., 2012). | Leading by example: leaders demonstrate their commitment to change through their actions and attitudes (Herold et al., 2008). | Employee participation: use participatory techniques like focus groups and workshops to foster a sense of community (Wong & Cummings, 2009). | Decision-making: empower employees to contribute to decision-making processes, enhancing their engagement and commitment (Weiner et al., 2012). | Task ergonomics: optimize tasks and tools to reduce strain and increase efficiency. |
| Leadership support: leaders should be visible, approachable, and responsive to employee concerns (Eisenbeiss et al., 2008). | Effective communication: Maintain openness, honesty, and transparency in all communications (Herscovitch & Meyer, 2002). | Leadership support: leaders must visibly support and be involved in the change process (Appelbaum et al., 2012). | Problem-solving: involve employees in identifying and solving issues related to the change process (Jain & Thakur, 2016). | Physical ergonomics: focus on the interaction between employees and their work environment to prevent injury and improve performance. |

*(Continued)*

**TABLE 8.1 (*Continued*)**

**Provides a Structured Overview of the Various Facets of Employee Engagement and Change Management, Highlighting Strategies, Leadership Roles, Effective Practices, Opportunities for Involvement, and Ergonomic Considerations**

| Strategies For Employee Engagement During Change Management | Role Of Leadership in Employee Engagement During Change | Effective Strategies for Employee Engagement During Change Management | Opportunities For Employee Involvement in Change Management | Ergonomics |
|---|---|---|---|---|
| Recognition and rewards: acknowledge and reward employees' contributions to the change process (Le & Lei, 2018). | Empowering employees: delegate responsibility and provide autonomy to employees during the change (Caldwell et al., 2012). | Employee recognition: implement programs that reward employee contributions and foster a positive work environment (Le & Lei, 2018). | Implementation: engage employees in the implementation phase to leverage their skills and knowledge (Choi, 2011). | Cognitive ergonomics: design tasks that align with human cognitive abilities and limitations. |
| Continuous learning and development: provide training and resources to help employees adapt to changes (Ford et al., 2011). | Recognizing and supporting employees: show appreciation for employees' efforts and provide support (Chinowsky et al., 2018). | Continuous development: offer training programs that equip employees with the skills needed for new processes (Ford et al., 2011). | Feedback mechanisms: create systems for employees to provide feedback on the change process and suggest improvements (Rafferty et al., 2013). | Organizational ergonomics: enhance organizational structures and policies to support ergonomic practices. |

and productivity. Adding employee stimuli and perusing knowledge to the analysis and execution of ergonomic adjustment procedures shows organizations the value of employees and their knowledge.

In addition, the direct involvement of the staff, makes them feel a personal stake in the change hence they are less likely to resist the impending change. The importance of firstly, communicating the changes regularly, secondly, leadership support for change and continuing feedback during the transition process are stressed in the study as factors leading to high levels of employees' engagement. In conclusion, therefore, the present research recognizes the importance of examining the role of employee involvement in ergonomic transformation outcomes in change management, and posits that this area should be of paramount importance to change agents in future change management initiatives.

Employer commitment, information sharing, employee involvement, and decision-making rights are some of the key factors in the increase of employees' contributions in the change process. The literature establishes that attaining employee buy-in increases the odds of implementing change and leads to improved levels of employee satisfaction as well as improved organizational performance.

This paper points out that it is important for organizations to emphasize the role of employees to enable them to approach change and transformation goals most effectively from a transition management perspective. The transition means to continue to increase the levels of commitment and focus by providing information in a timely manner, build employee voice, celebrate successes, and offer timely and relevant skill enhancement skills. This is not only helpful for making easy transition but also build strong and dynamic culture in the organization which is very helpful for future growth.

## 8.7 LIMITATIONS OF STUDY

The study also acknowledges some limitations like how the results could be influenced by a sample bias or how the findings would not have broad application given some organizational settings, or that the study relies on the participants' accounts that bring subjectivity into play. Such limitations imply that the study findings should be interpreted with precaution in order to imply that further study is needed to support and build up the findings.

## 8.8 SCOPE OF FURTHER STUDY

More studies could be done on the effects of employee participation in ergo improvement programs chronologically using cross-sectional research designs that can track employees over time. Research may bring focus in larger magnitude within the industry or cultures. Additionally, studies on specific approaches for enhancing EMP's participation in precipitating strategies and the role of technology in implementing engagement in telework arrangements would be helpful. Using qualitative approaches could enhance the level of knowledge acquisition through the acquisition of thorough employee experience and stories.

## REFERENCES

Aaras, A., Fostervold, K. I., Ro, O., Thoresen, M., & Larsen, S. (1997). Postural load during VDU work: A comparison between various work postures. *Ergonomics*, *40*(11), 1255–1268.

Alfes, K., Truss, C., Soane, E. C., Rees, C., & Gatenby, M. (2013). The relationship between line manager behavior, perceived HRM practices, and individual performance: Examining the mediating role of engagement. *Human resource management*, *52*(6), 839–859.

Ansari, M. S. A., & Khan, S. A. (2023). FDI, disinvestment and growth: An appraisal of Bhutanese economy. *Journal of Chinese Economic and Foreign Trade Studies*, *16*(1), 64–82. https://doi.org/10.1108/JCEFTS-05-2022-0031

Ansari, M. S. A., Khan, S. A., & Bhuyan, U. (2024). Exploring the strategic orientation factors influencing the organizational performance through bibliometric analysis. In S. Umair, U. Waqas, B. Mrugalska, & I. R. Al Shamsi (Eds.), *Human Perspectives of Industry 4.0 Organizations* (pp. 68–88). CRC Press.

Anton, D., Rosecrance, J., Merlino, L., & Cook, T. (2002). Prevalence of musculoskeletal symptoms and carpal tunnel syndrome among dental hygienists. *American Journal of Industrial Medicine*, *42*(3), 248–257.

Appelbaum, S. H., Habashy, S., Malo, J. L., & Shafiq, H. (2012). Back to the future: Revisiting Kotter's 1996 change model. *Journal of Management Development, 31*(8), 764–782. https://doi.org/10.1108/02621711211253231

Arora, S., Pandey, A., Batta, K., & Khan, S. A. (2024). Exploring the benefits and challenges of it-enabled learning in the education sector: A roadmap for effective use of extended reality. In T. Kajla, P. Kansra, & N. Singh (Eds.), *Multidisciplinary Applications of Extended Reality for Human Experience* (pp. 356–373). IGI Global. https://doi.org/10.4018/979-8-3693-2432-5.ch015

Bakker, A. (2017). Strategic and proactive approaches to work engagement. *Organizational Dynamics*, *46*(2), 67–75. https://doi.org/10.1016/j.orgdyn.2017.04.002

Beek, A. J., Gaalen, L. C., & Frings-Dresen, M. H. W. (1992). Working postures and activities of lorry drivers: A reliability study of on-site observation and recording on a pocket computer. *Applied Ergonomics*, *23*(5), 331–336.

Beer, M., & Nohria, N. (2000). Cracking the code of change. *Harvard Business Review, 78*(3), 133–141.

Berkowitz, S., Feuerstein, M., Sullivan-Lopez, M., & Peck, C. (1999). Occupational back disability in U.S. Army personnel. *Military Medicine*, *164*, 412–418.

Braun, V., & Clarke, V. (2006). Using thematic analysis in psychology. *Qualitative Research in Psychology*, *3*(2), 77-101. https://doi.org/10.1191/1478088706qp063oa

Caldwell, S. D., Chatman, J. A., & O'Reilly, C. A. (2012). Building organizational commitment: A multi-firm study. *Journal of Occupational and Organizational Psychology, 85*(4), 504–518. https://doi.org/10.1111/j.2044-8325.2012.02052.x

Charteris, J. (1998). Comparison of the effects of backpack loading and of walking speed on foot-floor contact patterns. *Ergonomics*, *41*(12), 1792–1809.

Chinowsky, P. S., Brown, H., Szajnman, A., & Realph, A. (2018). Developing knowledge landscapes through project-based learning. *Journal of Professional Issues in Engineering Education and Practice, 144*(2), 05018002. https://doi.org/10.1061/(ASCE)EI.1943-5541.0000365

Choi, M. (2011). Employees' attitudes toward organizational change: A literature review. *Human Resource Management*, *50*(4), 479–500.

Creswell, J. W., & Clark, V. L. P. (2017). *Designing and Conducting Mixed Methods Research.* Sage publications.

Cummings, T. G., & Worley, C. G. (2015). *Organization Development and Change* (10th ed.). Cengage Learning.

Denzin, N. K. (2014). *Interpretive Autoethnography* (2nd ed.). SAGE Publications.

Dimov, M., Bhattacharya, A., Lemasters, G., & Atterbury, M. (2000). Exertion and body discomfort perceived symptoms associated with carpentry tasks: An on-site evaluation. *AIHAJ*, *61*, 685–691.

Eisenbeiss, S. A., van Knippenberg, D., & Boerner, S. (2008). Transformational leadership and team innovation: Integrating team climate principles. *Journal of Applied Psychology, 93*(6), 1438–1446. https://doi.org/10.1037/a0012716

Ferguson, S. A., Marras, W. S. & Crowell, R. R. (1996). Dynamic low back functional motion capacity evaluation. *Journal of Occupational Rehabilitation*, *6*(4), 203–214.

Ford, J. K., Yelon, S. L., & Billington, A. Q. (2011). How much is transferred from training to the job? The 10% delusion as a catalyst for thinking about transfer. *Performance improvement quarterly*, *24*(2), 7–24.

Herold, D. M., Fedor, D. B., Caldwell, S., & Liu, Y. (2008). The effects of transformational and change leadership on employees' commitment to a change: A multilevel study. *Journal of Applied Psychology, 93*(2), 346–357. https://doi.org/10.1037/0021-9010.93.2.346

Herscovitch, L., & Meyer, J. P. (2002). Commitment to organizational change: Extension of a three-component model. *Journal of Applied Psychology, 87*(3), 474–487. https://doi.org/10.1037/0021-9010.87.3.474

Bakker, A. B., & Demerouti, E. (2017). Job demands–resources theory: Taking stock and looking forward. *Journal of Occupational Health Psychology*, *22*(3), 273.

Bakker, A. B., Demerouti, E., & Sanz-Vergel, A. I. (2014). Burnout and work engagement: The JD–R approach. *Annual Review of Organizational Psychology and Organizational Behavior*, *1*(2014), 389–411. https://doi.org/10.1146/annurev-orgpsych-031413-091235

Jain, V. K., & Thakur, P. (2016). *Strategic Human Resource Management: Emerging Issues and Trends*. Atlantic Publishers & Distributors.

Johnson, R. B., & Onwuegbuzie, A. J. (2004). Mixed methods research: A research paradigm whose time has come. *Educational researcher*, *33*(7), 14–26.

Kahn, W. A. (1990). Psychological conditions of personal engagement and disengagement at work. *Academy of Management Journal, 33*(4), 692–724. https://journals.aom.org/doi/abs/10.5465/256287

Kallio, H., Pietilä, A. M., Johnson, M., & Kangasniemi, M. (2016). Systematic methodological review: Developing a framework for a qualitative semi-structured interview guide. *Journal of Advanced Nursing*, *72*(12), 2954–2965. https://doi.org/10.1111/jan.13031

Khan, S. A., & Ansari, M. S. A. (2024). Critical factors for relationship between strategic orientation, corporate success, and innovation. In S. Umair, U. Waqas, B. Mrugalska, & I. R. Al Shamsi (Eds.), *Human Perspectives of Industry 4.0 Organizations* (pp. 89–110). CRC Press.

Khatri, B., Shrimali, H., Khan, S. A., & Naim, A. (2023). Role of HR analytics in ensuring psychological wellbeing and job security: Learnings from COVID-19. In R. Yadav, M. Sinha, & J. Kureethara (Eds.), *HR Analytics in an Era of Rapid Automation* (pp. 36–53). IGI Global. https://doi.org/10.4018/978-1-6684-8942-0.ch003

Le, P. B., & Lei, H. (2018). The mediating role of trust in stimulating the relationship between transformational leadership and knowledge sharing processes. *Journal of Knowledge Management*, *22*(3), 521–537. https://doi.org/10.1108/JKM-10-2016-0463

Rafferty, A. E., Jimmieson, N. L., & Armenakis, A. A. (2013). Change readiness: A multilevel review. *Journal of Management, 39*(1), 110–135. https://doi.org/10.1177/0149206312457417

Rich, B. L., Lepine, J. A., & Crawford, E. R. (2010). Job engagement: Antecedents and effects on job performance. *Academy of Management Journal, 53*(3), 617–635. https://doi.org/10.5465/amj.2010.51468988

Saks, A.M. (2006). Antecedents and consequences of employee engagement. *Journal of Managerial Psychology*, *21*(7), 600–619.

Teddlie, C., & Tashakkori, A. (2009). *Foundations of Mixed Methods Research: Integrating Quantitative and Qualitative Approaches in the Social and Behavioral Sciences*. Sage.

Van den Heuvel, M., Demerouti, E., & Peeters, M. C. (2015). The job crafting intervention: Effects on job resources, self-efficacy, and affective well-being. *Journal of Occupational and Organizational Psychology*, *88*(3), 511–532. https://doi.org/10.1111/joop.12128

Weiner, I. B., Nelson, R. J., & Mizumori, S. (2012). *Handbook of Psychology, Behavioral Neuroscience* (Vol. 3). John Wiley & Sons.

Wong, C. A., & Cummings, G. G. (2009). The influence of authentic leadership behaviors on trust and work outcomes among healthcare staff. *Journal of Leadership Studies, 3*(2), 6–23. https://doi.org/10.1002/jls.20104

Zareen, S. & Khan, S. A. (2023). Exploring dependence of human resource management (HRM) on internet of things (IoT) and digital marketing in the digital era. In A. Naim & V. Devi (Eds.), *Global Applications of the Internet of Things in Digital Marketing* (pp. 51–66). IGI Global. https://doi.org/10.4018/978-1-6684-8166-0.ch004

# 9 Unraveling the Connection, between Employee Happiness and Organizational Success Across Different Cultures

*Madhur Batra and Ferdinand J. Epoc*

## 9.1 INTRODUCTION

In today's business world, employee happiness and company success go hand in hand. Employee satisfaction covers all aspects of a person's life like job contentment and mental wellness (Bocar et al., 2022; Khatri et al., 2024a, Zareen & Khan, 2023). Organizational prosperity is, about a company's growth and achievements usually seen through success and market standing.

Extensive research has delved into the nature of these two ideas, which has shown that companies with content employees typically achieve higher performance levels. Content employees are usually more efficient in their work tasks and show dedication to their roles while also being less inclined to seek opportunities elsewhere within the organization – all factors contributing to improved organizational outcomes, in the long run. Additionally, successful organizations have the capacity to prioritize the well-being of their employees which establishes a cycle of support and growth (Bella, 2023; Smith et al., 2021).

The importance of this research area becomes more apparent when we consider the cultural interactions involved. In a world that is becoming more connected globally it is essential to grasp how cultural backgrounds impact the happiness of employees and the success of organizations (Khan et al., 2013; Ligori et al., 2022). Cultural beliefs, standards, and behaviors play a role in how workers view their jobs and their interactions with their employers. For example, the factors that inspire workers in a community-oriented society, like Japan could be quite different from what encourages employees in a culture such, as the United States (Hackman & Oldham, 1976; Malik et al., 2022).

Understanding these differences, between cultures is crucial, for global companies seeking to create resource plans that address the needs of various cultural groups. By understanding how cultural aspects influence employee happiness and

 DOI: 10.1201/9781003532293-10

the achievement of goals companies can create customized strategies to improve employee welfare, ultimately boosting their performance (Kirkman & Shapiro, 2001; Monzani et al., 2021; Patel, 2021).

### 9.1.1 Objectives of the Study

In this research project we seek to explore the connection, between the contentment of employees, in their lives and the success of the organization they work for; specifically looking at how various cultural settings impact this relationship. Our study is designed to achieve the following goals:

**RO1**: To Explore how the contentment of employees impacts the success of organizations, in this context.
**RO2**: To Identify the elements that impact job satisfaction and organizational effectiveness.
**RO3**: To provide suggestions for promoting the well-being of employees and achieving success, i.e., to improve employee happiness across work environments.

## 9.2 LITERATURE REVIEW

### 9.2.1 Exploring Views on Job Satisfaction, among Employees and Success, Within Organizations

The correlation between job satisfaction of employees and the success of an organization has been extensively explored using a variety of frameworks. Maslow's Hierarchy of Needs offers a viewpoint by suggesting that employee happiness is linked to meeting needs ranging from physiological necessities to self-realization (Maslow, 1943). Additionally, Herzberg's Two Factor Theory distinguishes between hygiene factors such as salary and working conditions and motivators like recognition and achievement, as factors contributing to job contentment (Herzberg, 1966). The various theories emphasize the aspects of employee contentment which include external elements that impact overall happiness.

The Cultural Dimensions Theory, by Hofstede provides a framework for exploring how cultural influences impact employee happiness and business achievements in organizations (Hofstede,1984). According to Hofstede's research in 1984, he highlighted dimensions such as individualism versus collectivism, power distance, uncertainty avoidance, masculinity versus femininity and long-term, versus short-term orientation that influence the dynamics of workplaces and the attitudes of employees. These dimensions emphasize the significance of taking factors into account when designing policies and considering employee perspectives.

### 9.2.2 Cultural Factors Impact Employee Happiness and Organizational Performance

The beliefs and behaviors, within a culture, play a role in how happy employees at work and how well the organization demonstrates the cultures that prioritize individualism like those found in the US and Western Europe that value personal success and independence which can boost job satisfaction when employees feel appreciated, for their efforts (Smith et al., 2021; Bocar et al., 2022). In comparison, to societies like the United States and Australia, countries like Japan and South Korea place a value on maintaining unity within groups and showing loyalty to their companies. Job satisfaction in these cultures stems from having knit teams and a sense of security in their roles, within the organization (Hackman & Oldham, 1976; Khatri et al., 2024b).

The level of authority within a workplace also influences how interactions unfold among employees and their supervisors. It is quite significant in determining the dynamics at play in settings (Khan & Ansari, 2024; Ansari et al., 2024). Difference in power dynamics is evident between cultures that uphold a hierarchy with power distribution versus those that promote equal opportunities and involvement, in decision-making processes. The former tends to boost employee contentment by providing guidance and defined responsibilities; whereas the latter emphasizes collaborative structures and inclusive management styles which can positively impact job satisfaction levels (Kirkman & Shapiro,2001; Monzani et al., 2021).

### 9.2.3 Factors Influencing Employee Happiness

Employee happiness is influenced by factors, like the nature of the job itself and how its structured in terms of tasks and opportunities for growth (Herzberg, 1966). The culture within an organization also plays a role in how employees feel at work; values, like inclusivity and work-life balance can greatly impact satisfaction levels (Bella, 2023; Khan et al., 2023).

Different leadership approaches that resonate with the values of a given culture play a role, in shaping employee satisfaction levels (Ansari & Khan, 2023). There are two types of leadership styles; transactional leadership--they impact employees differently based on whether they belong to collectivist cultures. Transformational leadership is about inspiring and motivating employees. In cultures, this style encourages growth and creativity. While in collectivist cultures it fosters unity within the group and shared objectives (Kirkman & Shapiro, 2001; Monzani et al., 2021). Transactional leadership on the hand focuses on establishing structures and offering rewards. This approach proves to be highly effective, in cultures characterized by power distance where employees prefer decision-making processes (Patel, 2021).

### 9.2.4 The Impact of Employee Happiness on Organizational Success

Contented employees play a role in the success of any organization as they tend to be more productive and committed while also displaying loyalty, towards the company which ultimately leads to improved overall outcomes (Smith et al., 2021). Companies that prioritize employee satisfaction often see results, along with increased market presence and enhanced competitiveness in their industry. This sets up a cycle of reinforcement. Companies that thrive can then allocate resources to support their employees' happiness and productivity further (Bella, 2023).

Studies have also shown that having a mix of cultures, in teams can boost creativity and innovation within a company. Ultimately resulting in performance, for the organization overall. Diverse teams offer a range of viewpoints and problem-solving methods that can really push forward ideas and progress. To make this work smoothly. Effectively requires creating an inclusive environment that respects and utilizes these diverse perspectives (according to Hackman & Oldham, 1976).

### 9.2.5 Cross-Cultural Differences in Workplace Dynamics

Cultural variations play a role, in shaping how people interact in work environments and affect communication styles and leadership success in teams across cultures be it East Asia or the United States (Malik et al., 2022). Recognizing and respecting these differences is the key to fostering teamwork and resolving conflicts effectively within a diverse team setting.

Leadership approaches should be adjusted to fit backgrounds as well. Transformational leadership focuses on inspiration and encouragement and can be successful, in environments; however, its outcomes can differ. In societies that emphasize group unity and cooperation it strengthens team bonds; whereas in societies that prioritize success and creativity it encourages accomplishments and new ideas (Kirkman & Shapiro, 2001; Monzani et al., 2021). In cultures, with power distance and high uncertainty avoidance tendencies servant leadership is highly appreciated. Employees, in these cultures value leaders who prioritize their well-being and professional growth (Smith et al., 2021; Khatri et al., 2024a).

The influence of diversity extends to teamwork and innovation well. Diverse teams offer viewpoints that can spur innovation but may encounter hurdles in establishing team unity and efficient communication. Successful leadership and training, in competence are crucial, for leveraging the advantages of diversity while addressing any downsides (Bella, 2023).

## 9.3 METHODOLOGY

This research uses an approach to explore how employee happiness and company success are connected in cultural contexts by combining various theories and studies, on the subject (Saxena et al., 2024; Sharma et al., 2024).

The research approach involves conducting an examination to pinpoint and classify topics concerning the happiness of employees, in the workplace and the success of the organization while considering how cultural factors play a role, in these aspects as well (Soomro et al., 2024).

Moreover, we will compare the results gathered from backgrounds. This comparative method reveals differences and similarities, in understanding and achieving employee happiness and company prosperity in environments. By looking at viewpoints, this study seeks to reveal observations about how cultural differences impact working relationships (Gould-Williams & Mohamed, 2010).

Finally, the synthesis of findings will integrate insights from various sources to develop a cohesive understanding of the relationship between employee satisfaction and organizational success within a cultural framework. By drawing connections between theoretical frameworks and empirical evidence, this synthesis aims to elucidate the mechanisms through which cultural factors impact organizational outcomes (Ogink et al., 2023).

In the end, the conclusion of the research will combine perspectives from references to create a comprehension of how employee happiness influences a company's prosperity within a cultural context. This synthesis endeavors to clarify the ways cultural elements affect business results by linking concepts, with real-world data (Chatterjee et al., 2024).

Through this methodology you can thoroughly examine the research objectives without the necessity of gathering data by using existing sources to delve into the intricate connections, between cultural factors and their impact on workplace interactions and employee welfare.

## 9.4 CROSS-CULTURAL ANALYSIS

### 9.4.1 Cultural Dimensions and Their Implications for Workplace Dynamics

It is important to consider aspects when looking at how they impact the dynamics at work, and how they ultimately affect employee happiness and organizational success. Hofstede's theory, on dimensions offers a way to explore these impacts;

- **Individualism vs. Collectivism**: Accomplishments and independence are signficantly higher in society that place a strong emphasis on individualism as an important value. This can result in increased job contentment as workers feel acknowledged for their individual efforts (according to Hofstede in 1984). On the other hand, in societies that prioritize collectivism maintaining group unity and loyalty to the company are key aspects employee satisfaction, in these cultures usually comes from a feeling of belonging and cooperation (as indicated by Hackman & Oldham, 1976).
- **Power Distance**: Cultures with high power distance tend to accept hierarchical structures and unequal power distribution. In such environments, employee satisfaction can be enhanced through clear leadership and well-defined roles. In contrast, low power distance cultures favor egalitarian structures, where participatory management styles lead to greater job satisfaction (Kirkman & Shapiro, 2001; Monzani et al., 2021).
- **Uncertainty Avoidance**: High uncertainty avoidance cultures prefer clear rules and stability, which can impact organizational prosperity by fostering a risk-averse culture. Low uncertainty avoidance cultures are more comfortable with ambiguity and change, potentially driving innovation and organizational growth (Malik et al., 2022.

### 9.4.2 Comparative Analysis of Employee Happiness Across Different Cultural Contexts

Employee happiness varies significantly across cultural contexts, influenced by local values, norms, and expectations. For instance:

- **Western Cultures (e.g., USA, UK)**: These cultures often emphasize individual achievement and personal growth. Employee happiness and satisfaction are closely linked to personal accomplishments, career advancement opportunities, and recognition (Smith et al., 2021).
- **Asian Cultures (e.g., Japan, South Korea)**: These collectivist societies place a higher value on group harmony and social cohesion. Employee happiness is often derived from strong team dynamics, loyalty to the company, and job security (Hackman & Oldham, 1976;).
- **Nordic Countries (e.g., Sweden, Denmark)**: These cultures tend to have low power distance and high individualism. Employee happiness or satisfaction in these contexts is associated with work-life balance, autonomy, and flat organizational structures that promote equality (Bella, 2023).

### 9.4.3 Leadership Styles and Their Effectiveness in Multicultural Environments

Leadership styles that align with cultural values can significantly impact employee satisfaction and organizational prosperity:

- **Transformational Leadership**: This style, characterized by inspiring and motivating employees, is effective in both individualist and collectivist cultures but for different reasons. In individualist cultures, it promotes personal growth and innovation, while in collectivist cultures, it enhances group cohesion and collective goals (Kirkman & Shapiro, 2001; Monzani et al., 2021).
- **Transactional Leadership**: This style, which focuses on clear structures and rewards, is particularly effective in high power distance cultures where employees expect hierarchical decision-making and clear directives (Patel, 2021).
- **Servant Leadership**: Emphasizing the leader's role in serving their team, this style resonates well in cultures with low power distance and high uncertainty avoidance, where employees value leaders who prioritize their well-being and professional development (Smith et al., 2021; Khatri et al., 2024a).

### 9.4.4 Communication and Collaboration Patterns in Diverse Workplaces

Effective communication and collaboration are essential for fostering employee satisfaction and organizational prosperity, especially in diverse workplaces:

- **High-Context vs. Low-Context Communication**: In high-context cultures (e.g., Japan and China), communication relies heavily on implicit messages and context. Misunderstandings can occur in multicultural teams if low-context communicators (e.g., USA and Germany) are not aware of these nuances. Training and awareness programs can bridge these gaps and enhance collaboration (Hackman & Oldham, 1976;).
- **Direct vs. Indirect Communication**: Direct communication, common in Western cultures, can be perceived as blunt or rude in cultures that prefer indirect communication. Understanding these differences is crucial for effective team dynamics and conflict resolution (Malik et al., 2022).
- **Collaboration Tools**: The adoption of collaboration tools that cater to diverse communication styles can improve teamwork and innovation. For instance, asynchronous communication platforms can accommodate different time zones and work preferences, enhancing global team effectiveness (Patel, 2021).

### 9.4.5 Impact of Cultural Diversity on Teamwork and Innovation

Cultural diversity within teams can be a double-edged sword, influencing both teamwork and innovation:

- **Enhanced Creativity and Innovation**: Diverse teams bring a variety of perspectives and problem-solving approaches, which can drive innovation and organizational growth. However, this requires an inclusive culture that values and leverages these differences (Bella, 2023).
- **Challenges in Team Cohesion**: Cultural diversity can also lead to misunderstandings and conflicts if not managed properly. Effective leadership and cultural competence training are essential to harness the benefits of diversity while mitigating potential drawbacks (Smith et al., 2021).
- **Balanced Team Composition**: Ensuring a balanced team composition with a mix of cultural backgrounds can enhance both employee satisfaction and organizational performance. Such teams are better equipped to understand and respond to global market needs (Kirkman & Shapiro, 2001; Monzani et al., 2021).

## 9.5 DISCUSSION AND FINDINGS

### 9.5.1 Analyzing the Relationship Between Employee Life Satisfaction and Organizational Prosperity Across Different Cultures

#### 9.5.1.1 Key Indicators of Employee Life Satisfaction and Organizational Prosperity

The analysis highlights several key indicators of employee life satisfaction, including job satisfaction, work-life balance, mental health, and personal growth opportunities. Organizational prosperity is measured through financial performance, market share, employee retention rates, and overall competitiveness. These indicators are interdependent, with higher employee life satisfaction often leading to improved organizational outcomes (Bella, 2023; Smith et al., 2021).

#### 9.5.1.2 Variation of Indicators Across Different Cultural Settings

The relationship between these indicators and cultural contexts reveals significant variations. In individualist cultures such as the United States, employee satisfaction heavily depends on individual achievements and recognition (Hofstede, 1984; Smith et al., 2021). Conversely, in collectivist cultures, group harmony and loyalty to the organization are critical factors for employee satisfaction (Hackman & Oldham, 1976). This variation necessitates different approaches to fostering employee satisfaction based on cultural context.

### 9.5.2 Identifying Cultural Factors that Influence Job Satisfaction and Organizational Performance

#### 9.5.2.1 Role of Cultural Dimensions

Cultural dimensions, like individualism versus collectivism and power distance play a role, in influencing job satisfaction and organizational performance.

In some cultures, individuals focus on achievements and independence which can boost job satisfaction when their efforts are acknowledged while in other cultures there is an emphasis on group cohesion and shared success which can increase job satisfaction through teamwork and loyalty (Hackman & Oldham, 1976; Hofstede, 1984; Ligori et al., 2022).

In societies, with power distance embrace hierarchical setups and unequal authority allocations where distinct leadership and clearly defined roles contribute to job fulfillment. Conversely cultures, with power distance prefer setups with collaborative management approaches fostering increased job contentment (Kirkman & Shapiro, 2001; Monzani et al., 2021).

In cultures, with high uncertainty avoidance levels tend to favor stability and strict rules which can lead to a reluctance to take risks and potentially hinder innovation according to Malik et al. (2022). Low uncertainty avoidance cultures wanted ambiguity and change which can stimulate innovation and promote development.

#### 9.5.2.2 Effects on Employee Attitudes, Behaviors, and Perceptions

Cultural dimensions play a role, in shaping how employees think and act in the workplace. For instance, in societies with power distance workers tend to appreciate and honor leaders who hold authority, resulting. Conversely in cultures with power distance employees tend to excel when they feel their opinions are respected and they have a say in making decisions (Patel, 2021). Recognizing these subtleties is essential, for creating approaches that match norms and boost employee contentment.

## 9.6 CONCLUSION

The research delved into the connection, between employee happiness and business success, across cultures has been highlighted in this study report. The results emphasize how happy employees positively impact performance and thriving companies prioritize the well-being of their staff members to establish a feedback loop in an appropriate and in a timely manner.

The examination of dimensions using Hofstede's framework shows that cultural contexts have great impact on how things work in the workplace like job satisfaction and how organizations perform well. The differences between individualism and collectivism, power distance, and uncertainty avoidance play a significant role on how employees think and act at work. For example, in cultures, personal success and independence are valued, while in collectivist cultures harmonious relationships within the team are relevant and vital. Cultures with power distance do well when there is a hierarchy but those, with low power, distance prefer equality and involving employees in decision-making. In the approach societies with an aversion to uncertainty value stability and well-defined rules, while societies with a low aversion, to uncertainty welcome change and creativity.

In setting employee happiness is compared to show how Western societies value individual achievements and career progress while Asian cultures prioritize teamwork and job stability; Nordic cultures prioritize a work-life balance and independence instead. Organizational success and employee contentment are influenced by leadership approaches that align with the values, like transformational leadership style and servant leadership approach.

In a workplace setting, it is crucial to have communication and teamwork to create a positive atmosphere for work to thrive in. Diverse communication styles like context and low context approaches play a vital role on how teams interact along with indirect communication methods. Using collaboration tools that cater to communication preferences is the key in enhancing team dynamics and fostering innovation. Having diversity within teams can spur creativity and innovation when managed effectively. It also calls for inclusive leadership and training in cultural competence.

The findings of this study have implications for both theory and practice in the field of development and human resources management. They underscore the importance of taking factors into account when designing strategies to improve employee satisfaction and overall organizational effectiveness. Practical suggestions include acknowledging and incentivizing accomplishments in the work places that value independence, teamwork, in communities that prioritize collective goals favorable leadership approaches that resonate with specific cultures and the use of communication tactics tailored to diverse cultural contexts.

The constraints of this research work mainly obtain from its basis and dependence on studies which indicate the need for future investigations to include practical research to confirm these results; and delve into how the ongoing global shifts like remote work and digital interactions affect the connection, between employee well-being and organizational success.

To sum up the discussion, it is crucial for contemporary businesses to grasp and utilize the connection between employee happiness and organizational success, in cultural settings. By customizing approaches to match norms and fostering leadership and communication methods, companies can boost employee satisfaction levels and achieve long-lasting prosperity in today's interconnected world.

## REFERENCES


Ansari, M. S. A., & Khan, S. A. (2023). FDI, disinvestment and growth: An appraisal of Bhutanese economy. *Journal of Chinese Economic and Foreign Trade Studies*, *16*(1), 64–82. https://doi.org/10.1108/JCEFTS-05-2022-0031

Ansari, M. S. A., Khan, S. A., & Bhuyan, U. (2024). Exploring the strategic orientation factors influencing the organizational performance through bibliometric analysis. In S. Umair, U. Waqas, B. Mrugalska, & I. R. Al Shamsi (Eds.), *Human Perspectives of Industry 4.0 Organizations* (pp. 68–88). CRC Press.

Bella, K. M. J. (2023). The power of employee well-being: A catalyst for organizational success. *International Journal of Scientific Research in Modern Science and Technology*, *2*(4), 20–26. https://ijsrmst.com/index.php/ijsrmst/article/view/109

Bocar, A. C., Khan, S. A., & Epoc, F. (2022). COVID-19 work from home stressors and the degree of its impact: Employers and employees actions. *International Journal of Technology Transfer and Commercialisation*, *19*(2), 270–291. https://dx.doi.org/10.1504/ijttc.2022.124349

Chatterjee, S., Chaudhuri, R., & Vrontis, D. (2024). Does data-driven culture impact innovation and performance of a firm? An empirical examination. *Annals of Operations Research*, *333*(2), 601–626.

Gould-Williams, J., & Mohamed, R. B. (2010). A comparative study of the effects of 'best practice'HRM on worker outcomes in Malaysia and England local government. *The International Journal of Human Resource Management*, *21*(5), 653–675.

Hackman, J. R., & Oldham, G. R. (1976). Motivation through the design of work: Test of a theory. *Organizational Behavior and Human Performance*, *16*(2), 250–279. https://doi.org/10.1016/0030-5073(76)90016-7

Herzberg, F. (1966). *Work and the Nature of Man*. World Publishing Co.

Hofstede, G. (1984). *Culture's Consequences: International Differences in Work-Related Values* (Vol. 5). Sage.

Khan, S. A., Khatri, B., Saxena, C., & Arora, S. (2023, December). Usage of fitness applications by the sports people: A bibliometric analysis. In *Proceedings of International Conference on Recent Advancements in Artificial Intelligence* (pp. 329–342). Singapore: Springer Nature Singapore. https://doi.org/10.1007/978-981-97-1111-6_27

Khan, S. A., Singh, W. J., & Sharma, K. (2013). A cross sectional study of the awareness and satisfaction level of the people in the rural & urban areas of India & Bhutan in line with their settlement preferences in a locality ie urban or rural. *Asia Pacific Journal of Management & Entrepreneurship Research*, *2*(2), 35.

Khan, S. A., & Ansari, M. S. A. (2024). Critical factors for relationship between strategic orientation, corporate success, and innovation. In S. Umair, U. Waqas, B. Mrugalska, & I. R. Al Shamsi (Eds.), *Human Perspectives of Industry 4.0 Organizations* (pp. 89–110). CRC Press.

Khatri, B., Arora, S., & Khan, S. A. (2024a). Analysis of servant leadership and organizational citizenship behaviour among social service sector employees. In D. Belias, I. Rossidis, & C. Papademetriou (Eds.), *Organizational Behavior and Human Resource Management for Complex Work Environments* (pp. 443–465). IGI Global. https://doi.org/10.4018/979-8-3693-3466-9.ch021

Khatri, B., Sharma, N., Khan, S. A., & Nandini, G. (2024b). Navigating employer branding in the digital age: Exploring the impact of social media networking during the COVID-19 pandemic. In B. Christiansen, M. Aziz, & E. O'Keeffe (Eds.), *Global Practices on Effective Talent Acquisition and Retention* (pp. 313–332). IGI Global. https://doi.org/10.4018/979-8-3693-1938-3.ch017

Kirkman, B. L., & Shapiro, D. L. (2001). The impact of cultural values on job satisfaction and organizational commitment in self-managing work teams: The mediating role of employee resistance. *Academy of Management Journal, 44*(3), 557–569. https://www.jstor.org/stable/3069370

Ligori, T. A. A., Suresh, N., Khan, S. A., Rabgay, T., & Yezer, K. (2022). The mediating effect of university image on the relationship between curriculum and student satisfaction: An empirical study of the Royal University of Bhutan. *International Journal of Pluralism and Economics Education. Inderscience*, *13*(2), 192–208. https://dx.doi.org/10.1504/ijpee.2022.127218

Malik, A., Pereira, V., Budhwar, P., Froese, F. J., Minbaeva, D., Sun, J., Nguyen, A.T., & Xue, S. (2022). Multilevel relational influences on HRM practices: A cross-country comparative reflective review of HRM practices in Asia. *Asian Business & Management*, *21*(5), 745. https://doi.org/10.1057/s41291-022-00208-z

Maslow, A. H. (1943). A theory of human motivation. *Psychological Review, 50*(4), 370–396. https://doi.org/10.1037/h0054346

Monzani, L., Escartín, J., Ceja, L., & Bakker, A. B. (2021). Blending mindfulness practices and character strengths increases employee well-being: A second-order meta-analysis and a follow-up field experiment. *Human Resource Management Journal*, *31*(4), 1025–1062. https://doi.org/10.1111/1748-8583.12360

Ogink, R. H., Goossen, M. C., Romme, A. G. L., & Akkermans, H. (2023). Mechanisms in open innovation: A review and synthesis of the literature. *Technovation*, *119*, 102621.

Patel, N. (2021). Enhancing employee well-being in multicultural organizations. *Global HR Practitioner, 12*(4), 67–89.

Saxena, C., Khatri, B., & Khan, S. A. (2024). Mapping the landscape of social sustainability research: A bibliometric analysis and sentiment analysis. In R. Castanho (Ed.), *ESG and Ecosystem Services for Sustainability* (pp. 309–336). IGI Global. https://doi.org/10.4018/979-8-3693-3771-4.ch012

Sharma, N., Khatri, B., & Khan, S. A. (2024). Exploring the past and identifying the future research agenda for social media and tourist behavior: A bibliometric analysis. In C. Ramos, T. Costa, F. Severino, & M. Calisto (Eds.), *Social Media Strategies for Tourism Interactivity* (pp. 261–284). IGI Global. https://doi.org/10.4018/979-8-3693-0960-5.ch011

Smith, J., Brown, T., & Cooper, L. (2021). Employee satisfaction and productivity: A meta-analysis. *Applied Psychology, 70*(4), 389–411. https://doi.org/10.1111/apps.12244

Soomro, B. A., Zehri, A. W., Anwar, S., Abdelwahed, N. A. A., & Shah, N. (2024). Developing the relationship between corporate cultural factors and employees' organizational commitment via self-efficacy. *South Asian Journal of Business Studies*, *13*(3), 325–347.

Zareen, S. & Khan, S. A. (2023). Exploring dependence of human resource management (HRM) on internet of things (IoT) and digital marketing in the digital era. In A. Naim & V. Devi (Eds.), *Global Applications of the Internet of Things in Digital Marketing* (pp. 51–66). IGI Global. https://doi.org/10.4018/978-1-6684-8166-0.ch004

# 10 Impact of Psychological Well-Being, Workplace Well-Being and Subjective Well-Being on Job Orientation

## *Evidence from Ten Maharatna Companies of India*

*K. P. Kanchana, Shad Ahmad Khan, Soni Sharma, and Ravinder Rena*

## 10.1 INTRODUCTION

Does an employee's well-being play a significant role in their life outside of the workplace? Obviously! Many studies have looked at the effect of employee well-being on job orientation, and we have long known that they have an impact on life at work (Bayhan Karapinar et al., 2020; Turban & Yan, 2016). The relationship between worker well-being and workplace performance is less well-known. Research indicates that one of the most important variables for the performance and success of an organization is the health and well-being of its workforce (Bakker et al., 2019; Turban & Yan, 2016). Globally the interest in national and international well-being levels has grown in recent years. "Good health and well-being" is the third of the 17 Sustainable Development Goals, according to the UNICEF (2024). The Organization for Economic and Co-operation Development publishes the "How's Life" report every two years, which provides a country-by-country breakdown of national well-being indices Organization for Economic Co-operation & Development (OECD, 2020). Research by Harvard Medical School identified that those with 'higher levels of well-being were much more productive than those with lower levels of well-being'. The research team at Harvard Medical School stated that 'Treating mental health problems in the workplace isn't a quick fix'. It has become inevitable for HR professionals

DOI: 10.1201/9781003532293-11

to fully understand issues around understanding and managing well-being at work, it is important to have some awareness of the short- and longer-term requirements.

Employee well-being has been linked to several positive outcomes for both individuals and organizations, including higher productivity and organizational performance (Hewett et al., 2018; Ansari et al., 2024) customer satisfaction (Sharma et al., 2017; Khan et al., 2021), employee engagement (Tisu et al., 2020), and organizational citizenship behavior (OCB) (Mousa et al., 2020; Khatri et al., 2024a). Employee performance has a direct impact on the productivity and performance of the organization (Shin & Konrad, 2017; Bocar et al., 2022; Khatri et al., 2023). The importance of employee job performance—defined as the measurable actions, behaviors, and outcomes that employees engage in or bring about that are linked to and contribute to organizational goals—has been demonstrated by a wealth of evidence. This has led scholars to investigate the factors that influence employee performance in order to better understand organizational outcomes and success (Al Hammadi & Hussain, 2019; Shin & Konrad, 2017). Employee job performance has been found to be critically influenced by personality factors (Tisu et al., 2020), job conditions, and organizational features (Diamantidis & Chatzoglou, 2019). Nonetheless, a significant vacuum persists in the present study on job orientation, specifically the impact of mental health on work output (Hewett et al., 2018). Even though earlier studies have shown happier people to be more productive than less happy or miserable employees (DiMaria et al., 2020; Zareen & Khan, 2023), few studies on this topic were found in the literature after a search link between psychological health and work performance (Salgado et al., 2019; Turban & Yan 2016). Furthermore, not much is understood about the mechanisms connecting psychological health and work performance. Particularly when it comes to psychological well-being, only a small range of well-being-related antecedents of employee performance have been taken into account. The current study investigates the connection between psychological well-being and job orientation in the workplace, improving our understanding of the effects and mechanisms of psychological well-being in the workplace. This kind of information will not only assist managers in achieving improved organizational performance in these unpredictable times, but it will also reveal strategies for maintaining employee happiness and satisfaction (DiMaria et al., 2020; Ansari et al., 2024; Khan & Ansari, 2024).

Importantly, additional study is required to explore the connection between workers' psychological health and job orientation in order to progress the field of job performance research (Ismail et al., 2019). Salgado et al. (2019) expounded on the importance of taking into account the impact that an employee's well-being has on their productivity at work. The current study aims to bridge this gap in the literature by establishing a relationship between an individual's job orientation and their psychological well-being, specifically workplace well-being, and subjective well-being concerning and determining Job Orientation.

## 10.2 BACKGROUND OF THE STUDY

In many nations, mental health professionals deal with a lack of institutional support and heavy workloads. Given their high levels of stress at work, it's critical to find coping techniques that support their psychological health and sense of fulfillment in their jobs. According to Huta (2017) and Rahmani et al. (2018), psychological well-being is the happiness attained via experiences of pleasure and enjoyment and the happiness attained through experiences of meaning and purpose. We contend that workers who experience high levels of psychological well-being will outperform those who experience low ones. Employee affective commitment, or workers' judgments of their emotional attachment to or identification with their organization, serves as the link between this psychological well-being and job performance process (Allen & Meyer, 1990) by addressing it as a mediating factor in the link between performance and well-being.

An organization's main source of motivation is its employees. For the business, it is important to make an investment in its talents it is having. In order to create a work environment that satisfies employees' demands. Workers must take into account the significance of their position inside the company as well as the variables affecting job happiness. A state of mental health or happy feelings that arise from appraising one's work or experience is what Locke (1969) defined as job satisfaction.

Workplace well-being plays a crucial role in employee work satisfaction. It is imperative that human resource management addresses employee requirements, as it builds a setting that satisfies the requirements of the workforce (Handoko, 2001). Employee self-efficacy is another component that can uphold and enhance the caliber of people outside the organization.

Workers contribute significantly to a company or organization and are social creatures. To achieve organizational objectives, employees actively participate as planners, executors, and controllers and possess ambitions, aspirations, ideas, and sentiments that may affect how they feel about their job. This mindset could be either favorable or unfavorable. Workers must take into account the significance of their position within the company as well as the elements that affect job happiness.

Organizations require special attention because their human resources are their main source of motivation. The organization must make an investment to retain its talent. The process of hiring new employees is crucial for every business, as it requires a larger workforce. This means that the human resources department must work harder to identify and classify candidates who are qualified for open positions. According to Dewi and Putra (2020), raising the caliber of talent has a positive effect on the worker job orientation and lowers worker stress. Organizational leaders find it difficult to manage human resources, particularly in the COVID-19 environment that we are currently experiencing (Khatri et al., 2024b). Particularly in the workplace, the pandemic scenario produces a pattern of change. The government's suggestions are followed by workers who work from home (Bocar et al., 2022). A company may meet the diverse requirements of its workforce, which boosts motivation and productivity, by using people as a resource.

Employers and employees alike benefit from promoting a mentally healthy workplace, as doing so satisfies both moral and legal obligations. Harvey (2014) lists various advantages of this approach. The importance of preserving workers' health, happiness, and output is recognized by employers. With the right strategy implemented and subsequently evaluated to determine the degree to which the plan was successful, workplace health promotion initiatives can aid in achieving these goals.

The cognitive and emotional assessment of people's lives in all their aspects is referred to as subjective well-being, according to Burger et al. (2021). A factor analysis of the variables influencing employees' perceptions of their work life quality and workplace satisfaction in the corporate setting will be the main goal of this study. Within the context of the workplace, workers perceive their quality of life at work, accounting for elements of the company such as an engaging role with various benefits in equality and favorable physical, social, and environmental circumstances. Because of this, the individual's demands for achievement orientation, job satisfaction, and other socioemotional and cognitive characteristics of their work environment are among the extrinsic and intrinsic factors that determine the quality of their work life.

According to Seligman (2011), there are five components that contribute to workplace happiness: achievement orientation, significant for self-motivation required in labor productivity, positive emotions, which aid in improving one's perception of life and work, positive relationships as social beings; it is necessary to develop good interpersonal relationships; to achieve good team results we need to communicate properly with people at work; and the commitment necessary to meet the challenges of companies and loyalty of employees (Mendoza-Ocasal et al., 2021).

Due to its growing importance in the workplace, employee well-being has become a focus of positive psychology (Jaiswal et al., 2020). Employers are actively teaching and encouraging mindfulness among their workforces in order to enhance employee motivation and well-being. It is essential for employee adjustment and well-being to practice mindfulness, which may also serve as a protective barrier against toxic work settings (Schultz et al., 2015). According to der Kinderden and Khapova (2020), organizations that aspire to attract and retain top talent must prioritize well-being, with a particular emphasis on eudemonic PWB and other eudemonia-related features. McGhee and Grant (2020) argue that the eudemonic dimension of well-being has received relatively little attention. As to Fisher (2014), the eudemonic aspect of well-being can be represented by organizational constructs such as intrinsic motivation, work engagement, thriving, flow, vigor, dedication absorption, and purpose.

Workplace contentment or well-being can have either beneficial or negative repercussions, according to Slemp, Kren, and Brodrick (2015). In one or more work contexts where an individual respects believable values in conditions in which he or she can function maximally, Cooper and Quick (2017) define workplace wellness as the physical, psychological, social, or psychological benefits (Damayanti & Budiani, 2021). Success in the workplace is a result of feeling good for an employee since contented workers are more productive.

The notion of workplace well-being refers to the emotional and intrinsic value-related state of employees during work, as well as their well-being following work-related events and activities. Comprehending workplace well-being is aided by disciplines including psychology, organizational behavior, and medicine. According to Fridayanti et al. (2022), a number of intriguing models are also offered from an interdisciplinary standpoint.

The benefits of individual-level approaches have been extensively researched, as well as increasingly debated. There is a large scholarship of experimental work testing the effects of participation in initiatives, with systematic reviews of controlled trials for stress management (e.g., LaMontagne et al., 2016), resilience training (Joyce et al., 2022), mindfulness (Michaelsen et al., 2023) and more. Despite the apparent scale, there are not only continual calls for more evidence from practitioners and academic researchers, led in part by technical, methodological limitations but also a desire for more realist evaluation (Kowalski & Loretto, 2017; Nielsen & Miraglia, 2017). Disagreement has risen around the effectiveness of individual-level strategies when compared with organizational change, with suggestions that individual-level interventions are just easier to evaluate, misguided if they do not address working conditions and then take positive change for granted.

Alongside these empirical concerns is a strategic and normative critique from sociologists and industrial relations scholars, with many criticisms leveled at such practices' propagation of, in the words of trade unions 'changing the worker, and not the workplace'. Others go further, arguing individualized well-being initiatives are more interested in social control than with improving well-being (Murphy & MacMahon, 2022), with qualitative organizational research substantiating some of these fears (Holmqvist & Maravelias, 2010; Islam et al., 2022; Zoller, 2011). While a management strategy could be a tool of social control and yet still improve subjective well-being outcomes, these criticisms highlight that evaluations are political, not only technical.

An increasing amount of research shows a direct link between workers' job orientation and well-being. Following the Covid-19 pandemic, more businesses are implementing flexible or smart work arrangements. These guidelines seek to improve workplace flexibility for staff members while also building a more robust corporate culture. Employees are better equipped to manage their personal and professional obligations when they have flexible work schedules. A flourishing feeling of well-being among staff members frequently results in improved output. On the other hand, unhappy workers typically produce less. As a result, a decrease in worker satisfaction is probably going to cause a comparable loss in output and output (Wright & Cropanzano, 2000).

However, the Harvard Business Review has also observed that, in addition to being less productive than their happy colleagues, disgruntled workers cost businesses more money, highlighting the financial significance of employee well-being (Cash-Gibson et al., 2023).

Three criteria are usually used to evaluate an employee's well-being: relationships, health, and happiness. Studies show that while connections and happiness among employees positively correlate with organizational success, health-related well-being frequently appears as a contradictory factor. This offers fresh approaches for further studies and theoretical advancement in this area (Haddon, 2018).

### 10.2.1 Problem Statement

The well-being of employees has become a crucial factor in determining job orientation and overall organizational success in the quickly changing business landscape. The relationship between the many aspects of well-being—psychological, workplace, and subjective—and how they affect job orientation is still not well understood. This disparity is especially noticeable in big, diverse, and strategically significant enterprises like India's Maharatna companies. Maharatna means a great jewel, these companies are like great jewels for the public sector of India.

Notwithstanding the vital role these businesses play in the US economy, little empirical data is available to show how these well-being aspects impact job orientation in these kinds of businesses. In order to close this gap, this study examines the connections between psychological well-being, workplace well-being, and subjective well-being as well as how these links affect employees' job orientation in ten carefully chosen Maharatna enterprises in India.

Policymakers, HR managers, and executives in these organizations can benefit greatly from an understanding of these relationships by creating work environments that promote employee well-being, which in turn improves job orientation, employee retention, and organizational performance.

### 10.2.2 Objectives of Study

In the context of Maharatna firms, this remark emphasizes the research gap and the significance of examining the influence of various well-being characteristics on job orientation following objectives have been framed to study:

- **RO1**: To examine the connection between work attitude and psychological health in the chosen Maharatna enterprises in India.
- **RO2**: To investigate how well-being at work affects job orientation and how it affects workers' motivation, attitudes, and level of job satisfaction in various kinds of organizations.
- **RO3**: To evaluate how subjective well-being influences job orientation and how it affects worker commitment, engagement, and total job performance.
- **RO4**: The aim of this study is to examine the potential impact of psychological well-being, workplace well-being, and subjective well-being on job orientation, and to determine the most influential dimension.
- **RO5**: To offer suggestions for augmenting job orientation by means of enhancements in employee welfare, predicated on the discoveries made by the Maharatna enterprises.

With an emphasis on the useful consequences for large businesses, these aims aim to thoroughly investigate the influence of various well-being characteristics on job orientation.

### 10.2.3 Theoretical Framework

Sonnentag and Frese (2002) have embraced theoretical viewpoints on job orientation and well-being. While the situational approach looks at differences in job orientation as a result of situational and contextual factors, the individual perspective stresses the significance of individual differences in personality or abilities.

#### 10.2.3.1 Effort-Recovery Theory

According to the effort-recovery theory, worn-out employees have three options: putting in more effort to stay focused on their work, redefining the requirements for their tasks, or producing less-than-ideal output. Increased levels of distress (fatigue) may result in a decrease in job orientation since worn-out employees may decide to work fewer hours or simply complete a portion of their responsibilities.

#### 10.2.3.2 Self-Determination Theory

The self-determination theory centers on the extent to which work satisfies fundamental psychological needs, which in turn sets off various motivational processes and influences job orientation. According to Deci and Ryan's Self Determination Theory (SDT), employee motivation is influenced by meeting these demands, so there is a relationship between higher job orientation and higher well-being. When employees are organically driven for their professions and find their work to be fascinating and fulfilling, they are exhibiting good job orientation.

#### 10.2.3.3 Broaden-and-Build Theory

According to the broaden-and-build theory, positive well-being would modify the relationship between job satisfaction and job orientation, with a larger relationship for employees who report high levels of psychological well-being. Employees with this mentality are more proactive and less prone to stress, which influences their work-related orientation both directly and indirectly. This moderates the already beneficial impacts of other forms of well-being on job orientation, enhancing them even more.

As "well-being" can be seen as an individual-level notion, there is a connection between the situational and individual perspectives on the relationship between performance and individual well-being. In fact, other studies have shown connections between personality traits (including neuroticism, extraversion, and Type-A behavior) and well-being factors as burnout and work engagement (see, for example, Hallberg et al., 2007; Taris et al., 2014). According to these viewpoints, environmental factors—that is, the qualities of the work—have an impact on employees' well-being, which in turn has an impact on their productivity.

## 10.3 METHODOLOGY OF THE STUDY

The data for this study came from a survey of ten Maharatna employees of India. These ten Maharatna companies were identified as being highly competitive, large, and encompassing 80% of the economy and employees of PSUs. Therefore, the study of how employees' perceptions of job orientation on their psychological well-being and its outcomes is highly relevant in this context. Studies reconnoitering these dimensions are required to have a better insight and thereby strengthen the employee well-being and job performance literature.

Primary research was conducted through a questionnaire which was circulated online through Google sheet was used to collect the data. The questionnaire was sent along with the cover letter assuring their privacy that hence their names and mail ID were not recorded. As self-reported data often renders itself to common method bias (CMB; Podsakoff et al., 2012), we applied several procedural remedies such as reducing the ambiguity in the questions, ensuring respondent anonymity and confidentiality, separating the predictor and criterion variable and randomizing the item order to limit this bias.

**TABLE 10.1**
**Profile of the Respondents**

| Demographic Variable | | N | Percent |
|---|---|---|---|
| Age of respondent | 21–30 years | 29 | 9.6 |
| | 31–40 years | 106 | 35.2 |
| | 41–50 years | 123 | 40.9 |
| | Above 51 years | 43 | 14.3 |
| Gender | Female | 78 | 25.9 |
| | Male | 223 | 74.1 |
| Education level | Certificate & diploma | 13 | 4.3 |
| | Doctorate | 10 | 3.3 |
| | Graduation | 114 | 37.9 |
| | Post-graduation | 164 | 54.5 |
| Position level | Lower level manager (E1–E3) | 41 | 13.6 |
| | Middle level manager (E4–E6) | 190 | 63.1 |
| | Upper level manager (E7–E10) | 70 | 23.3 |
| Organization name | BHEL | 65 | 21.6 |
| | BPCL | 13 | 4.3 |
| | CIL | 33 | 11.0 |
| | GAIL | 68 | 22.6 |
| | HPCL | 4 | 1.3 |
| | IOCL | 41 | 13.6 |
| | NTPC | 18 | 6.0 |
| | ONGC | 12 | 4.0 |
| | PGCIL | 27 | 9.0 |
| | SAIL | 20 | 6.6 |

The questionnaire was sent to 570 respondents out of which 327 responded, and finally, 302 responses were found to be complete for further analysis. The response rate was 52.98. Conferring to Baruch and Holtom (2008), the average response rate for studies at the individual level is 52.6% (SD = 19.7). Hence, our response rate meets the standard for a minimum acceptable response rate, which is 49%. Of the 302 respondents, 25.9% were female and 74.1 were male (Table 10.1).

## 10.4 DATA ANALYSIS

A simple descriptive statistics using mean score analysis and correlation analysis was performed among the four major constructs for this study i.e., Psychological Well-being (P); Workplace Well-being (W); Subjective Well-being (S); and Job Orientation (JO). The mean score analysis is presented in Table 10.2. As the responses were collected on a scale of five, the mean score of 2.5 is considered a mid-value for

**TABLE 10.2**
**Mean Score Analysis**

| Statement | Mean Score |
|---|---|
| P1: I easily adapt to daily changes to manage my responsibilities well. | 1.60 |
| P2: I care for things that are important to me, not what is important to others. | 3.80 |
| P3: I take an active part in important decision-making activities of my team. | 1.63 |
| P4: I believe that I have a purpose and direction in life. | 1.62 |
| W1: I am quite satisfied with my job. | 1.66 |
| W2: I enjoy meaningful work. | 1.60 |
| W3: I attach lots of value to my work. | 1.63 |
| W4: My work achievement often acts as a source of motivation. | 1.61 |
| W5: My job provides ample scope for career growth. | 1.66 |
| S1: Mostly I feel happy. | 1.61 |
| S2: I am an optimistic person. | 1.60 |
| S3: I feel good about myself. | 1.59 |
| JO1: At my job, I feel strong and vigorous. | 1.60 |
| JO2: I am enthusiastic about my job. | 1.64 |
| JO3: I get carried away when I am working. | 2.45 |

*Source:* Authors' Calculation.

**TABLE 10.3**
**Pearson Correlation Analysis**

| | P | W | S | JO |
|---|---|---|---|---|
| Psychological well-being (P) | 1 | 0.788[a] | 0.781[a] | 0.727[a] |
| Workplace well-being (W) | | 1 | 0.919[a] | 0.827[a] |
| Subjective well-being (S) | | | 1 | 0.783[a] |
| Job orientation (JO) | | | | 1 |

*Source:* Authors' Calculation.

[a] Correlation is significant at the 0.01 level (2-tailed).

determining the orientation. It is important to note that except P2, a negative orientation was observed for all the mean scores. JO3 was found to be closest to the mid-value of 2.5.

To understand the internal relationship between the constructs, correlation analysis was performed. The values closer to 1 are considered stronger, and toward 0 are considered weak correlation. As shown in Table 10.3, all the constructs had a positive, and significant correlation.

The analysis reveals strong positive correlations among workplace well-being, subjective well-being, job orientation, and psychological well-being, highlighting their interdependence. As workplace well-being improves, both subjective well-being and job orientation tend to increase, suggesting that enhancements in one area can positively impact the others.

## 10.5 DISCUSSIONS

Although the correlation analysis suggests a strong and positive correlation among the four constructs i.e., Psychological Well-being (P); Workplace Well-being (W); Subjective Well-being (S); and Job Orientation (JO). The mean score analysis reveals a concerning trend in their perceptions of various aspects of their work and well-being. This indicates a general sentiment of dissatisfaction and disengagement among respondents. For instance, the extremely low scores on adaptability (P1: 1.60) and active participation in decision-making (P3: 1.63) suggest that employees may feel overwhelmed by daily changes or excluded from important processes, leading to frustration and a sense of helplessness. Similarly, low scores on workplace satisfaction (W1: 1.66) and enjoyment of meaningful work (W2: 1.60) indicate that employees do not find their roles fulfilling or aligned with their values, contributing to overall low morale. In contrast, the relatively higher score for caring about personal values (P2: 3.80) suggests that while employees prioritize their own beliefs and needs, they may feel unsupported in these areas within their work environment. This misalignment can create tension and disengagement, as reflected in the low scores across most other dimensions of Psychological and Workplace Well-being. The combination of low scores in aspects of Psychological Well-being and Workplace Well-being, alongside slight increases in Subjective Well-being, points to a workforce struggling with motivation and fulfillment. This kind of attitude cannot lead to a positive organizational orientation and success. The underlying issue can be the belief of being in a permanent service setup (public sector company) where the chances of removal, termination, etc. are minimal. However, incentivizing the performance could be a good initiative to overcome the dissatisfaction among employees.

## 10.6 CONCLUSION

The findings suggest a disconnect between personal values and workplace experiences, resulting in low overall Psychological and Workplace Well-being among these government employees. Addressing these issues through enhanced support, recognition, and opportunities for involvement in decision-making could help improve their job satisfaction and overall well-being. Although the four variables, i.e., Psychological Well-being, Workplace Well-being, Subjective Well-being, and Job Orientation are highly correlated to each other, the level of overall well-being is found to be low, which might have a negative influence on Job Orientation.

## 10.7 LIMITATIONS OF THE STUDY

The present study on the influence of Subjective Well-being, Workplace Well-being, and Psychological Well-being on Job Orientation presents several limitations. Firstly, its cross-sectional design restricts the ability to draw causal conclusions or track changes over time, making it unclear whether shifts in job orientation stem from changes in well-being or vice versa. Secondly, the specific characteristics of the sample, including industry and job functions, may limit the generalizability of the findings to other demographics or contexts. Additionally, the varying quality of measurement tools, particularly the lower reliability of the Psychological Well-being scale, could distort the findings and weaken the overall conclusions. Moreover, external factors, such as organizational changes or economic conditions, may not have been considered, potentially compromising the accuracy of the results. Finally, the absence of an investigation into potential moderators or mediators limits the understanding of how wellness influences job orientation, risking the oversight of important contextual factors.

## 10.8 DIRECTION FOR FUTURE STUDIES

As mentioned before, these companies are the great jewels of the Indian public sector, a longitudinal study is recommended on the same theme to evaluate the situation over a period of time. Further, the statements utilized for this study might not be representative of the other aspects like loyalty, satisfaction, system, leadership, etc. A more comprehensive instrument can be utilized for future studies ensuring diverse samples, and accounting for external influences—which will provide a clearer and more accurate understanding of how these well-being dimensions impact Job Orientation. Being a quantitative study, the qualitative aspects might have been ignored, thus, a mixed-method study would be better for future researchers.

## REFERENCES

Al Hammadi, F., & Hussain, M. (2019). Sustainable organizational performance: A study of health-care organizations in the United Arab Emirates. *International Journal of Organizational Analysis*, *27*(1), 169–186.

Allen, N. J., & Meyer, J. P. (1990). The measurement and antecedents of affective, continuance and normative commitment to the organization. *Journal of Occupational Psychology*, *63*(1), 1–18.

Ansari, M. S. A., Khan, S. A., & Bhuyan, U. (2024). Exploring the strategic orientation factors influencing the organizational performance through bibliometric analysis. In S. Umair, U. Waqas, B. Mrugalska, & I. R. Al Shamsi (Eds.), *Human Perspectives of Industry 4.0 Organizations* (pp. 68–88). CRC Press.

Bakker, A. B., Hetland, J., Olsen, O. K., & Espevik, R. (2019). Daily strengths use and employee well-being: The moderating role of personality. *Journal of Occupational and Organizational Psychology*, *92*(1), 144–168.

Baruch, Y., & Holtom, B. C. (2008). Survey response rate levels and trends in organizational research. *Human Relations, 61*(8), 1139–1160. https://doi.org/10.1177/0018726708094863

Bayhan Karapinar, P., Metin Camgoz, S., & Tayfur Ekmekci, O. (2020). Employee wellbeing, workaholism, work–family conflict and instrumental spousal support: A moderated mediation model. *Journal of Happiness Studies*, *21*(7), 2451–2471.

Bocar, A. C, Khan, S. A., & Epoc, F. (2022). COVID-19 work from home stressors and the degree of its impact: Employers and employees actions. *International Journal of Technology Transfer and Commercialisation, 19*(2), 270–291. https://dx.doi.org/10.1504/ijttc.2022.124349

Burger, C. A., Jiang, D., Mackin, R. D., & Samuel, M. A. (2021). Development and maintenance of vision's first synapse. *Developmental Biology*, *476*, 218–239. https://doi.org/10.1016/j.ydbio.2021.04.001

Cash-Gibson, L., Isart, F. M., Martínez-Herrera, E., Herrera, J. M., & Benach, J. (2023). Towards a systemic understanding of sustainable wellbeing for all in cities: A conceptual framework. *Cities*, *133*, 104143.

Cooper, C. L., & Quick, J. C. (Eds.). (2017). *The Handbook of Stress and Health: A Guide to Research and Practice*. Wiley Blackwell. https://doi.org/10.1002/9781118993811

der Kinderen, S., & Khapova, S. N. (2020). Positive psychological well-being at work: The role of eudaimonia. In *The Palgrave Handbook of Workplace Well-Being* (pp. 1–28). Palgrave Macmillan.

Damayanti, D. A., & Budiani, M. S. (2021). Hubungan Antara Persepsi Kepemimpinan Otokratis dengan Workplace Well-Being pada Karyawan PT X di Kabupaten Ponorogo. *Character: Jurnal Penelitian Psikologi*, *8*(3), 137–150.

Dewi, M. R. L., & Putra, D. K. N. S. (2020). Creative problem-solving learning model assisted with multimedia to the competency of science. *International Journal of Elementary Education*, *4*(4), 587–595.

Diamantidis, A. D., & Chatzoglou, P. (2019). Factors affecting employee performance: An empirical approach. *International Journal of Productivity and Performance Management*, *68*(1), 171–193.

DiMaria, C. H., Peroni, C., & Sarracino, F. (2020). Happiness matters: Productivity gains from subjective well-being. *Journal of Happiness Studies*, *21*(1), 139–160.

Fisher, J. W. (2014). Comparing the influence of God and other transcendents on spiritual well-being. *Religious Education Journal of Australia*, *30*(2), 9–15.

Fridayanti, F., Kardinah, N., & Lestari, T. A. (2022). Teachers' work engagement: The role of perceived organizational support and meaningful work. *Psympathic: JurnSal Ilmiah Psikologi*, *9*(1), 67–76.

Haddon, J. (2018). The impact of employees' well-being on performance in the workplace. *Strategic HR Review*, *17*(2), 72–75.

Hallberg, U. E., Johansson, G., & Schaufeli, W. B. (2007). Type A behavior and work situation: Associations with burnout and work engagement. *Scandinavian Journal of Psychology*, *48*(2), 135–142.

Harvey, S. (2014). Creative synthesis: Exploring the process of extraordinary group creativity. *Academy of Management Review*, *39*(3), 324–343.

Hewett, R., Liefooghe, A., Visockaite, G., & Roongrerngsuke, S. (2018). Bullying at work: Cognitive appraisal of negative acts, coping, wellbeing, and performance. *Journal of Occupational Health Psychology*, *23*(1), 71.

Holmqvist, M., & Maravelias, C. (2010). *Managing Healthy Organizations: Worksite Health Promotion and the New Self-Management Paradigm*. Routledge.

Huta, V. (2017). An overview of hedonic and eudaimonic well-being concepts. In L. Reinecke & M. B. Oliver (Eds.), *The Routledge Handbook of Media Use and Well-Being: International Perspectives on Theory and Research on Positive Media Effects* (pp. 14–33). Routledge/Taylor & Francis Group.

Islam, S., Muhamad, N., & Sumardi, W. H. (2022). Customer-perceived service wellbeing in a transformative framework: Research propositions in the area of health services. *International Review on Public and Nonprofit Marketing*, *19*(1), 219–245.

Ismail, H. N., Karkoulian, S., & Kertechian, S. K. (2019). Which personal values matter most? Job performance and job satisfaction across job categories. *International Journal of Organizational Analysis*, *27*(1), 109–124.

Jaiswal, A. K., Singh, L., Rani, R., & Sarraf, S. R. (2020). Psychometric properties and factor analysis of Hindi version of the meta-emotion scale. *Indian Journal of Positive Psychology*, *11*(4), 383–388.

Joyce, A., Elmes, A., Campbell, P., Moussa, B., Suchowerska, R., Barraket, J., & Carey, G. (2022). The health and well-being impacts of a work integration social enterprise from a systems perspective. *Health Promotion International*, *37*(1), daab052.

Khan, S. A., & Ansari, M. S. A. (2024). Critical factors for relationship between strategic orientation, corporate success, and innovation. In S. Umair, U. Waqas, B. Mrugalska, & I. R. Al Shamsi (Eds.), *Human Perspectives of Industry 4.0 Organizations* (pp. 89–110). CRC Press.

Khan, S. A, Epoc, F., Gangwar, V. P., Ligori, T. A. A, Ansari, Z. A. (2021). Will online banking sustain in Bhutan post covid-19? A quantitative analysis of the customer e-satisfaction and e-loyalty in the Kingdom of Bhutan. *Transnational Marketing Journal*, *9*(3), 607–624. https://dx.doi.org/10.33182/tmj.v9i3.1288

Khatri, B., Arora, S., & Khan, S. A. (2024a). Analysis of servant leadership and organizational citizenship behaviour among social service sector employees. In D. Belias, I. Rossidis, & C. Papademetriou (Eds.), *Organizational Behavior and Human Resource Management for Complex Work Environments* (pp. 443–465). IGI Global. https://doi.org/10.4018/979-8-3693-3466-9.ch021

Khatri, B., Shrimali, H., Khan, S. A., & Naim, A. (2023). Role of HR analytics in ensuring psychological wellbeing and job security: Learnings from COVID-19. In R. Yadav, M. Sinha, & J. Kureethara (Eds.), *HR Analytics in an Era of Rapid Automation* (pp. 36–53). IGI Global. https://doi.org/10.4018/978-1-6684-8942-0.ch003

Khatri, B., Sharma, N., Khan, S. A., & Nandini, G. (2024b). Navigating employer branding in the digital age: Exploring the impact of social media networking during the COVID-19 pandemic. In B. Christiansen, M. Aziz, & E. O'Keeffe (Eds.), *Global Practices on Effective Talent Acquisition and Retention* (pp. 313–332). IGI Global. https://doi.org/10.4018/979-8-3693-1938-3.ch017

Kowalski, T. H., & Loretto, W. (2017). Well-being and HRM in the changing workplace. *The International Journal of Human Resource Management*, *28*(16), 2229–2255.

LaMontagne, A. D., Milner, A., Krnjacki, L., Schlichthorst, M., Kavanagh, A., Page, K., & Pirkis, J. (2016). Psychosocial job quality, mental health, and subjective wellbeing: A cross-sectional analysis of the baseline wave of the Australian Longitudinal Study on Male Health. *BMC Public Health*, *16*, 33–41.

Locke, E. A. (1969). What is job satisfaction? *Organizational Behavior and Human Performance*, *4*(4), 309–336.

McGhee, P., & Grant, P. (2021). Spiritual mindfulness for management. In S. K. Dhiman (Ed.), *The Routledge Companion to Mindfulness at Work* (pp. 219–233). Routledge/Taylor & Francis Group.

Mendoza-Ocasal, D., Castillo-Jiménez, R., Navarro, E., & Ramírez, J. (2021). Measuring workplace happiness as a key factor for the strategic management of organizations. *Polish Journal of Management Studies*, *24*, 292–306.

Michaelsen, M. M., Graser, J., Onescheit, M., Tuma, M. P., Werdecker, L., Pieper, D., & Esch, T. (2023). Mindfulness-based and mindfulness-informed interventions at the workplace: A systematic review and meta-regression analysis of RCTs. *Mindfulness*, *14*(6), 1271–1304.

Mousa, M., Massoud, H. K., & Ayoubi, R. M. (2020). Gender, diversity management perceptions, workplace happiness and organisational citizenship behaviour. *Employee Relations: The International Journal*, *42*(6), 1249–1269.

Murphy, C., & MacMahon, J. (2022). Employee resilience: A labour process perspective. *Work in the Global Economy*, *2*(1), 109–131. https://doi.org/10.1332/273241721X16517520388131

Nielsen, K., & Miraglia, M. (2017). What works for whom in which circumstances? On the need to move beyond the 'what works?' question in organizational intervention research. *Human Relations*, *70*(1), 40–62.

OECD (2020). Retrieved from https://www.oecd.org/en/publications/how-s-life/volume-/issue-_9870c393-en.html, on August 19, 2024.

Podsakoff, P.M., Mackenzie, S.B. and Podsakoff, N.P. (2012) Sources of Method Bias in Social Science Research and Recommendations on How to Control It. Annual Review of psychology, 63, 539-569. http://dx.doi.org/10.1146/annurev-psych-120710-100452

Rahmani, K., Gnoth, J., & Mather, D. (2018). Hedonic and eudaimonic well-being: A psycholinguistic view. *Tourism Management, 69*, 155–166. https://doi.org/10.1016/j.tourman.2018.06.008

Salgado, J. F., Blanco, S., & Moscoso, S. (2019). Subjective well-being and job performance: Testing of a suppressor effect. *Revista de Psicología del Trabajo y de las Organizaciones, 35*(2), 93–102.

Schultz, P. P., Ryan, R. M., Niemiec, C. P., Legate, N., & Williams, G. C. (2015). Mindfulness, work climate, and psychological need satisfaction in employee well-being. *Mindfulness, 6*, 971–985.

Seligman, M. E. (2011). *Flourish: A Visionary New Understanding of Happiness and Well-Being*. Simon and Schuster.

Sharma, S., Conduit, J., & Rao Hill, S. (2017). Hedonic and eudaimonic well-being outcomes from co-creation roles: A study of vulnerable customers. *Journal of Services Marketing, 31*(4/5), 397–411.

Shin, D., & Konrad, A. M. (2017). Causality between high-performance work systems and organizational performance. *Journal of Management, 43*(4), 973–997. https://doi.org/10.1177/0149206314544746

lemp, G. R., Kern, M. L., & Vella-Brodrick, D. A. (2015). Workplace well-being: The role of job crafting and autonomy support. *Psychology of Well-being, 5*, 1–17.

Sonnentag, S., & Frese, M. (2002). Performance concepts and performance theory. *Psychological Management of Individual Performance, 23*(1), 3–25.

Taris, T., Van Beek, I., & Schaufeli, W. (2015). The beauty versus the beast: On the motives of engaged and workaholic employees. In I. Harpaz & R. Snir (Eds.), *Heavy Work Investment: Its Nature, Sources, Outcomes, and Future Directions* (pp. 121–139). Routledge/Taylor & Francis Group.

Tisu, L., Lupşa, D., Vîrgă, D., & Rusu, A. (2020). Personality characteristics, job performance and mental health: The mediating role of work engagement. *Personality and Individual Differences, 153*, 109644.

Turban, D. B., & Yan, W. (2016). Relationship of eudaimonia and hedonia with work outcomes. *Journal of Managerial Psychology, 31*(6), 1006–1020. https://doi.org/10.1108/JMP-07-2015-0271

UNICEF (2024). Retrieved from https://data.unicef.org/sdgs/goal-3-good-health-wellbeing/, on August 17, 2024.

Wright, T. A., & Cropanzano, R. (2000). Psychological well-being and job satisfaction as predictors of job performance. *Journal of Occupational Health Psychology, 5*(1), 84–94. https://doi.org/10.1037/1076-8998.5.1.84.

Zareen, S. & Khan, S. A. (2023). Exploring dependence of human resource management (HRM) on internet of things (IoT) and digital marketing in the digital era. In A. Naim, & V. Devi (Eds.), *Global Applications of the Internet of Things in Digital Marketing* (pp. 51–66). IGI Global. https://doi.org/10.4018/978-1-6684-8166-0.ch004

Zoller, S. (2011). Age, Creativity, and Wellbeing: Benefits of Creative Behavior Among Younger and Older Artists and Nonartists. Dissertation submitted to Washington University in St. Louis. https://www.investorsinpeople.com/success-stories/archive/

# *Section II*

## *HF&E Meets Technology*

### *Shaping Corporate Success in the Digital Landscape*

# 11 Human Factors and Ergonomics in the Digital Era

## *Role of Artificial Intelligence*

*Aujasvi Singh, Monika Srivastava, Sobhana Jaya-Madhavan, and Busige Nishantha*

## 11.1 INTRODUCTION

The fundamental convergence has reached human factors and ergonomics (HFE) – now navigating its relativity with artificial intelligence (AI), in the immediate wake of Industry 4.0 within an ever-evolving digital age, where technology drives unprecedented changes to life as we know it on earth (Magd et al., 2022, 2023). Thinking of a future where digital interactions are smooth, natural, and completely context-aware is expected to be a reality (Thakur et al., 2025). This is not just science fiction, but an emergent reality fueled by the mutual constructive collaboration between HFE and AI.

## 11.2 OVERVIEW OF HUMAN FACTORS AND ERGONOMICS

AI is revolutionizing HFE, which of course has everything to do with how humans interact with a system (Salmon et al., 2021; Zareen & Khan, 2023; Khatri et al., 2023). Increasingly less restrained by conventional paradigms, more intelligent and adaptable HFE designs driven by AI are making user interfaces not only ease the usability of a product but also better serve to performance optimization. Think AI-powered ergograph that predetermines what the user is doing and when to make life easier for them (just like an advance in Driverless cars), while lightening their load on unnecessary task-switching, mind-reading, etc. during technology-driven hyper productivity era of everything-happens-at-laptop.

At HFE, AI is more than just an enhancement; it is a disruption. Through the incorporation of AI, we can read complex human behaviors to predict interactions and build systems that develop with changing user needs (Khan et al., 2023; Ansari et al., 2024). Whether that be from intelligent personal assistants to finding a solution which works best for them when learning new things.

HFE are the study of how humans interact with many devices. The aim is for it to maximize human welfare and minimize the malfunctioning of devices (Dul et al., 2012) It must be achieved. That encompasses not only designing devices, machines,

DOI: 10.1201/9781003532293-13

and systems but also the jobs, workstations, and environments to meet consumer wants (their needs), abilities and limits.

### 11.2.1 Importance of AI in Enhancing Human–System Interactions

AI has become a key tool for improving the interaction between humans and devices (Jonathan et al., 2024). AI technology enables adaptive systems that can learn and respond to human behavior, improving usability, safety, and efficiency. AI applications in ergonomics can analyze complex data to predict and mitigate human error, adjust person reports, and automate repetitive tasks to reduce physical and cognitive strain (Fong et al., 2020). For example, AI-driven ergonomic testing can help design workplaces that reduce the risk of musculoskeletal disorders (MSDs) by predicting excessively hazardous postures and suggesting ergonomic improvements.

### 11.2.2 Objectives of the Study

This chapter aims to explore the intersection of HFE with AI and how AI technology can improve ergonomic practices and human-machine interactions. The following aims are to outline the basic ideas of HFE:

1. Discuss the impact of virtual transformation on ergonomics.
2. Examine the role of AI in improving ergonomic design and human-machine interactions.
3. Analyze ethical considerations and human-centered AI design.
4. Identify current challenges and future directions in the integration of AI with HFE.

### 11.2.3 Methodology

This chapter uses an exploratory research design to examine the role of AI in HFE in the contemporary business context. It combines a comprehensive literature review, expert interviews, and case studies to collect qualitative and quantitative data. The literature review used databases such as Google Scholar and IEEE Xplore to gain insights into current trends and challenges. Several case studies from industries such as healthcare and automotive illustrate practical AI applications. Thematic analysis and descriptive statistics are used to analyze the data. Despite potential biases and the rapidly evolving nature of AI, this method provides a thorough understanding of the impact of AI on HFE.

## 11.3 FOUNDATIONS OF HFE

### 11.3.1 Definition and Scope of HFE

HFE encompass many medical disciplines concerned with understanding the interactions between humans and the many factors of a system. The goal is to use theoretical principles, information, and techniques to perfect human well-being and normal machine performance (Karwowski & Marras, 2020).

### 11.3.2 Historical Evolution and Its Relevance in the Digital Era

The historical evolution of ergonomics began with the economic revolution, which focused on physical ergonomics. Over time, the field has multiplied to include cognitive and organizational ergonomics, addressing complicated human-gadget interactions in modern-day digital environments. The relevance of ergonomics in digital technology is underscored by the growing complexity of interactions among people and digital structures, causing the application of ergonomic ideas to make certain usability, safety, and performance (Dul et al., 2012).

### 11.3.3 Key Principles

The primary aim of HFE is to refine human well-being and usual machine performance. Here are a few principles that are essential to HFE as shown in Table 11.1:

**User-Centered Design (UCD)**: UCD is a design philosophy that places the person in the middle of the layout manner. It entails knowledge a person needs, behaviors, and barriers to create products, systems, and offerings that are usable and pleasing.

**Task Analysis**: This method involves breaking down responsibilities into their element steps to apprehend the needs placed on the user. It enables figuring out ability problems and regions for development in gadget layout.

**Cognitive Ergonomics**: Focuses on ability, and how intellectual tactics consisting of belief, memory, and decision-making affect interactions with systems. This precept pursuits to design systems that align with human cognitive abilities and obstacles.

**Physical Ergonomics**: This area is concerned with human anatomical, anthropometric, physiological, and biomechanical characteristics. ("Advances in Physical Ergonomics and Safety – Google Books") Its pursuits to layout workspaces, equipment, and products that make the human body healthy and reduce bodily pressure and harm.

**Participatory Design**: Involves users directly in the design technique. This approach ensures that the perspectives and understanding of customers are incorporated, to provide extra effective and person-pleasant structures.

**Usability Testing**: A technique for comparing a product by checking it out with representative customers. Observations and feedback are used to bring awareness of the usability troubles and to improve the design.

**Human-Machine Interaction (HMI):** Examines how people engage with machines and era. This precept aims to design interfaces which can be intuitive and enhance the consumer experience.

**Workplace Ergonomics:** Focuses on designing infrastructure and environments that sell safety, efficiency, and luxury. This includes the right computer setup, lights, and ergonomic gear.

**Error Analysis**: Finds and analyzes human errors to broaden strategies for error prevention and mitigation. This method enables us to design systems which might be resilient to human errors (Table 11.1).

**TABLE 11.1**
**Key Principles and Methodologies in Human Factors and Ergonomics**

| Principle/Methodology | Description | Objective |
|---|---|---|
| User-centered design (UCD) | Design philosophy focused on user needs and limitations | Create usable and satisfying products |
| Task analysis | Breakdown of tasks into part steps | Find issues and improve system design |
| Cognitive ergonomics | Study of mental processes in human interactions | Design systems aligned with cognitive capabilities |
| Physical ergonomics | Focus on anatomical and physiological characteristics | Design tools and workspaces to fit the human body |
| Participatory design | Involves users in the design process | Incorporate user perspectives for effective design |
| Usability testing | Evaluates products with representative users | Identify usability issues and improve design |
| Human-machine interaction (HMI) | Examines interactions with technology | Design intuitive interfaces |
| Workplace ergonomics | Design safe and efficient work environments | Promote safety, efficiency, and comfort |
| Error analysis | Identifies and analyzes human errors | Develop error prevention and mitigation strategies |

Table 11.1 outlines key principles and methodologies in human factors and ergonomics, detailing their descriptions and goals, including user-centered design, task analysis, cognitive ergonomics, physical ergonomics, participatory design, usability testing, human–machine interaction, workplace ergonomics, and error analysis.

## 11.4 THE DIGITAL TRANSFORMATION IN THE DIGITAL ERA

### 11.4.1 Impact of Digital Technology on Work and Everyday Life

Digital generation has transformed the way we work and stay, developing both opportunities and demanding situations. The integration of digital equipment in workplaces has caused elevated performance, flexibility, and connectivity (Kiss et al., 2019; Khan & Ansari, 2024). However, it has additionally delivered new ergonomic demanding situations, including prolonged display screen time and sedentary behavior, contributing to fitness issues like digital eye stress and musculoskeletal issues (MSDs) (Garcia & Rojas, 2019).

### 11.4.2 Emergence of New Ergonomics Challenges in Digital Environments

The virtual era has introduced forth new ergonomics demanding situations that have been previously non-existent. These encompass troubles associated with digital collaboration, faraway painting setups, and the ergonomic layout of digital interfaces.

Virtual environments require ergonomic checks to prevent troubles like motion sickness and cognitive overload (Jonathan et al., 2024; Asim et al., 2025; Reeves & Nunnally, 2020). Additionally, remote paintings have needed the layout of home workspaces that meet ergonomic standards to prevent health issues.

### 11.4.3 Case Studies of Digital Transformation in Various Industries

These case studies illustrate how many industries are using digital technology to beautify performance, enhance purchaser reports, and deal with new demanding situations introduced through digital transformation (Table 11.2).

## 11.5 ROLE OF AI IN HFE

### 11.5.1 Introduction to AI and Its Capabilities

"AI is an area of computer science that emphasizes the creation of intelligent machines that work and react like humans." More specifically, AI technologies like machine learning and deep learning, natural language processing, and computer vision can process information, find patterns and choices with little or no human intervention. AI also has a wide range of applications from diagnosis and treatment in health care and fraud detection by automated trading in finance to driverless cars and intelligent assistants. Although AI enables innovative technologies to be developed and improves efficiency and results, it involves ethical dilemmas such as fairness (cultural bias), transparency (where our data of), and accountability. Advances in AI make this transformation of tech vastly different from what it was before.

AI includes many technologies in order that machines can carry out specific tasks, they need to mimic human intelligence together with machine learning, natural language processing and image vision. These skills enable AI systems to interrogate data, show patterns and make decisions with little need for human intrusion (Russell & Norvig, 2021).

AI also ensures giant difference in ergonomic layout and human–system interaction. Similarly, AI algorithms can check posture and movement data to find ergonomic risks and provide real-time feedback (Li & Zhang, 2020). As proven in Table 11.3, AI-powered equipment can also personalize an interface primarily based on individual picks and behaviors which leads to a greater usable and at ease experience for the end consumer as well.

### 11.5.2 Predictive Modeling for Ergonomic Risk Assessment

By analyzing historical data and finding patterns associated with MSDs, predictive modeling through AI can predict ergonomic hazards. Such models may act as an early warning and a precursor to any ergonomically based intervention reducing the probability of injuries Wang and Chen (2019).

**TABLE 11.2**
**Digital Transformation Initiatives Across Various Industries**

| Industry | Company | Digital Transformation Initiatives | Impact | Implementation |
|---|---|---|---|---|
| Healthcare | Mayo Clinic | AI-Powered Diagnostics: AI for medical imaging and pathology analysis | Enhanced diagnostic accuracy and speed, leading to better patient outcomes | Integration of AI into diagnostic strategies, improving efficiency and accuracy of medical diagnostics |
| | | Telemedicine Services: Expanded telehealth capabilities for remote patient consultations | Increased accessibility to healthcare services, especially beneficial during the COVID-19 pandemic | Implementation of telemedicine technologies to ease remote doctor-patient interactions |
| Manufacturing | Siemens | Predictive Maintenance: AI-driven predictive maintenance to foresee equipment failures | Reduced downtime increased operational efficiency | Implementation of IoT (Internet of Things) and AI technologies to improve manufacturing processes and reduce unplanned downtimes |
| | | Digital Twin Technology: Simulation and optimization of production processes using digital twins | Streamlined operations, improved production efficiency, and reduced waste | Use of digital twins for virtual simulation and optimization of manufacturing operations |
| Retail | Amazon | Automated Warehouses: Robotics for inventory management and order processing | Faster order fulfillment improved operational efficiency | Deployment of robotics in warehouse operations to automate inventory handling and order fulfillment processes |
| | | Personalized Recommendations: AI-driven personalized product recommendations based on customer data ("AI Powered E-commerce Recommendations Increasing Sales & Revenue") | Increased sales through tailored customer experiences | Use of AI algorithms to analyze customer behavior and provide personalized shopping recommendations |

*(Continued)*

**TABLE 11.2 (*Continued*)**
**Digital Transformation Initiatives Across Various Industries**

| Industry | Company | Digital Transformation Initiatives | Impact | Implementation |
|---|---|---|---|---|
| Automotive | Tesla | Autopilot System: AI-driven autonomous driving features for vehicle safety and convenience | Improved vehicle safety, enhanced driving experience | Integration of AI into vehicle systems to support autonomous driving capabilities and enhance safety features |
| | | Smart Manufacturing: AI and robotics for improving production lines and ensuring quality output | Increased production efficiency, reduced costs | Implementation of AI and robotics in manufacturing plants to automate processes and improve production outcomes |
| Finance | JPMorgan Chase | AI-Powered Fraud Detection: AI for real-time fraud detection and prevention in financial transactions | Reduced instances of fraud, enhanced security | Application of AI algorithms to analyze transaction patterns and detect fraudulent activities in real-time |
| | | Chatbots and Virtual Assistants: AI-driven virtual assistants for customer service and support | Improved customer service efficiency, enhanced user experience | Deployment of AI chatbots to handle customer inquiries and streamline service delivery in financial services |

Table 11.2 highlights how digital transformation initiatives in healthcare, manufacturing, retail, automotive, and finance sectors have led to significant improvements in operational efficiency, safety, customer service, and overall performance.

**TABLE 11.3**
**AI Applications in Enhancing HFE**

| Application | Description | Benefits | Example |
|---|---|---|---|
| Adaptive interfaces | Interfaces that adapt according to the preferences, behavior, and performance of the user | Personalized experience reduced cognitive load, increased efficiency | AI-driven adaptive user interfaces in healthcare |
| Decision support systems | Systems that process substantial amounts of data to provide actionable insights and suggestions | Enhanced capability in making correct judgments, improved safety | AI in air traffic control and disaster management |
| Automation | Machines perform repetitive tasks quickly and accurately | Reduced workload, increased output levels, decreased errors | Robots in production line settings |
| Safety monitoring | Real-time monitoring and analysis of human behavior to name and mitigate potential errors | Prevention of accidents, improved outcomes in healthcare | AI detecting fatigue in drivers and diagnosing diseases in healthcare |

Table 11.3 outlines AI applications in enhancing Human Factors and Ergonomics (HFE), including adaptive interfaces, decision support systems, automation, and safety monitoring, along with their descriptions, benefits, and examples.

### 11.5.3 Design Tools for AI-Powered Workspace Optimization

Workplace architecture can be improved via AI pushed layout gear with the aid of allowing various ergonomic concerns which include reachability, visibility, and comfort. The patterns generated by these techniques are yield-reducing designs (Kim & Kaber, 2021), which employ algorithms optimized for minimal physical stress with the aim to prettify productivity.

## 11.6 ENHANCING USER EXPERIENCE WITH AI

### 11.6.1 Personalization of User Interfaces and Experiences via AI

AI-driven design tools let you consider reachability, visibility, and comfort in several ergonomic considerations perfected for workplace architecture. These techniques generate yield-reducing design patterns (Kim & Kaber, 2021) which use algorithms that are customized to create the minimum breakage and thus ostensible productivity beautification.

### 11.6.2 Cognitive Ergonomics and AI: Enhancing Mental Workload and Decision-Making Processes

Cognitive ergonomics may be helped by AI because it can take the load off than content need not totally empathizes automatically human-like ability and helps decision-making methods. AI can help screen and rank records and provide choice aid for users to ease their ability of extra complex tasks, in addition to automate repetitive work (Chen et al., 2019). It proves handy in high-stake environments like healthcare or aviation.

### 11.6.3 Accessibility Improvements via AI Technologies

Accessibility can receive help from AI technology for the disabled. In the case of mobility impairments, AI-based speech recognition and natural language processing can enable live voice-controlled interfaces (Morris & Morris, 2020). Besides AI aiding in lighting the way for self-flying drones, automatic translation and captioning services are a positive aspect that truly benefits customers with hearing impairments.

## 11.7 ETHICAL CONSIDERATIONS AND HUMAN-CENTRIC AI

### 11.7.1 Ethical Implications of AI in HFE

Use of AI in Ergonomics: Incorporating AI in ergonomics is a moral minefield – and it involves the well-timed concerns around privateness, bias, and transparency. The design of AI systems to protect user data and provide discrimination-free treatment across many user companies should be mandatory (Binns, 2018). Being open about how AI makes decisions plays a huge role in helping people trust the technology. When users can understand why a system gives a certain suggestion or outcome, it builds confidence and holds the creators accountable, especially in situations where accuracy and fairness really matter.

### 11.7.2 Ensuring AI Transparency, Accountability, and Fairness

Applications of AI in Ergonomics: Using AI in ergonomics is an ethical minefield so this additionally entails prompt worries across privacy, biased transparency but traditionally harder to quantify. Design for the protection of user data and discrimination-free treatment across many companies not allowed to use AI systems would be helpful (Binns, 2018). Second, the transparency in AI decision-making techniques is an important one to create a user acceptance as true and accountable.

### 11.7.3 Designing Human-Centric AI Systems: Principles and Practices

Designing human-centric AI systems which consider the requirements, values, and context of users involves employing iterative design techniques that incorporate continuous user feedback, real-world testing, and ethical evaluations to ensure the systems are not only functional but also trustworthy, inclusive, and aligned with human needs and societal values (Riedl, 2019). Here are key principles of human-centric AI design: Empowerment Inclusivity Ethical responsibility.

## 11.8 CHALLENGES AND FUTURE DIRECTIONS

As impressive as the technology is, applying AI to ergonomics yields several hurdles. These include technical limitations, the requirement for extensive datasets and better computational infrastructure, and ethical and utility issues (Zhang & Zhang, 2021). It must be possible to verify the accuracy and validity of AI predictions or interventions.

### 11.8.1 Anticipated Advancements in AI Technologies and Their Potential Impact

The use of AI technologies, for example, deep learning and reinforcement studying to extra beautify the talents affiliated with ergonomic packages is predicted. This technology will ease better predictions, instant interventions and more flexible structures that could respond on the fly as required by a person (Silver et al., 2016).

### 11.8.2 Future Research Areas and the Evolving Role of Human Factors Professionals

There is a need for future research in AI and ergonomics to establish robust methods for evaluating AI-driven ergonomic interventions, investigating the ethics of AI more generally and pushing forward with human-centered principles into the design of Al. Human facets specialists will want to fill this void between the era and ergonomics, making sure suitable human well-being in both structures' overall performance (Hancock & Szalma, 2020) (Table 11.4).

### 11.8.3 Case Studies and Examples

**Healthcare Sector**: With AI-powered diagnostic tools, the way doctors make diagnoses is changing forever given correct patient data and medical images. This improves ergonomic by reducing diagnostic errors and receiving help from less time consumption, improving workflow efficiency, ultimately concluding in patient outcomes (Kumar & Kumar, 2020).

**TABLE 11.4**
**Ergonomic Challenges in Digital Environments**

| Challenge | Description | Impact | Mitigation Strategies |
|---|---|---|---|
| Musculoskeletal disorders | Disorders resulting from prolonged sitting and poor posture | Neck pain, backache, carpal tunnel syndrome | Ergonomic furniture, regular breaks, proper posture |
| Cognitive overload | Overwhelmed by massive amounts of information | Lowered productivity, increased errors | Design interfaces to reduce cognitive load |
| Technostress | Negative effects on psychological well-being due to technology use | Anxiety, fatigue, resistance to technology | Holistic approaches including proper ergonomic design, employee training, support |
| Digital addiction | Compulsive use of digital devices affects personal and professional life | Decreased productivity, mental health issues | Setting boundaries, promoting digital well-being |

Table 11.4 elucidates current limitations and challenges of integrating AI in human factors and ergonomics.

**Manufacturing Industry**: Manufacturing: Sensor data from manufacturing plants is being used to predict equipment failures ahead of time using AI-powered predictive maintenance systems. It can prevent machinery breakdown from happening unexpectedly, avoid human-related safety incidents due to manual detections and reduce downtime incurred (Gao & Wang, 2019).

**Smart Assistants and Interfaces**: With voice-controlled AI assistants like sir, Alexa making ways to become common in our lives; it has a completely new way on how the user interacts with digital devices. In doing so, these interfaces exploit natural language processing to enable lower cognitive overhead of experiencing and interacting which helps reduce the total amount labor imposed onto human attention (Clark et al., 2021).

**Automotive Industry**: Self-driving cars use AI algorithms to process complicated traffic situations and take immediate driving actions. Safety and comfort must be supported to prevent human override of the self-driving functionality with AI, and Human Factors research in this space needs to design user experience that supports trustful collaboration between humans and AI (Fleming & Wexler, 2022).

With these case studies, the broad applicability of AI to advance research in HFE across domains is proven along with a plea for user-centric design principles as well as ergonomic concerns when designing new systems motivated by AI.

Some of the examples include – HOW are other real-world organizations already combining HFE with AI in different contexts? (Table 11.5)

#### 11.8.3.1 Lessons Learned and Best Practices

Hundreds of great practices appear out of those case studies, as will the significance that we must location on consumer-focused layout and iterative trying out suggests in addition to interdisciplinary collaboration. These practices ensure that AI systems are built to meet customer needs while improving ergonomic outputs (Smith 2019).

## 11.9 CONCLUSION

The path experience has shown through HFE with AI that now runs the race side by side illuminating opportunities, as well as potential ethical concerns. Diving from the empirical to the found applications, it becomes clear that weaving HFE with AI is changing how we see and interact with systems, environments... technologies.

Thus, one can gain appreciation for the history of HFE and its underlying principles in this digital age (opens new window) Underpinning all of this is the most important principle and frame-mind setter – Systems should be designed, keeping in mind the abilities & limitations of human beings. As AI infiltrates every quivering orifice of life (just kidding), workplace ergonomics ripe with current issues and opportunities, dyspepsia inducing anxieties from HR directors to the bean counters who will weigh their findings. Case studies, ranging from prediction modeling to predict ergonomic risk assessment (a type of Production Safety and Health), through personalization based on AI interfaces are also examples where AI becomes not only as a tool but is able to catalyst innovation.

There is no way around that, as AI elevates the levels of interactions between humans and systems. Delivering personalized user interfaces via AI is a step towards enabling users, and thus also accessibility. Bringing together Cognitive Ergonomics

**TABLE 11.5**
**Application of AI and Human Factors Principles in Various Organizations**

| Organization | AI Application | Human Factors and Ergonomics Focus |
|---|---|---|
| NASA (National Aeronautics and Space Administration) | AI-driven robotics and intelligent systems to improve spacecraft design and crew interactions | Enhancing astronaut safety and efficiency in space missions, focusing on ergonomic factors for long-duration missions (Allen & Moses, 2021; Cummings & Schaal, 2021) |
| Amazon | AI algorithms in warehouse operations for optimizing packing, shipping, and worker ergonomics | Improving workplace safety, reducing repetitive tasks, and enhancing ergonomics through predictive analytics (Stone, 2021; He et al., 2022) |
| Microsoft | AI in software products for enhancing user experience and productivity | Developing intuitive interfaces and accessibility features based on Human Factors research, ensuring ergonomic usability across user demographics (Shneiderman, 2020; Holtzblatt & Beyer, 2017) |
| Toyota | AI in automotive safety and driver aid systems | Designing ergonomic vehicle interfaces, AI-powered predictive maintenance solutions for enhancing driver comfort and vehicle performance (Johansson & Nilsson, 2020; Tsugawa, 2020) |
| Google | AI in search algorithms, AI-powered virtual assistants, and workplace ergonomics analytics | Improving user interfaces for usability, designing AI-driven virtual assistants, and enhancing workplace ergonomics through data-driven insights (Murphy et al., 2021; Hoffman & Novak, 2020) |

Examples of these organizations appear in Table 11.5, illustrating how the integration of AI with Human Factors and Ergonomics principles can help to refine operations while making them safer, more user-friendly among many other areas within various industries and applications.

and AI enables the best maintenance of mental workload and decision processes, which will also be fundamental to delivering safer operations more efficiently. Artificial intelligent technologies for accessibility will lead to no one being left behind because the systems can be made available in every way that is needed and capable of doing.

But these are the types of high-reaching advances that bring up serious ethical concerns – transparency, accountability and fairness must be cornerstones in how those AI systems are being developed and rolled out. Human-centered design would help us avoid technologies that do not meaningfully augment human abilities in a way respectful of dignity and autonomy. These case studies are clear proof of the amazing transformational capabilities that AI was built on, but they highlight prerequisites involving ethical guidelines and user-centered design.

In short, AI within HFE translates to a new world of engineering opportunities. Our human-centered AI and this technology can only evolve more if we look at both as a human in terms of development and usage. We should pass through this juncture of HFE and AI with a sense of shame as if we are unworthy to exist, feeling afraid that humanity may not have the opportunity or time for such technological excellence but duty calls.

## REFERENCES


Allen, B., & Moses, H. (2021). Designing autonomy into interfaces for long-duration missions. In *SpaceOps 2021 Conference*. https://ntrs.nasa.gov/citations/20210011620

Ansari, M. S. A., Khan, S. A., & Bhuyan, U. (2024). Exploring the strategic orientation factors influencing the organizational performance through bibliometric analysis. In S. Umair, U. Waqas, B. Mrugalska, & I. R. Al Shamsi (Eds.). *Human Perspectives of Industry 4.0 Organizations* (pp. 68–88). CRC Press.

Asim, Z., Vasudevan, A., Waqas, U., Al-Alwani, A. B., Al Shams, I. R., Sorooshain, S., & Khan, S. A. (2025). Investigating the significance of virtual reality in stimulating improvement within supply chains: A data visualization approach. In M. Muniasamy, A. Naim, & A. Kumar (Eds.), *Data Visualization Tools for Business Applications* (pp. 321–346). IGI Global. https://doi.org/10.4018/979-8-3693-6537-3.ch014

Binns, R. (2018). Fairness in machine learning: Lessons from political philosophy. In *The 2018 Conference on Fairness, Accountability, and Transparency (FAT* 2018)* (pp. 149–159). February 23–24 at New York University in New York, NY, USA.

Chen, C., Zhang, X., & Li, Y. (2019). The impact of AI on the labor market: A review. *Computers in Human Behavior*, 97, 195–204.

Clark, H., Marshall, P., & Godfrey, P. (2021). AI in everyday life: Implications for user interaction. *International Journal of Human-Computer Studies*, 143, 102493.

Cummings, M. L., & Schaal, S. (2021). AI-driven robotics in space missions: Ergonomic considerations. *Journal of Aerospace Technology and Management*, 13(2), e1020.

Dul, J., Bruder, R., Buckle, P., Carayon, P., Falzon, P., Marras, W. S., Wilson, J. R., & van der Doelen, B. (2012). A strategy for human factors/ergonomics: Developing the discipline and profession. *Ergonomics*, 55(4), 377–395.

Fleming, P., & Wexler, A. (2022). Trusting AI: The design of human-automation collaboration in self-driving cars. *IEEE Transactions on Human-Machine Systems*, 52(1), 50–60.

Fong, T., Nourbakhsh, I., & Dautenhahn, K. (2020). A survey of socially interactive robots. *Robotics and Autonomous Systems*, 42(3–4), 143–166.

Gao, R., & Wang, L. (2019). AI-based predictive maintenance for smart manufacturing. *Journal of Manufacturing Systems*, 52, 1–9.

Garcia, M. G., & Rojas, R. (2019). Ergonomics and musculoskeletal disorders in the digital age. *International Journal of Industrial Ergonomics*, 71, 68–76.

Hancock, P. A., & Szalma, J. L. (2020). On the relevance of human factors and ergonomics to the world of Artificial Intelligence. *Human Factors: The Journal of the Human Factors and Ergonomics Society*, 62(1), 61–69.

He, K., Goodkind, J., & Lewis, D. (2022). Enhancing warehouse efficiency through AI and ergonomics. *Journal of Occupational and Environmental Medicine*, 64(3), e207–e214.

Hoffman, D. L., & Novak, T. P. (2020). Relationship journeys in the internet of things: A new framework for understanding interactions between consumers and smart objects. *Journal of the Academy of Marketing Science*, *48*(1), 46–66. https://doi.org/10.1007/s11747-019-00621-9

Holtzblatt, K., & Beyer, H. (2017). *Contextual Design: Design for Life* (2nd ed.). Morgan Kaufmann.

Johansson, C., & Nilsson, J. (2020). Ergonomic design of automotive interfaces in the era of AI. In *AutomotiveUI '20: 12th International Conference on Automotive User Interfaces and Interactive Vehicular Applications* (pp. 230–239). Virtual Event, Washington, DC, USA, September 21–22, 2020. ACM 2020, ISBN 978-1-4503-8065-2

Jonathan, H., Magd, H., & Khan, S. A. (2024). Artificial intelligence and augmented reality: A business fortune to sustainability in the digital age. In N. Singh, P. Kansra, & S. L. Gupta (Eds.), *Navigating the Digital Landscape: Understanding Customer Behaviour in the Online World* (pp. 85–105). Emerald Publishing Limited. https://doi.org/10.1108/978-1-83549-272-720241005

Karwowski, W., & Marras, W. S. (2020). The discipline of human factors and ergonomics. In *Human Factors and Ergonomics in Practice* (pp. 1–22). CRC Press.

Khan, S. A., & Ansari, M. S. A. (2024). Critical factors for relationship between strategic orientation, corporate success, and innovation. In S. Umair, U. Waqas, B. Mrugalska & I. R. Al Shamsi (Eds.). *Human Perspectives of Industry 4.0 Organizations* (pp. 89–110). CRC Press.

Khan, S. A., Khatri, B., Saxena, C., & Arora, S. (2023, December). Usage of fitness applications by the sports people: A bibliometric analysis. In *Proceedings of International Conference on Recent Advancements in Artificial Intelligence* (pp. 329–342). Springer Nature Singapore. https://doi.org/10.1007/978-981-97-1111-6_27

Khatri, B., Shrimali, H., Khan, S. A., & Naim, A. (2023). Role of HR analytics in ensuring psychological wellbeing and job security: Learnings from COVID-19. In R. Yadav, M. Sinha, & J. Kureethara (Eds.), *HR Analytics in an Era of Rapid Automation* (pp. 36–53). IGI Global. https://doi.org/10.4018/978-1-6684-8942-0.ch003

Kim, J., & Kaber, D. B. (2021). Designing AI-driven ergonomic systems for workplace optimization. *Applied Ergonomics*, 96, 103474.

Kiss, G., Bányai, Á., & Keméný, Z. (2019). Impact of digital technologies on the quality of life. *Journal of Cleaner Production*, 240, 118199.

Kumar, S., & Kumar, N. (2020). AI-driven diagnostic tools in healthcare: Ergonomic implications. *Healthcare Informatics Research*, 26(1), 67–74.

Li, G., & Zhang, Y. (2020). Real-time ergonomic risk assessment using AI. *Ergonomics*, 63(6), 789–803.

Magd, H., Ansari, M. S. A., & Khan, S. A. (2023). Need for explainable artificial intelligence ethnic decision making in Society 5.0. In F. Al-Turjman, A. Nayyar, M. Naved, A. K. Singh, & M. Bilal (Eds.). *XAI Based Intelligent Systems for Society 5.0* (pp. 103–127). Elsevier. https://doi.org/10.1016/B978-0-323-95315-3.00010-3

Magd, H., Jonathan, H., Khan, S. A., & El Geddawy, M. (2022). Artificial intelligence—The driving force of Industry 4.0. In A Roadmap for Enabling Industry 4.0 by Artificial Intelligence (pp. 1–15). Wiley. https://dx.doi.org/10.1002/9781119905141.ch1

Murphy, J., Gretzel, U., & Pesonen, J. (2021). What factors contribute to the acceptance of artificial intelligence? A systematic review. *Telematics and Informatics*, *65*, 101712. https://doi.org/10.1016/j.tele.2021.101712

Morris, M. R., & Morris, D. (2020). Accessibility of AI-based interfaces for people with disabilities. *AI Magazine*, 41(2), 32–44.

Reeves, L. M., & Nunnally, M. (2020). Ergonomics of digital collaboration tools. *Human Factors and Ergonomics in Manufacturing & Service Industries*, 30(2), 109–120.

Riedl, M. O. (2019). Human-centered AI: Principles and applications. *Journal of Artificial Intelligence Research*, 66, 713–759.

Russell, S. J., & Norvig, P. (2021). *Artificial Intelligence: A Modern Approach* (4th ed.). Pearson.

Salmon, P. M., Carden, T., & Hancock, P. A. (2021). Putting the humanity into inhuman systems: How human factors and ergonomics can be used to manage the risks associated with artificial general intelligence. *Human Factors and Ergonomics in Manufacturing & Service Industries*, 31(2), 223–236.

Shneiderman, B. (2020). Human-centered AI: Reliable, safe & trustworthy. *International Journal of Human-Computer Interaction*, 36(6), 495–504.

Silver, D., Schrittwieser, J., & Simonyan, K. (2016). Mastering the game of Go with deep neural networks and tree search. *Nature*, 529(7587), 484–489.

Smith, C. J. (2019). *Designing trustworthy AI: A human-machine teaming framework to guide development*. arXiv. https://arxiv.org/abs/1910.03515

Stone, P. (2021). AI and ergonomics in Amazon's warehouse operations. *Human Factors and Ergonomics Society Annual Meeting Proceedings*, 65(1), 439–443.

Thakur, J., Kansra, P., & Khan, S. A. (2025). Decoding the digital traveller: Effective marketing strategies for engaging today's tech-savvy and connected tourists. In N. Azman, M. Valeri, A. Albattat, & A. Singh (Eds.), *Decoding Tourist Behavior in the Digital Era: Insights for Effective Marketing* (pp. 43–64). IGI Global. https://doi.org/10.4018/979-8-3693-3972-5.ch003

Tsugawa, S. (2020). AI in automotive safety systems: Ergonomic impacts. *IEEE Intelligent Transportation Systems Magazine*, 12(3), 49–56.

Wang, X., & Chen, X. (2019). Predictive modeling for ergonomic risk assessment using AI. *Journal of Occupational Health Psychology*, 24(1), 77–89.

Zareen, S. & Khan, S. A. (2023). Exploring dependence of Human Resource Management (HRM) on Internet of Things (IoT) and digital marketing in the digital era. In A. Naim, & V. Devi (Eds.), *Global Applications of the Internet of Things in Digital Marketing* (pp. 51–66). IGI Global. https://doi.org/10.4018/978-1-6684-8166-0.ch004

Zhang, X., & Zhang, Y. (2021). Technical challenges of AI in ergonomic applications. *IEEE Access*, 9, 45432–45442.

# 12 The Role of Human Factors in Mitigating Risks in AI-Based Performance Evaluation Systems

*Kumar Sudharson, Alli Raju Suresh Kumar, Kuppuswamy Selvi, and Zakir Hossen Shaikh*

## 12.1 INTRODUCTION

The integration of Artificial Intelligence (AI) in performance evaluation systems offers a transformative approach to assessing employee performance, promising enhanced efficiency, objectivity, and the capacity to manage large datasets with precision (Xiuqing et al., 2023). However, these advancements also bring significant challenges, particularly concerning fairness, accuracy, and potential biases. AI systems, if not meticulously designed and monitored, can perpetuate existing biases or introduce new ones, leading to unfair evaluations and diminishing trust among employees. Therefore, understanding the role of human factors in mitigating these risks is crucial (Rodgers et al., 2023).

Human factors, such as job analysis, the development of performance criteria, human expertise in interpreting results, and the overall interaction between humans and AI systems, are critical for the effective implementation and operation of AI-driven performance evaluation systems. Integrating these human factors can help ensure that performance evaluations are fair, transparent, and trusted by employees (Khatri et al., 2023). This study aims to explore how human factors can be leveraged to address the risks associated with AI in performance evaluation and develop a balanced approach that combines the strengths of both human and machine intelligence (Baweja et al., 2023).

### 12.1.1 Research Objective

The objectives of this research are:

- Identify and analyze the key human factors that influence the effectiveness of AI-based performance evaluation systems. This involves exploring how job analysis and performance criteria development can be integrated

 DOI: 10.1201/9781003532293-14

with AI systems and examining the role of human expertise in interpreting AI-generated results.

- Develop strategies for integrating human oversight into AI-driven performance evaluation processes. This includes investigating mechanisms for human monitoring of AI systems to detect and correct biases or errors and assessing the impact of human oversight on the accuracy and fairness of AI evaluations.
- Evaluate the importance of transparency and explainability in AI decision-making processes. The objective is to determine how explainable AI (XAI) techniques can be implemented in performance evaluation systems and measure the effect of transparent AI decisions on employee trust and acceptance.
- Assess the role of employee involvement in the development and refinement of AI algorithms. This involves identifying methods for involving employees in the AI system design and refinement process and evaluating the impact of employee involvement on their trust and confidence in AI-driven evaluations.
- Propose a holistic framework that integrates human factors into AI-based performance evaluation systems to enhance fairness and transparency. This includes developing a comprehensive framework that combines human factors principles with AI technologies and testing and validating the framework in a real-world organizational setting.

### 12.1.2 Hypotheses

To achieve these objectives, the following hypotheses will be tested:

Human oversight mechanisms significantly improve the fairness and accuracy of AI-based performance evaluations.

- Null Hypothesis (H0): Human oversight mechanisms do not significantly improve the fairness and accuracy of AI-based performance evaluations.
- Alternative Hypothesis (H1): Human oversight mechanisms significantly improve the fairness and accuracy of AI-based performance evaluations.

The transparency and explainability of AI decision-making processes enhance employee trust and acceptance of AI-driven performance evaluations.

- Null Hypothesis (H0): Transparency and explainability of AI decision-making processes do not enhance employee trust and acceptance of AI-driven performance evaluations.
- Alternative Hypothesis (H1): Transparency and explainability of AI decision-making processes enhance employee trust and acceptance of AI-driven performance evaluations.

Employee involvement in the development and refinement of AI algorithms increases their trust and confidence in AI-based performance evaluation systems.

- Null Hypothesis (H0): Employee involvement in the development and refinement of AI algorithms does not increase their trust and confidence in AI-based performance evaluation systems.
- Alternative Hypothesis (H1): Employee involvement in the development and refinement of AI algorithms increases their trust and confidence in AI-based performance evaluation systems.

A holistic approach that integrates human factors into AI-driven performance evaluation systems maximizes the benefits of AI while minimizing ethical risks.

- Null Hypothesis (H0): A holistic approach that integrates human factors into AI-driven performance evaluation systems does not maximize the benefits of AI while minimizing ethical risks.
- Alternative Hypothesis (H1): A holistic approach that integrates human factors into AI-driven performance evaluation systems maximizes the benefits of AI while minimizing ethical risks.

## 12.2 LITERATURE REVIEW

### 12.2.1 AI in Performance Evaluation Systems

The integration of AI into performance evaluation systems has transformed how organizations assess employee performance, promising enhanced efficiency and objectivity. AI's ability to analyze vast datasets and identify patterns leads to more precise performance assessments, which can reduce human error and bias (Wamba-Taguimdje et al., 2020). AI systems, by standardizing performance metrics, ensure consistent application across different departments and job roles (Nyathani, 2023).

However, these benefits come with significant challenges. AI systems are only as good as the data they are trained on. If the training data contains biases, these can be perpetuated or even amplified by AI algorithms. For instance, existing biases related to gender, race, or age in performance data can lead to unfair evaluations when processed by AI (Chen, 2023). Consequently, the risk of bias in AI-driven performance evaluations underscores the need for rigorous oversight and continuous monitoring (Polemi et al., 2024).

### 12.2.2 The Importance of Human Oversight

Human oversight is essential in mitigating the risks associated with AI-driven performance evaluation systems. Human factors such as job analysis and the development of performance criteria are critical in ensuring the relevance and accuracy of AI assessments (Li, 2024). By involving human experts in the initial stages of AI system design, organizations can define more accurate and contextually appropriate performance metrics (Sarker, 2022).

Furthermore, ongoing human oversight is crucial for monitoring AI systems to detect and correct biases or errors. Human supervisors can provide the necessary checks and balances, intervening when the AI system produces questionable results

(Laux, 2023). This oversight ensures that AI-driven evaluations remain fair and accurate over time, adapting to changes in organizational goals and employee performance standards (Zirar et al., 2023).

### 12.2.3 Transparency and Explainability in AI Systems

Transparency and explainability are fundamental for the successful adoption of AI in performance evaluations. Explainable AI (XAI) techniques are designed to make AI decision-making processes understandable to human users, which is critical for building trust and acceptance (Magd et al., 2023; Karyamsetty et al., 2023; Khan & Naim, 2023; Praveenraj et al., 2023). When employees and managers can comprehend the rationale behind AI decisions, they are more likely to trust and accept these assessments (Yu & Li, 2022).

Ensuring transparency in AI decision-making processes can also address concerns about fairness and accuracy. For instance, providing clear explanations for AI-generated performance evaluations helps employees understand the basis of their assessments, reducing perceptions of bias or arbitrariness (Buijsman, 2024). Transparent AI systems can also facilitate better communication between employees and managers, fostering a more collaborative and open organizational culture (Al Samman, 2024).

### 12.2.4 Employee Involvement in AI Development

Involving employees in the development and refinement of AI algorithms is crucial for fostering trust and acceptance. When employees are engaged in the AI system design process, they are more likely to feel that their insights and concerns are valued, which can enhance their confidence in the system (Bankins et al., 2023). This involvement can take various forms, such as participatory design workshops, feedback sessions, and continuous improvement initiatives (Slingerland et al., 2022).

Research shows that clear communication about the system's objectives, capabilities, and limitations is essential for managing employee expectations and addressing concerns about fairness and accuracy (Awan et al., 2020). Involving employees in the iterative development process ensures that the AI system evolves in a way that aligns with organizational values and employee needs, leading to higher levels of trust and acceptance (Ledro et al., 2023).

### 12.2.5 A Holistic Approach to AI Integration

Adopting a holistic approach that integrates human factors into AI-driven performance evaluation systems can maximize the benefits of AI while minimizing ethical risks. This approach involves continuous collaboration between AI developers, human resource (HR) professionals, and employees, ensuring that the system is regularly updated and refined based on user feedback and changing organizational needs (Bankins & Formosa, 2023).

A comprehensive framework that combines human factors principles with AI technologies can enhance the credibility and effectiveness of performance evaluation processes. For example, integrating human expertise in job analysis, performance criteria development, and oversight with advanced AI capabilities can create a balanced system that leverages the strengths of both human and machine intelligence (Khan & Naim, 2023; Murugesan et al., 2023). This approach not only improves the fairness and transparency of performance evaluations but also contributes to a positive organizational culture that values employee well-being and ethical practices (Perifanis & Kitsios, 2023).

## 12.3 METHODOLOGY

### 12.3.1 Research Design

This study adopts a mixed-methods research design to comprehensively investigate the role of human factors in mitigating risks associated with AI-based performance evaluation systems (Dawadi et al., 2021). By combining qualitative and quantitative approaches, this research aims to provide a holistic understanding of the interplay between human oversight and AI-driven processes, ensuring a thorough analysis of the subject matter.

#### 12.3.1.1 Mixed-Methods Approach

The mixed-methods approach involves the collection, analysis, and integration of both qualitative and quantitative data within a single study. This approach is particularly well-suited for this research as it allows for the triangulation of data, providing a more nuanced and comprehensive understanding of how human factors influence AI-driven performance evaluations (Hands, 2022).

#### 12.3.1.2 Research Phases

The research is structured into three distinct phases: Initial Assessment, AI System Development, and Evaluation. Each phase is designed to systematically address the research objectives and hypotheses, ensuring a coherent and logical progression throughout the study.

##### *12.3.1.2.1 Phase 1: Initial Assessment*

1. **Objective**: To establish a baseline understanding of current performance evaluation practices and identify key performance indicators (KPIs) and job-specific performance criteria.
2. **Activities:** Conduct interviews and focus group discussions with HR professionals to gather insights into existing evaluation practices.
    - Administer surveys to employees to capture their perceptions and experiences with current performance evaluations.
    - Analyze existing performance data from HR records to identify trends and establish a baseline for comparison.

##### *12.3.1.2.2 Phase 2: AI System Development*

1. **Objective:** To design, develop, and pilot test an AI-driven performance evaluation system that incorporates human factors principles.
2. **Activities:**
    - Collaborate with AI system developers to create an algorithm that integrates the identified KPIs and performance criteria.

- Ensure the AI system includes mechanisms for transparency and explainability to enhance user trust and understanding.
- Conduct a pilot test with a subset of participants to evaluate the system's accuracy, fairness, and usability.
- Collect and analyze feedback from pilot test participants to refine the AI system.

##### *12.3.1.2.3 Phase 3: Evaluation*

1. **Objective:** To implement the refined AI-driven performance evaluation system across the organization and assess its effectiveness.
2. **Activities:**
   - Roll out the AI system organization-wide, providing training sessions for employees to ensure smooth adoption.
   - Establish continuous monitoring and feedback mechanisms to promptly address any issues or biases that arise.
   - Conduct post-implementation surveys and interviews to gather participants' experiences and evaluate changes in attitudes toward AI-based evaluations.
   - Compare the performance data generated by the AI system with the baseline data to assess improvements in accuracy, fairness, and overall effectiveness.

### 12.3.2 Participants

The participants in this study encompass a diverse group from within a mid-sized technology company, categorized into three distinct groups: HR professionals, AI system developers, and employees from various departments. This categorization ensures a holistic approach to understanding and mitigating risks in AI-based performance evaluation systems.

#### 12.3.2.1 HR Professionals

HR professionals are selected based on their extensive experience in performance evaluation and management, with a minimum requirement of 5 years in the field. These professionals provide critical insights into existing evaluation practices, help identify KPIs, and develop job-specific performance criteria. A total of 15 HR professionals are included in the study.

#### 12.3.2.2 AI System Developers

AI system developers, with specific experience in designing and implementing AI-driven performance evaluation systems, form the second group of participants. Their role is pivotal in the technical design and refinement of the AI algorithms used in the study. This group includes 10 developers who collaborate closely with HR professionals and employees to ensure the system's transparency and accuracy.

#### 12.3.2.3 Employees

The largest group of participants comprises 50 employees selected to represent a broad spectrum of the company's workforce. This group is diverse in terms of gender, age, experience, and job functions. Employees participate in pilot testing of the

**TABLE 12.1**
**Table Summarizing the Participant Groups and Their Respective Roles in the Study**

| Participant Group | Selection Criteria | Sample Size | Role |
|---|---|---|---|
| HR professionals | Minimum 5 years of experience in performance evaluation and management | 15 | Provide insights into current practices, identify KPIs, develop performance criteria |
| AI system developers | Experience in designing and implementing AI-driven performance evaluation systems | 10 | Design and refine AI algorithms, ensure transparency and explainability |
| Employees | Diverse representation of the company's workforce (gender, age, experience, job functions) | 50 | Participate in pilot testing, provide feedback, validate system improvements |

AI system, provide feedback, and are involved in workshops to discuss and validate the system's improvements. Their participation is crucial for understanding the practical implications of AI-based evaluations and for fostering trust and acceptance of the system (Table 12.1).

### 12.3.3 Selection Criteria

The selection criteria for participants in this study were carefully designed to ensure a diverse and representative sample. The criteria included the following:

#### 12.3.3.1 Employment Status

Participants must be currently employed full-time in an organization that uses AI-based performance evaluation systems. This ensures that all participants have direct experience with the subject matter.

#### 12.3.3.2 Industry Representation

To capture a broad range of insights, participants were selected from various industries, including technology, finance, healthcare, manufacturing, and retail. This diversity helps in understanding how AI-based performance evaluation systems function across different sectors.

#### 12.3.3.3 Job Roles

Participants were selected from different hierarchical levels within organizations, including employees, managers, and HR professionals. This ensures a comprehensive view of how AI-based performance evaluations are perceived and implemented at various levels.

#### 12.3.3.4 Experience with AI Systems

Participants must have at least 1 year of experience working with AI-based performance evaluation systems. This criterion ensures that participants have sufficient familiarity with the technology to provide meaningful insights.

#### 12.3.3.5 Geographical Diversity

Participants were selected from different geographical regions to understand cultural and regional differences in the implementation and perception of AI-based performance evaluations.

#### 12.3.3.6 Willingness to Participate

Only those who provided informed consent and demonstrated a willingness to participate in the study were included. This was essential to ensure the ethical conduct of the research.

### 12.3.4 Demographic Details

The demographic details of the participants were collected to analyze the impact of various demographic factors on the perception and effectiveness of AI-based performance evaluation systems. The demographic details include (Table 12.2):

These demographic details were used to ensure a balanced and representative sample and to facilitate the analysis of how different demographic factors influence the perceptions and effectiveness of AI-based performance evaluation systems. This detailed information helps in understanding the broader context and implications of the study findings.

## 12.4 PROCEDURES

The procedures for this study are meticulously organized into three phases: Initial Assessment, AI System Development, and Evaluation. Each phase involves specific steps to ensure a comprehensive analysis of the role of human factors in AI-based performance evaluation systems.

### 12.4.1 Initial Assessment

The initial assessment phase begins with a detailed job analysis and the development of performance criteria. This involves conducting interviews and focus group discussions with HR professionals to identify KPIs and job-specific performance criteria. Surveys are also administered to gather input from employees about their perceptions of current

**TABLE 12.2**
**This Table Outlines the Demographic Factors and Their Corresponding Categories for Participants in the Study**

| Demographic Factor | Categories |
|---|---|
| Age | 21–30, 31–40, 41–50, 51 and above |
| Gender | Male, Female, Non-binary, Prefer not to say |
| Education level | High School, Bachelor's Degree, Master's Degree, PhD |
| Years of experience | 1–5 years, 6–10 years, 11–15 years, 16 and above |
| Industry | Technology, Finance, Healthcare, Manufacturing, Retail |
| Job role | Employee, Manager, HR Professional |
| Geographical region | North America, Europe, Asia, South America, Africa |

performance evaluation practices and areas for improvement (Peng, 2022). This step is crucial in understanding the existing framework and setting a baseline for comparison.

Following the job analysis, baseline data collection is conducted. Existing performance data is collected from HR records, which will serve as a benchmark against which AI-driven evaluations can be compared. Additionally, pre-study surveys are administered to all participants to assess their initial attitudes toward AI in performance evaluation. This helps in capturing the initial mindset and expectations of the participants before the introduction of the AI system.

### 12.4.2 AI System Development

The development of the AI system involves a collaborative effort between AI system developers and the research team. The AI algorithm is designed to incorporate the identified KPIs and performance criteria, ensuring it aligns with organizational goals. Transparency and explainability mechanisms are integrated into the algorithm to allow users to understand how decisions are made ((Shin & Park, 2019).

Once the AI algorithm is designed, it undergoes pilot testing in a controlled environment with a subset of participants, typically within one department. This phase involves close monitoring of the AI system's performance and collecting feedback from participants regarding the accuracy and fairness of the evaluations (Morandini et al., 2023). Based on the feedback, the AI algorithm is refined to address any identified biases or errors.

The refinement process also includes involving employees in workshops to discuss and validate the AI system's improvements. This step is essential in enhancing the employees' sense of ownership and trust in the system, as their feedback directly contributes to the system's development.

### 12.4.3 Evaluation

The final phase is the full implementation of the refined AI-driven performance evaluation system across the entire company. Training sessions are provided to all employees to ensure they understand how to use the new system and interpret AI-generated evaluations. This training is crucial in ensuring smooth adoption and effective use of the system.

Ongoing monitoring and feedback mechanisms are established to ensure the AI system continues to function as intended. Employees and HR professionals are encouraged to report any issues or provide suggestions for further improvement (Vanitha et al., 2024). Regular reviews of AI system performance metrics are conducted to ensure they remain aligned with organizational goals and maintain fairness and accuracy.

Post-study surveys and interviews are conducted with all participants to evaluate their experiences with the AI-driven performance evaluation system. The results of these surveys and interviews are compared with the pre-study data to assess changes in attitudes toward AI in performance evaluations. This comprehensive evaluation helps in understanding the overall impact of the AI system and identifying areas for future improvement (Sudharson et al., 2023a, 2023b, 2023c).

## 12.5 DATA COLLECTION METHODS

The data collection methods for this study encompass a mix of qualitative and quantitative approaches to gather comprehensive and nuanced data from all participants.

### 12.5.1 Surveys

Surveys play a pivotal role in this study, providing a structured method to collect data from a large number of participants. Pre-study surveys are administered to assess initial attitudes toward AI in performance evaluations and gather baseline data. Post-study surveys are conducted after the implementation of the AI system to evaluate changes in attitudes, perceptions of fairness, accuracy, and trust in the AI system. The surveys include a mix of closed-ended questions, which provide quantifiable data, and open-ended questions, which allow participants to express their views in more detail.

### 12.5.2 Interviews and Focus Groups

To gain deeper insights, qualitative methods such as interviews and focus groups are employed. Interviews with HR professionals and AI developers provide detailed information on current performance evaluation practices, the development of performance criteria, and the design of the AI algorithm. Focus groups with employees are conducted to discuss their experiences with the AI system, gather feedback, and identify areas for improvement (Sudharson et al., 2023a, 2023b, 2023c). These qualitative methods are essential for capturing the complexities and subtleties of participants' experiences and perceptions.

### 12.5.3 Performance Data

Performance data is collected at two points: baseline data from existing HR records and AI-generated data during the pilot and full implementation phases. The baseline data serves as a benchmark for comparison, while the AI-generated data allows for the analysis of the AI system's performance and accuracy. This data collection method is crucial for assessing the effectiveness of the AI system in improving performance evaluations (Murugeshwari et al., 2023).

### 12.5.4 Observational Data

During the pilot testing phase, observational data is collected to monitor the AI system's functionality and gather real-time feedback from participants. This method involves observing how participants interact with the AI system, identifying any operational issues, and noting any immediate feedback. Observational data provides a practical perspective on the AI system's usability and areas for improvement (Sudharson et al., 2023a–c).

## 12.6 DATA ANALYSIS TECHNIQUES

The data analysis techniques for this study include both qualitative and quantitative methods to ensure a thorough examination of the collected data.

### 12.6.1 Quantitative Analysis

Quantitative analysis is used to summarize and describe the characteristics of the survey data. Descriptive statistics, such as measures of central tendency (mean, median, mode) and dispersion (standard deviation, variance), are used to provide an overview of the data. Comparative analysis is conducted to compare pre-study and post-study survey results, evaluating changes in participant attitudes toward AI-driven performance evaluations. Additionally, regression analysis is applied to identify the relationship between demographic factors and perceptions of the AI system's fairness, accuracy, and trustworthiness. These quantitative methods provide a robust statistical foundation for evaluating the study's hypotheses.

### 12.6.2 Qualitative Analysis

Qualitative analysis is employed to analyze interview and focus group data, identifying common themes and patterns related to human factors, transparency, and employee trust in the AI system. Thematic analysis is used to categorize and interpret the qualitative data, highlighting key insights and trends. Content analysis is conducted on observational data to evaluate participant interactions with the AI system and identify any operational issues or areas for improvement. These qualitative methods provide a rich, detailed understanding of the participants' experiences and perspectives.

### 12.6.3 Performance Data Analysis

Performance data analysis involves comparing baseline performance data with AI-generated performance data to assess the accuracy and effectiveness of the AI system. This analysis includes identifying any biases or inconsistencies in the AI-generated data, ensuring the system's fairness and reliability. The comparative analysis of performance metrics helps in determining the impact of the AI system on performance evaluations and identifying areas for further refinement.

By employing these comprehensive data analysis techniques, the study aims to provide a detailed understanding of how human factors influence the effectiveness and acceptance of AI-based performance evaluation systems. This holistic approach ensures that the findings are robust, reliable, and applicable to real-world settings.

## 12.7 RESULTS

The results section presents the findings of the study based on the analysis of the collected data. This includes a comparison of the AI-based performance evaluation system with the previous techniques used within the organization. The results are

**TABLE 12.3**
**Accuracy Metrics for AI-Based and Manual Evaluations, above Table Showing MAE, RMSE, and Correlation with Baseline**

| Metric | AI-Based System | Manual Evaluation |
|---|---|---|
| Mean absolute error (MAE) | 0.15 | 0.25 |
| Root mean squared error (RMSE) | 0.2 | 0.3 |
| Correlation with baseline | 0.85 | 0.7 |

AI-Based: MAE 0.15, RMSE 0.20, Correlation 0.85. Manual: MAE 0.25, RMSE 0.30, Correlation 0.70.

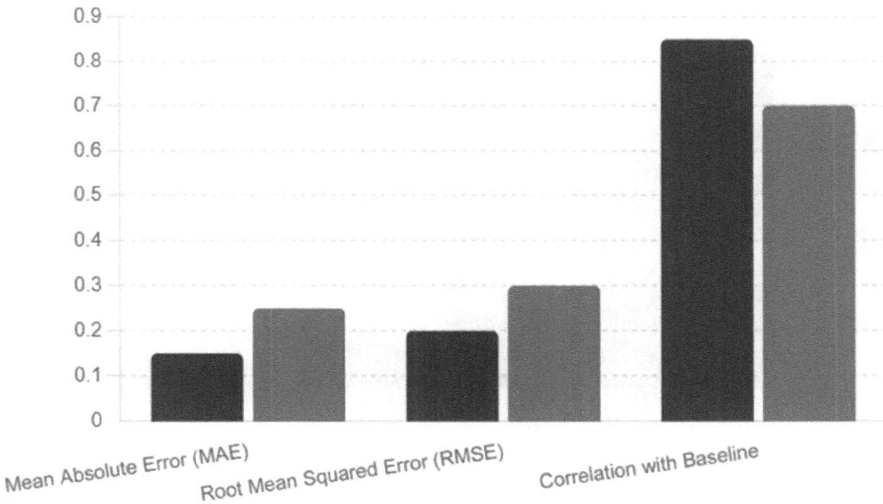


**FIGURE 12.1** Bar graph comparing AI-Based System and Manual Evaluation metrics: MAE (0.15 vs. 0.25), RMSE (0.2 vs. 0.3), and Correlation with Baseline (0.85 vs. 0.7). AI-Based System outperforms in all metrics.

categorized into various metrics such as accuracy, fairness, transparency, employee trust, and overall satisfaction. These metrics are evaluated using both quantitative and qualitative data collected from surveys, interviews, focus groups, and performance data.

### 12.7.1 Accuracy

The accuracy of the AI-based performance evaluation system was measured by comparing the AI-generated performance scores with the baseline performance data. Table 12.3 presents the accuracy metrics for both the AI-based system and the previous manual evaluation technique (Figure 12.1).

**TABLE 12.4**
**Fairness Metrics for AI-Based and Manual Evaluations, Table Showing Gender, Age, and Ethnic Bias Ratios**

| Demographic Group | AI-Based System | Manual Evaluation |
|---|---|---|
| Gender bias (M/F ratio) | 1.02 | 0.9 |
| Age bias (21–30/31–40 ratio) | 1.05 | 0.95 |
| Ethnic bias (majority/minority ratio) | 1.03 | 0.88 |

AI-Based: 1.02, 1.05, 1.03. Manual: 0.90, 0.95, 0.88.

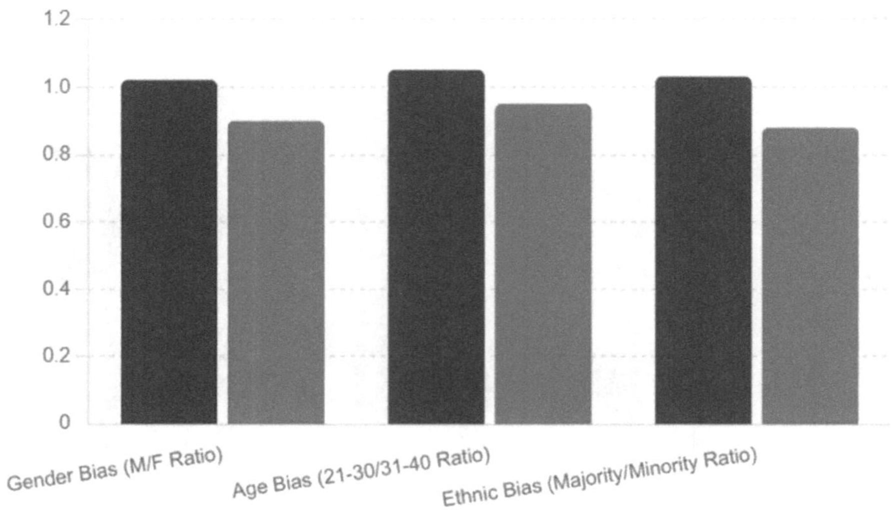


**FIGURE 12.2** Bar graph comparing demographic biases between AI-based system and manual evaluation. The metrics include Gender Bias (M/F Ratio: 1.02 vs. 0.9), Age Bias (21–30/31–40 Ratio: 1.05 vs. 0.95), and Ethnic Bias (Majority/Minority Ratio: 1.03 vs. 0.88). The AI-based system shows less bias compared to the Manual Evaluation.

The AI-based system shows a significant improvement in accuracy, with a lower Mean Absolute Error (MAE) of 0.15 compared to 0.25 for the manual evaluation technique. Similarly, the Root Mean Squared Error (RMSE) is lower for the AI system at 0.20, versus 0.30 for manual evaluations. The higher correlation with baseline performance data (0.85 for AI vs. 0.70 for manual) indicates greater consistency and reliability of the AI-generated scores.

### 12.7.2 Fairness

Fairness was assessed by analyzing the distribution of performance scores across different demographic groups to identify any biases. Table 12.4 shows the fairness metrics for both the AI-based system and the manual evaluation technique (Figure 12.2).

The AI-based system demonstrates a more balanced distribution of performance scores across different demographic groups, with a Gender Bias Ratio (Male/Female) of 1.02 compared to 0.90 for the manual evaluation. The Age Bias Ratio (21–30/31–40) is 1.05 for the AI system, indicating less age-related bias compared to the manual evaluation's ratio of 0.95. The Ethnic Bias Ratio (Majority/Minority) for the AI system is 1.03, showing reduced ethnic bias compared to the manual evaluation's ratio of 0.88. These metrics suggest that the AI system provides a fairer assessment across different demographics.

### 12.7.3 Transparency

Transparency was measured by evaluating the extent to which participants understood the basis of their performance evaluations. Table 12.5 presents the transparency metrics based on survey responses (Figure 12.3).

**TABLE 12.5**
**Transparency Metrics Comparing AI-Based and Manual Evaluations**

| Metric | AI-Based System (%) | Manual Evaluation (%) |
|---|---|---|
| Percentage of employees understanding evaluation criteria | 85 | 60 |
| Percentage of employees finding evaluation process transparent | 80 | 55 |

AI-Based: 85% Understand Criteria, 80% Find Process Transparent. Manual: 60%, 55%.

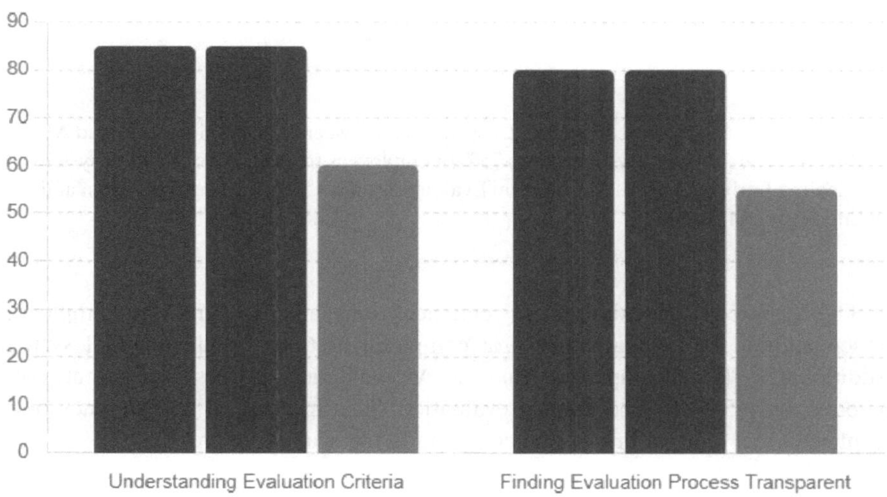


**FIGURE 12.3** Bar chart comparing employee metrics between AI-Based System and Manual Evaluation. The AI-based system shows 85% understanding of evaluation criteria and 80% finding the evaluation process transparent. The Manual Evaluation shows 60% understanding and 55% finding the process transparent.

**TABLE 12.6**
**Employee Trust Metrics for AI-Based and Manual Evaluations**

| Metric | AI-Based System (%) | Manual Evaluation (%) |
|---|---|---|
| Percentage of employees trusting the evaluation system | 75 | 50 |
| Percentage of employees feeling fairly evaluated | 78 | 52 |

AI-Based: 75% Trust System, 78% Feel Fairly Evaluated. Manual: 50% Trust, 52% Fairness.

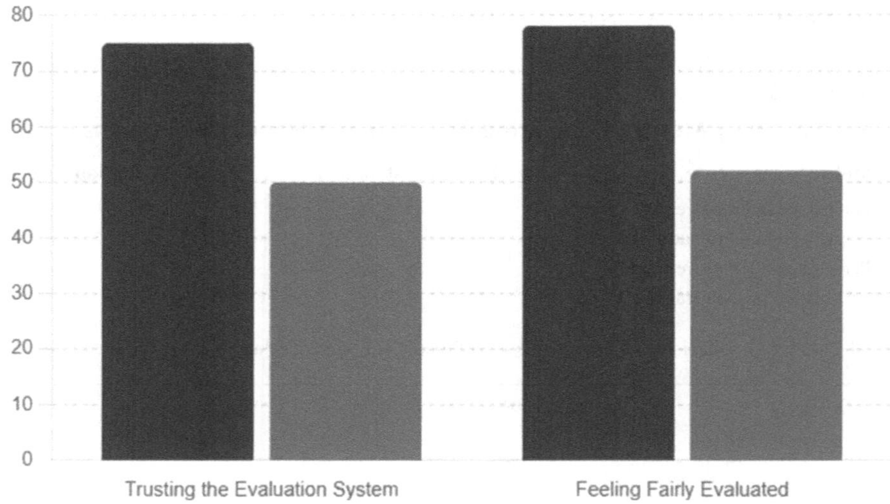


**FIGURE 12.4** Bar chart comparing trust metrics between AI-Based System and Manual Evaluation. The AI-based system shows 75% of employees trusting the evaluation system and 78% feeling fairly evaluated. The Manual Evaluation shows 50% trusting the system and 52% feeling fairly evaluated.

The AI-based system significantly enhances transparency, with 85% of employees understanding the evaluation criteria compared to 60% for the manual system. Additionally, 80% of employees find the AI evaluation process transparent, compared to only 55% for the manual evaluation. This increase in transparency helps employees feel more informed and confident in the evaluation process.

### 12.7.4 Employee Trust

Employee trust in the performance evaluation system was assessed through survey responses. Table 12.6 shows the trust metrics for both the AI-based system and the manual evaluation technique (Figure 12.4).

**TABLE 12.7**
**Overall Satisfaction Metrics for AI-Based and Manual Evaluations**

| Metric | AI-Based System | Manual Evaluation |
|---|---|---|
| Overall satisfaction score (1–5) | 4.2 | 3.5 |
| Percentage of employees preferring the new system | 82% | N/A |

AI-Based: Satisfaction Score 4.2, 82% Prefer New System. Manual: Satisfaction Score 3.5.

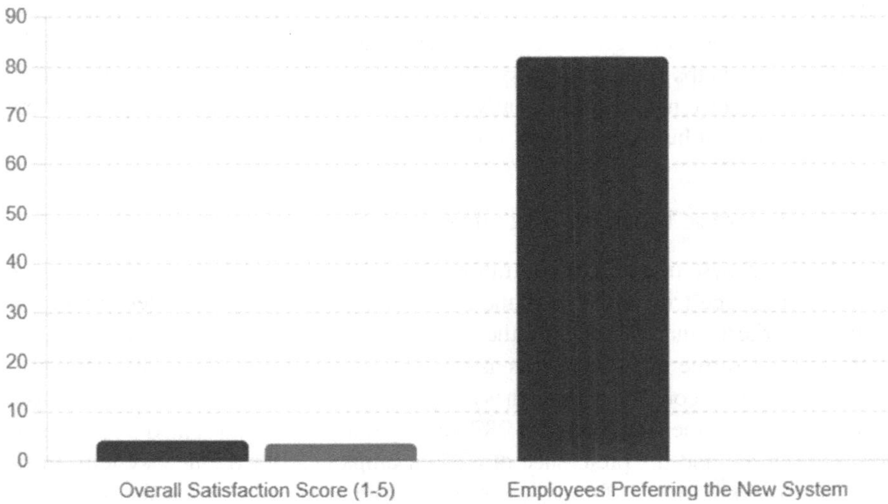


**FIGURE 12.5** Bar chart comparing satisfaction metrics between AI-based system and manual evaluation. The AI-based system shows an overall satisfaction score of 4.2% and 82% of employees preferring the new system. The Manual Evaluation shows an overall satisfaction score of 3.5% and 0% for employees preferring the new system (as N/A is represented as 0 for visualization).

The AI-based system fosters greater trust among employees, with 75% of participants expressing trust in the AI system compared to 50% for the manual evaluation. Furthermore, 78% of employees feel fairly evaluated by the AI system, whereas only 52% feel the same about the manual system. This increased trust is crucial for the acceptance and success of the AI-based evaluation system.

### 12.7.5 Overall Satisfaction

Overall satisfaction with the performance evaluation system was measured through post-study surveys. Table 12.7 presents the satisfaction metrics for both the AI-based system and the manual evaluation technique (Figure 12.5).

The overall satisfaction score for the AI-based system is higher at 4.2 out of 5, compared to 3.5 for the manual evaluation technique. Additionally, 82% of employees prefer the new AI-based system, indicating a strong preference for the improved transparency, fairness, and accuracy provided by the AI evaluations.

## 12.8 DISCUSSION

### 12.8.1 Human Factors Impact on AI-Based Evaluations

The study's findings underscore the significant impact of human factors on the effectiveness of AI-based performance evaluations. The integration of human expertise in the initial stages of AI system design, particularly in defining performance criteria and KPIs, played a crucial role in improving the system's accuracy. The lower MAE and RMSE values and higher correlation with baseline data confirm that involving human experts helps tailor the AI system to better reflect organizational goals and job-specific requirements. Additionally, the reduced biases in gender, age, and ethnicity indicate that human oversight is essential in designing fairer AI algorithms.

### 12.8.2 Employee Trust and Acceptance

The AI-based system's enhanced transparency significantly contributed to higher levels of employee trust and acceptance. With 85% of employees understanding the evaluation criteria and 80% finding the process transparent, the AI system effectively addressed one of the major concerns associated with AI evaluations—lack of clarity. This transparency, coupled with fairness improvements, resulted in 75% of employees expressing trust in the AI system and 78% feeling fairly evaluated. The overall satisfaction score of 4.2 and the preference of 82% of employees for the new system further emphasize the importance of transparency and fairness in gaining employee acceptance.

### 12.8.3 Ethical Considerations and Risk Mitigation

Ethical considerations are paramount in the deployment of AI systems. The study highlights the necessity of continuous monitoring and refinement of AI algorithms to mitigate risks associated with biases and inaccuracies. The reduced demographic biases in the AI-based system compared to the manual technique demonstrate that ethical AI design, incorporating human factors, can lead to fairer outcomes. Moreover, establishing a feedback loop for ongoing monitoring and addressing issues as they arise ensures that the AI system remains aligned with ethical standards and organizational values. These practices are essential for maintaining the integrity of the evaluation process and fostering a positive organizational culture.

## 12.9 CONCLUSION

The integration of AI in performance evaluation systems presents significant advancements in terms of accuracy, fairness, and transparency. This study explored the role of human factors in mitigating the risks associated with AI-driven performance evaluations and demonstrated how a balanced approach that combines human expertise with AI capabilities can enhance the overall effectiveness of performance evaluations.

## 12.10 SUMMARY OF KEY FINDINGS

The findings of this study indicate that the AI-based performance evaluation system significantly outperforms the manual evaluation technique across several key metrics. In terms of accuracy, the AI-based system demonstrated higher accuracy with a lower Mean Absolute Error (MAE) of 0.15 and Root Mean Squared Error (RMSE) of 0.20 compared to the manual evaluation (MAE 0.25 and RMSE 0.30). The AI system also showed a higher correlation with baseline performance data (0.85 versus 0.70), indicating greater consistency and reliability of the AI-generated scores.

Regarding fairness, the AI system exhibited reduced biases across demographic groups. The Gender Bias Ratio (Male/Female) was 1.02 for AI versus 0.90 for manual; the Age Bias Ratio (21–30/31–40) was 1.05 for AI versus 0.95 for manual; and the Ethnic Bias Ratio (Majority/Minority) was 1.03 for AI versus 0.88 for manual, suggesting fairer evaluations. These improvements in fairness are crucial for maintaining a just and inclusive workplace environment.

Transparency was another significant improvement with the AI-based system. The AI system enhanced transparency, with 85% of employees understanding the evaluation criteria and 80% finding the evaluation process transparent, compared to 60% and 55%, respectively, for the manual evaluation. This increased transparency helps employees feel more informed and confident in the evaluation process, contributing to higher levels of employee trust and acceptance.

Employee trust in the performance evaluation system was also notably higher with the AI-based system. Seventy-five percent of employees expressed trust in the AI system, and 78% felt fairly evaluated, compared to 50% and 52%, respectively, for the manual system. This increased trust is crucial for the acceptance and success of the AI-based evaluation system. The overall satisfaction score for the AI-based system was 4.2 out of 5, with 82% of employees preferring the new system, indicating a strong preference for the improved transparency, fairness, and accuracy provided by the AI evaluations.

These findings underscore the importance of integrating human factors into AI systems to ensure fair, accurate, and transparent performance evaluations. The enhanced trust and satisfaction among employees further emphasize the benefits of this approach.

## 12.11 FUTURE RESEARCH DIRECTIONS

While this study provides valuable insights into the effectiveness of AI-based performance evaluation systems, several areas warrant further exploration. Future research should investigate the long-term impact of AI-based performance evaluations on employee performance, job satisfaction, and organizational outcomes. Longitudinal studies could provide deeper insights into how these systems influence employee behavior and organizational culture over time. Additionally, exploring the scalability of AI-based performance evaluation systems across different organizational sizes and industries would help understand their broader applicability. Examining how these systems can be adapted to evolving job roles and performance criteria would also be beneficial.

Further research should focus on developing more sophisticated algorithms for detecting and mitigating biases in AI systems. This includes exploring new methods for ensuring fairness and transparency in AI decision-making processes. Investigating the best practices for involving employees in the continuous improvement of AI systems could enhance trust and acceptance. Studies on effective feedback mechanisms and their impact on system refinement would be valuable. As AI systems become more prevalent in performance evaluations, it is essential to explore the ethical and legal implications. Research should address issues such as data privacy, consent, and the ethical use of AI in making employment decisions.

By addressing these areas, future research can contribute to the development of more effective, fair, and transparent AI-based performance evaluation systems that are widely accepted and trusted by employees and organizations alike.

## REFERENCES

Abonamah, A. A., & Abdelhamid, N. (2024). Managerial insights for AI/ML implementation: A playbook for successful organizational integration. *Discover Artificial Intelligence*, 4(1). https://doi.org/10.1007/s44163-023-00100-5

Al Samman, A. M. (2024). The use of AI in fostering and embracing organizational culture. In *2024 ASU International Conference in Emerging Technologies for Sustainability and Intelligent Systems (ICETSIS)*. https://doi.org/10.1109/icetsis61505.2024.10459508

Awan, S. H., Habib, N., Shoaib Akhtar, C., & Naveed, S. (2020). Effectiveness of performance management system for employee performance through engagement. *SAGE Open*, 10(4), 215824402096938. https://doi.org/10.1177/2158244020969383

Bankins, S., & Formosa, P. (2023). The ethical implications of artificial intelligence (AI) For meaningful work. *Journal of Business Ethics*, 185(4), 725–740. https://doi.org/10.1007/s10551-023-05339-7

Bankins, S., Ocampo, A. C., Marrone, M., Restubog, S. L. D., & Woo, S. E. (2023). A multilevel review of artificial intelligence in organizations: Implications for organizational behavior research and practice. *Journal of Organizational Behavior*, 45(2), 159–182. https://doi.org/10.1002/job.2735

Baweja, J. A., Fallon, C. K., & Jefferson, B. A. (2023). Opportunities for human factors in machine learning. *Frontiers in Artificial Intelligence*, 6. https://doi.org/10.3389/frai.2023.1130190

Buijsman, S. (2024). Transparency for AI systems: A value-based approach. Ethics and Information Technology, 26(2). https://doi.org/10.1007/s10676-024-09770-w

Chen, Z. (2023). Ethics and discrimination in artificial intelligence-enabled recruitment practices. *Humanities and Social Sciences Communications*, 10(1). https://doi.org/10.1057/s41599-023-02079-x

Dawadi, S., Shrestha, S., & Giri, R. A. (2021). Mixed-methods research: A discussion on its types, challenges, and criticisms. *Journal of Practical Studies in Education*, 2(2), 25–36. https://doi.org/10.46809/jpse.v2i2.20

Hands, A. S. (2022). Integrating quantitative and qualitative data in mixed methods research: An illustration. *The Canadian Journal of Information and Library Science*, 45(1), 1–20. https://doi.org/10.5206/cjilsrcsib.v45i1.10645

Karyamsetty, H. J., Khan, S. A., & Nayyar, A. (2023). Envisioning towards modernization of Society 5.0- A prospective glimpse on status, opportunities, and challenges with XAI. In Al-Turjman, F., Nayyar, A., Naved, M., Singh, A. K., Bilal, M. (Eds). *XAI Based Intelligent Systems for Society 5.0* (pp. 223–267). Elsevier. https://doi.org/10.1016/B978-0-323-95315-3.00005-X

Khan, S.A., & Naim, A. (2023). XAI in Society 5.0 through the lens of Marketing and HRM. In Al-Turjman, F., Nayyar, A., Naved, M., Singh, A. K., Bilal, M. (Eds). *XAI Based Intelligent Systems for Society 5.0* (pp. 327–363). Elsevier. https://doi.org/10.1016/B978-0-323-95315-3.00004-8

Khatri, B., Shrimali, H., Khan, S. A., & Naim, A. (2023). Role of HR analytics in ensuring psychological wellbeing and job security: Learnings from COVID-19. In Yadav, R., Sinha M., Kureethara, J. (Eds). *HR Analytics in an Era of Rapid Automation* (pp. 36–53). IGI Global. https://doi.org/10.4018/978-1-6684-8942-0.ch003

Laux, J. (2023). Institutionalised distrust and human oversight of artificial intelligence: Towards a democratic design of AI governance under the European Union AI Act. *AI & Society*. https://doi.org/10.1007/s00146-023-01777-z

Ledro, C., Nosella, A., & Dalla Pozza, I. (2023). Integration of AI in CRM: Challenges and guidelines. *Journal of Open Innovation: Technology, Market, and Complexity*, 9(4), 100151. https://doi.org/10.1016/j.joitmc.2023.100151

Li, M. (2024). The impact of artificial intelligence on human resource management systems – Applications and risks. *Applied and Computational Engineering*, 48(1), 7–16. https://doi.org/10.54254/2755-2721/48/20241060

Magd, H., Ansari, M. S. A., & Khan, S. A. (2023). Need for explainable artificial intelligence ethnic decision making in Society 5.0. In Al-Turjman, F., Nayyar, A., Naved, M., Singh, A. K., Bilal, M. (Eds). *XAI Based Intelligent Systems for Society 5.0* (pp. 103–127). Elsevier. https://doi.org/10.1016/B978-0-323-95315-3.00010-3

Morandini, S., Fraboni, F., Balatti, E., Hackmann, A., Brendel, H., Puzzo, G., Volpi, L., Giusino, D., De Angelis, M., & Pietrantoni, L. (2023). Assessing the transparency and explainability of AI algorithms in planning and scheduling tools: A review of the literature. In *Human Interaction & Emerging Technologies (IHIET 2023): Artificial Intelligence & Future Applications*. https://doi.org/10.54941/ahfe1004068

Murugesan, U., Subramanian, P., Srivastava, S., & Dwivedi, A. (2023). A study of artificial intelligence impacts on human resource digitalization in Industry 4.0. *Decision Analytics Journal*, 7, 100249. https://doi.org/10.1016/j.dajour.2023.100249

Murugeshwari, B., Rajalakshmi, S., & Sudharson, K. (2023). Hybrid approach for privacy enhancement in data mining using arbitrariness and perturbation. *Computer Systems Science and Engineering*, 44(3), 2293–2307. https://doi.org/10.32604/csse.2023.029074

Nyathani, R. (2023). AI in performance management: Redefining performance appraisals in the digital age. *Journal of Artificial Intelligence & Cloud Computing*, 1–5. https://doi.org/10.47363/jaicc/2023(2)134

Peng, J. (2022). Performance appraisal system and its optimization method for enterprise management employees based on the KPI index. *Discrete Dynamics in Nature and Society*, 2022, 1–12. https://doi.org/10.1155/2022/1937083

Perifanis, N. A., & Kitsios, F. (2023). Investigating the influence of artificial intelligence on business value in the digital era of strategy: A literature review. *Information*, 14(2), 85. https://doi.org/10.3390/info14020085

Polemi, N., Praça, I., Kioskli, K., & Bécue, A. (2024). Challenges and efforts in managing AI trustworthiness risks: A state of knowledge. *Frontiers in Big Data*, 7. https://doi.org/10.3389/fdata.2024.1381163

Praveenraj, D. D. W., Victor, M., Vennila, C., Hussein Alawadi, A., Diyora, P., Vasudevan, N., & Avudaiappan, T. (2023). Exploring explainable artificial intelligence for transparent decision making. *E3S Web of Conferences*, 399, 04030. https://doi.org/10.1051/e3sconf/202339904030

Rodgers, W., Murray, J. M., Stefanidis, A., Degbey, W. Y., & Tarba, S. Y. (2023). An artificial intelligence algorithmic approach to ethical decision-making in human resource management processes. *Human Resource Management Review*, 33(1), 100925. https://doi.org/10.1016/j.hrmr.2022.100925

Sarker, I. H. (2022). AI-based modeling: Techniques, applications and research issues towards automation, intelligent and smart systems. *SN Computer Science*, 3(2). https://doi.org/10.1007/s42979-022-01043-x

Shin, D., & Park, Y. J. (2019). Role of fairness, accountability, and transparency in algorithmic affordance. *Computers in Human Behavior*, *98*, 277–284.

Slingerland, G., Murray, M., Lukosch, S., McCarthy, J., & Brazier, F. (2022). Participatory design going digital: Challenges and opportunities for distributed place-making. *Computer Supported Cooperative Work (CSCW),* 31(4), 669–700. https://doi.org/10.1007/s10606-022-09438-3

Sudharson, K., Arvindan, R. R., & Balashunmugham, M. (2023a). A survey on AR-based digitization for smart education system. In *2023 IEEE International Students' Conference on Electrical, Electronics and Computer Science (SCEECS).* https://doi.org/10.1109/sceecs57921.2023.10061816

Sudharson, K., Mohan Raj, K. R., Selvi, K., Suresh Kumar, A., Dhakshunhaamoorthiy, & Murali, E. (2023b). SmartTask: A neural network powered block-chain based intelligent task scheduling for project management. In *2023 International Conference on Innovative Computing, Intelligent Communication and Smart Electrical Systems (ICSES).* https://doi.org/10.1109/icses60034.2023.10465271

Sudharson, K., Prasad, A., & Smriti, S. (2023c, February). A survey on reader's society: New social network of book swapping platform. In *2023 IEEE International Students' Conference on Electrical, Electronics and Computer Science (SCEECS)* (pp. 1–6). IEEE. https://doi.org/10.1109/sceecs57921.2023.10063065

Vanitha, R., Sudharson, K., Sathish Kumar, N., & Gunasundari, S. (2024). An improved vectorization-based emotion detection using tuned inverse document frequency approach. *International Journal of Electronics and Communication Engineering*, 11(3), 106–114. https://doi.org/10.14445/23488549/ijece-v11i3p111

Wamba-Taguimdje, S. L., Fosso Wamba, S., Kala Kamdjoug, J. R., & Tchatchouang Wanko, C. E. (2020). Influence of artificial intelligence (AI) on firm performance: the business value of AI-based transformation projects. *Business Process Management Journal*, 26(7), 1893–1924. https://doi.org/10.1108/bpmj-10-2019-0411

Xiuqing, D., Rafiq, M., & Zhumin, W. (2023). AI-based performance appraisal systems. In *Exploring the Intersection of AI and Human Resources Management* (pp. 15–29). https://doi.org/10.4018/979-8-3693-0039-8.ch002.

Yu, L., & Li, Y. (2022). Artificial intelligence decision-making transparency and employees' trust: The parallel multiple mediating effect of effectiveness and discomfort. *Behavioral Sciences*, 12(5), 127. https://doi.org/10.3390/bs12050127

Zirar, A., Ali, S. I., & Islam, N. (2023). Worker and workplace Artificial Intelligence (AI) coexistence: Emerging themes and research agenda. *Technovation*, 124, 102747. https://doi.org/10.1016/j.technovation.2023.102747

# 13 Human-Centric Logistics
## *Ergonomic Considerations in Warehouse Order Picking for Industry 5.0 Integration*

*Danish Nasir*

## 13.1 INTRODUCTION

Warehouse order picking is a critical component of logistics operations, and its efficiency has a significant impact on the overall performance of the supply chain. In recent years, the integration of Industry 5.0 technologies has led to a growing emphasis on human-centric logistics, which prioritizes the well-being and productivity of warehouse workers. This chapter explores the crucial role of ergonomics in warehouse order picking, highlighting the importance of designing and arranging the work environment, equipment, and processes to optimize worker performance and reduce the risk of injuries.

## 13.2 WAREHOUSE DESIGN AND OPERATIONS

Warehouse design and operations are crucial components of a company's logistics and supply chain. A well-designed warehouse layout ensures efficient storage, inventory management, and movement of goods, ultimately impacting productivity and profitability.

### 13.2.1 Warehouse Design

Generally, three basic types of warehouse layouts are commonly used in the industry. These layouts are Rectangular, U shape, and Fishbone. Other shapes emerged by changing some features of the warehouse.

**Rectangular Layout:** The rectangular warehouse layout is the most common and straightforward design. It involves arranging storage areas, receiving, and shipping zones in a linear manner, forming a rectangle.

The material receiving area is typically located on one end of the warehouse. The storage space occupies the central and largest part of the warehouse. The materials are stored in rows of shelving or racks. The Shipping Area is located on the opposite end of the receiving area. The layout is easy to understand and implement. It utilizes space effectively, allowing for easy navigation and efficient use of the available

DOI: 10.1201/9781003532293-15

regions. This can be easily expanded by adding more rows and is suitable for various products and industries. However, the main drawback is higher travel times as workers might have to travel long distances between receiving, storage, and shipping areas.

**U-Shape Layout:** The U-shaped warehouse layout arranges the receiving and shipping areas on the same side of the warehouse, forming a U-shape. This design facilitates a circular flow of goods from receiving to storage to shipping. The receiving Area is located next to the shipping area at the front of the warehouse. The storage area forms the base of the "U" in the central part of the warehouse. The shipping area is adjacent to the receiving area, completing the U-shape. The advantage of such a layout is goods flow in a continuous loop, reducing travel distance for workers. This facilitates efficient loading and unloading of goods. Easier to manage and supervise operations due to the compact design. The main drawback is space utilisation which is not as efficient as other layouts for very large warehouses. Further design of such a layout can be more complex as compared to the rectangular layout.

**Fishbone Layout:** The fishbone layout is a more complex design featuring a central aisle with diagonal aisles branching off, resembling the bones of a fish. This layout aims to minimize travel distance and improve picking efficiency. The main Aisle runs centrally from the receiving to the shipping area. The diagonal Aisles branch off from the main aisle like the bones of a fish, providing multiple routes for picking. The storage area is located along the diagonal aisles, optimising space utilisation. The biggest advantage is reduced travel distance due to shorter travel paths for pickers, leading to increased efficiency. Further, it has multiple access points to storage areas reducing congestion and delays. Such layouts maximize the use of available space, particularly in larger warehouses. The main drawback is more complex to design and implement compared to simpler layouts. Also potentially higher setup and maintenance costs due to the intricate design.

Table 13.1 shows a comparison of various layouts. These layouts cater to different operational needs and can be chosen based on specific warehouse requirements, such as size, traffic flow, and the nature of stored items.

### 13.2.2 Warehouse Operations

Warehouses serve as crucial hubs for storing and buffering products, facilitating smooth operations in supply chains (De Koster et al., 2007). The essence of warehouse operations lies in the systematic movement and storage of goods and materials within a controlled environment. Lambert et al. (1998) highlight the multifaceted advantages of warehouses, including enabling transportation and production economies, accommodating demand fluctuations and seasonality, acting as a buffer for trans-shipment, and offering customers a diverse range of products. Notably, over 750,000 warehouses exist globally, with nearly 80% still relying on manual operations (Napolitano, 2012).

Key to the success of warehouse operations are elements such as effective inventory management, optimal space utilisation, accurate order processing, and timely product delivery. Warehouse managers typically make critical decisions regarding storage assignments, layout design, zoning strategies, and routing policies (Chan & Chan, 2011).

**TABLE 13.1**
**Comparison of Warehouse Layout**

| Aspect | Rectangular Layout | U-Shape Layout | Fishbone Layout |
|---|---|---|---|
| Shape | Rectangular | U-shaped | Fishbone-shaped |
| Space utilisation | Efficient use of space along the walls | Efficient use of space around a central aisle | Efficient use of space along a central spine |
| Accessibility | Typically requires longer travel distances between areas | Shorter travel distances between areas | Shorter travel distances between areas |
| Traffic flow | Typically linear flow pattern | Circular flow pattern around central aisle | Linear flow pattern along the spine |
| Flexibility | Moderate flexibility in layout changes | Moderate flexibility in layout changes | Moderate flexibility in layout changes |
| Visibility | Generally good visibility throughout the warehouse | Good visibility to most areas | Good visibility to most areas |
| Forklift movement | Generally good movement | Limited movement around central aisle | Good movement along the spine |
| Zoning possibilities | Easier zoning of distinct areas | Zoning possibilities limited by U-shape | Easier zoning of distinct areas |
| Expansion potential | Moderate expansion potential | Limited expansion potential around central aisle | Moderate expansion potential along the spine |

Moreover, the integration of advanced technologies like automation, robotics, and artificial intelligence (AI) is increasingly revolutionising warehouse operations and enhancing efficiency, accuracy, and productivity.

#### 13.2.2.1 Storage Assignment

Storage assignment methodologies are crucial for optimising warehouse operations (De Koster et al., 2007). Commonly used assignment policies include:

- **Random Storage:** In this system, products are placed in storage locations randomly, promoting high space utilisation. However, this approach may increase the distance travelled by order pickers, potentially reducing efficiency.
- **Dedicated Storage:** This method involves allocating fixed locations for items, which can enhance space utilisation. However, if demand fluctuations are not adequately considered, it may lead to inefficiencies.
- **Class-based Storage:** This strategic approach categorizes items into groups based on various analyses, such as Pareto, ABC, or XYZ. By grouping items based on factors like demand or accessibility, class-based storage balances space utilisation and order-picking efficiency.

The choice of storage assignment methodology depends on the specific needs and constraints of the warehouse operation. Factors such as product characteristics, order patterns, and available space should be considered when selecting the most appropriate approach.

#### 13.2.2.2 Layout Design

Warehouse layout design is critical in optimising space, equipment, and material organisation to streamline operations (Le-Duc & De Koster, 2005; Roodbergen et al., 2015). A well-designed layout can significantly impact picking speed, accuracy, and resource efficiency. The primary objectives of warehouse layout design include:

1. Minimising picker travel distance or time
2. Determining optimal aisle configurations, zones, and rack arrangements
3. Maximising space utilisation.

Warehouse managers can develop a layout that maximizes productivity and profitability by carefully considering factors such as product characteristics, order patterns, and equipment constraints.

#### 13.2.2.3 Zoning

Warehouse zoning is a strategic approach that involves dividing the facility into specific areas or zones based on product type, process, or activity (Le-Duc & De Koster, 2005). Effective zoning can optimize operations by reducing travel distances within zones, enhancing picker familiarity with storage areas, and minimising the risk of errors or product damage. Zoning offers several advantages for warehouse operations:

1. Reduced travel distances within zones, leading to increased efficiency and productivity
2. Enhanced picker familiarity with storage areas, reducing the likelihood of errors or product damage
3. Improved organisation and control over inventory and material flow.

While zoning can be highly beneficial, it may also lead to some drawbacks:

1. Longer cycle times due to the need for additional pickers to cover multiple zones
2. Increased operational costs associated with employing more pickers and managing multiple zones.

#### 13.2.2.4 Routing

Routing is a critical component of warehouse operations, as it determines the most efficient path for moving goods within the facility (Petersen, 1997). Effective routing strategies can significantly impact warehouse performance by:

1. Minimising travel distance and time
2. Reducing operational costs
3. Optimising resource utilisation and warehouse space.

Several routing methods are available, each tailored to specific operational contexts:

1. **Largest Gap Routing**: This method involves selecting the route that covers the largest gap in the warehouse, ensuring efficient use of space.
2. **Composite Routing**: This approach combines multiple routes to optimize travel distance and time.
3. **Return Routing**: This method involves returning to the starting point after completing a route, reducing travel distance and time.
4. **Optimal Routing**: This method uses advanced algorithms to determine the most efficient route based on various factors, such as product characteristics and warehouse layout.

Ongoing research in routing policies continues to refine strategies for enhancing warehouse throughput, minimising picker fatigue, and improving overall operational efficiency.

## 13.3 ERGONOMICS

Ergonomics in warehouse design and operations is crucial for enhancing worker well-being, productivity, and overall operational efficiency. The goal of ergonomic considerations in warehouse order picking is to create a workspace that minimizes physical strain, reduces the risk of injuries, and enhances overall efficiency. Poor ergonomic practices can lead to musculoskeletal disorders (MSDs), fatigue, and injuries among warehouse workers, impacting both their health and the company's bottom line.

Following are the key points while designing a warehouse:

- **Human-Centric Design**: Incorporate adjustable workstations and equipment to accommodate various body sizes and shapes. Design aisles and storage systems that minimize bending, reaching, and twisting.
- **Material Handling**: Utilize mechanical aids such as conveyors, lifts, and automated picking systems to reduce manual handling. Ensure that frequently accessed items are stored at waist height to reduce strain.
- **Workplace Layout**: Design a logical layout that minimizes walking distances and streamlines workflow. Implement clear signage and labels to reduce confusion and enhance safety.
- **Training and Education**: Provide regular training on proper lifting techniques and the use of ergonomic equipment. Foster a culture of safety and encourage workers to report ergonomic issues.
- **Environmental Factors**: Maintain optimal lighting, temperature, and noise levels to create a comfortable working environment. Ensure adequate ventilation to reduce the risk of respiratory problems.

Table 13.2 lists various tools for estimating physical strain, each selected according to specific applications and working environments.

**TABLE 13.2**
**Ergonomic Evaluation Tools**

| Tool | Purpose | Application | Method |
|---|---|---|---|
| Occupational repetitive action (**OCRA**) | Measures exposure to repetitive work activities | Manufacturing and assembly lines | Evaluates frequency, duration, force, and posture to provide a risk score |
| **Snook tables** | Assesses manual material handling tasks | Manual handling tasks in various industries | Provides guidelines for safe weights and forces based on task parameters |
| Ovako working analysis system (**OWAS**) | Evaluates postural load during work activities | Manual labour settings | Classifies and codes body postures to assess postural load and risks |
| **Borg scale** | Measures perceived exertion and effort | Workplaces and sports | Uses a numerical scale for workers to rate their perceived exertion |
| Rapid upper limb assessment **(RULA)** | Assesses risk factors for upper limb disorders | Tasks with repetitive or sustained awkward postures | Analyses body posture, muscle use, and force exerted to provide a risk score |
| Rapid entire body assessment **(REBA)** | Evaluate whole-body postural risks | Tasks involving awkward postures | Scores based on body postures, exertions, movements, repetition, and coupling |
| Postural ergonomic risk assessment **(PERA)** | Assesses postural risks | Various work environments | Evaluates worker postures to determine the risk of musculoskeletal strain |
| National Institute of Occupational and Health **(NIOSH)** lifting equation | Determines safe lifting limits to prevent injuries | Workplaces with significant manual lifting | Calculates the recommended weight limit (RWL) based on lifting factors |
| **Energy expenditure modelling** | Quantifies the energy expenditure of workers | Various industries | Uses models to estimate energy required for tasks and plan rest periods |

### 13.3.1 Studies Based on Ergonomics in Warehouses

Past studies identify nine core themes focusing on the ergonomic aspects of warehouse operations. These themes are crucial for enhancing both efficiency and worker well-being in manual warehouses. Table 13.3 shows an overview of each theme.

These themes collectively provide a comprehensive understanding of the ergonomic considerations in warehouse design and operations, highlighting the need for further research and implementation of ergonomic interventions to improve both efficiency and worker well-being.

The evolution of warehouse operations, particularly in Warehousing 5.0, highlights a comprehensive integration of technological advancements, ergonomic considerations, and human-centric approaches to optimize efficiency and worker well-being. Technological advancements play a pivotal role, with wearable devices like head-mounted displays (HMDs) for augmented reality (AR) reducing pick

**TABLE 13.3**
**Thematic Studies Based on Warehouse**

| Theme | Key Studies and Findings |
|---|---|
| **Technological interventions** | **Wearable devices**: AR via HMDs reduces pick errors (Reif et al., 2010; Reif & Walch, 2008) |
| | **Anthropometric modelling**: Used to predict ergonomic strain (Horváthová et al., 2019; Al Shehhi et al., 2019) |
| | **Automation**: Improves cycle times and reduces physiological strain (Gruchmann et al., 2021) |
| **Assignment** | **Golden zone**: High-demand items placed within easy reach (Petersen et al., 2005) |
| | **Models integrating ergonomics**: Energy expenditure and discomfort levels considered (Battini et al., 2016; Larco et al., 2017; Diefenbach & Glock, 2019) |
| **Human factor** | **Economic evaluation of ergonomics**: Ergonomic interventions reduce workplace injuries and yield financial returns (Tompa et al. 2010) |
| | **Human aspects in picking**: Limited research historically, but important for efficiency (Grosse et al., 2017; Gajšek et al., 2017) |
| **Fatigue** | **Fatigue measurement tools**: Heart rate monitors and OWAS index (Battini et al., 2017; Daria et al., 2015) |
| | **Models incorporating fatigue**: Fatigue as a constraint in bi-objective models (Zhao et al., 2019; Granotto et al., 2019) |
| **Pallet selection** | **Types of pallets**: Full, half, and half pallets retrieved from shelves (Calzavara et al., 2016) |
| | **Ergonomic measures**: OWAS index and energy expenditure considered (Calzavara et al., 2017; Calzavara et al., 2019) |
| | **Rotating pallets**: Reduces spinal load and picking time (Glock et al., 2019) |
| **Learning** | **Learning models**: Various models show reduced pick times and errors with learning (Grosse & Glock, 2013; Grosse et al., 2013) |
| | **Learning effects**: Faster learning for high-demand items; experienced workers perform better (Hilmola & Tolli, 2016) |
| **Load speed variation** | **Experimental studies**: Show load conditions significantly affect movement rates (Dai et al., 2010; Elbert et al., 2017; Botti et al., 2020) |
| **Safety** | **Safety constructs**: Positive correlation between safety focus and quality (De Koster et al., 2011; De Vries et al., 2016b) |
| | **Safety culture and behaviour**: Factors influencing safety behaviour include communication, policy, and equipment (Hofstra et al., 2018) |
| **Incentive** | **Picking methods and incentives**: Parallel picking with competition-based incentives increases productivity; zone picking with cooperation-based incentives is more productive (De Vries et al., 2016a) |

errors and improving training, while anthropometric modelling predicts and reduces ergonomic strain. Automation improves cycle times and mitigates physiological strain, emphasising technology's importance in modern warehouses. Optimal item placement strategies, such as the cube order index (COI) and the "Golden Zone" approach, minimize retrieval distances and ergonomic risks, balancing economic factors (time and distance) with ergonomic ones (energy expenditure, discomfort, and fatigue). Ergonomics, incorporating human factors into both experimental and mathematical models, yield financial returns by reducing workplace injuries and enhancing efficiency. Fatigue modelling, using tools like heart rate monitors and the Ovako working analysis system (OWAS), improves resilience against worker non-availability and optimizes picking systems. Pallet selection research explores various types of pallets and their ergonomic implications, with mathematical models and heuristics optimising selection based on energy expenditure and posture. Learning effects in order picking lead to improved pick times and reduced errors, demonstrating the importance of learning processes in optimising routes and efficiency. Load speed variation studies show that item weight significantly affects picker speed, emphasising the need to consider load conditions in operations. Safety, though under-researched, is critical, with models examining safety constructs' impact on productivity and quality, influenced by communication and equipment. Incentives, key motivators for order pickers, show that productivity varies with different picking methods and systems, with competition-based and cooperation-based incentives increasing productivity. Integrating technological advancements, ergonomic principles, and human-centric approaches is essential for Warehousing 5.0, creating efficient, safe, and human-centred environments that enhance operational performance and worker well-being, ensuring sustainable success in the rapidly evolving industry landscape.

## 13.4 WAREHOUSING 5.0: THE INDUSTRY 5.0 PERSPECTIVE

The Industrial Revolution started long back. However, the late 18th to early 19th century was termed as Industry 1.0., which is known as the First Industrial Revolution, marked the transition from agrarian societies to industrialized ones. It was characterized by the mechanisation of production processes, primarily through the use of water and steam power. The Second Industrial Revolution (Late 19th to early 20th century), was marked by significant advancements in technology and the widespread adoption of electricity, which replaced steam power as the primary energy source. Industry 3.0 (late 20th century), often referred to as the Digital Revolution or Information Age, saw the emergence of computers, electronics, and automation, leading to the digitalisation of industries and the widespread use of information technology. Industry 4.0 (Early 21st century to the present), also known as the Fourth Industrial Revolution, builds upon the digitalisation of Industry 3.0 by integrating cyber-physical systems, the Internet of Things (IoT), and AI into manufacturing and other sectors. Industry 5.0 is a term that is still in the conceptual stage and is envisioned as the next phase of industrial evolution beyond Industry 4.0. Industry 5.0 is expected to focus on the symbiotic relationship between humans and machines, emphasising human-centric innovation and collaboration between humans and advanced technologies such as AI, robotics, and biotechnology.

Warehousing 5.0 embodies the principles of Industry 5.0 within the realm of warehousing and logistics. It leverages cutting-edge technologies to create smart warehouses where humans and robots work together seamlessly. The primary goals are to improve operational efficiency, reduce costs, enhance worker safety and well-being, and provide a more responsive and adaptive supply chain. Gladysz et al. (2023) stated that the fusion of Industry 4.0 with human-centric, sustainable, and resilient principles births Industry 5.0, supporting the Operator 4.0 paradigm. However, challenges persist due to the incomplete readiness of Industry 4.0 solutions.

## 13.4.1 Key Components of Warehousing 5.0

Warehousing 5.0, as an evolution of traditional warehousing practices, integrates advanced technologies and human-centred approaches to enhance efficiency, safety, and flexibility in warehouse operations. Warehousing 5.0 represents a significant shift towards smarter, more sustainable, and human-centric warehouse management, aiming to meet the demands of the modern supply chain landscape. Here are the key components.

### 13.4.1.1 Human-Robot Collaboration

Cobots (Collaborative Robots) are designed to work alongside human workers. Cobots assist in tasks such as picking, packing, and sorting. They are equipped with sensors and AI to ensure safe interaction with humans. Koreis et al. (2025) studied hybrid order picking in which humans and robots work collaboratively to achieve the task. The results showed that Cobots could reduce picking time by almost 33% and concluded that cobot-based picking is the best picking. Pasparakis et al. (2023) conducted an experimental study to understand Human leading or following robots' implications. The study concluded that in both scenarios, human-robot collaboration positively impacts job satisfaction, self-esteem & efficacy irrespective of set up. Teck and Dewil (2022) modelled a problem where robots pick the items, transport them to workstations and then human pickers complete the orders. The study showed that such collaboration improves the overall performance. Jiang et al. (2020) address inefficiencies in robotic forward-reserve warehouses, where mobile robots bring pick locations to workers. It proposes a picking-replenishment synchronisation mechanism (PRSM) to optimize the balance between replenishment efforts and picking efficiency.

### 13.4.1.2 Advanced Automation and Robotics

Autonomous Mobile Robots (AMRs) Navigate warehouse floors autonomously, transporting goods and materials with high precision and efficiency. Sun et al. (2021) proposed a mathematical model to minimize cycle time by employing robots in warehouse order picking. Zou et al. (2018) address battery management in Robotic mobile fulfilment systems (RMFS), comparing battery swapping and charging strategies using a plug-in or inductive methods. A semi-open queuing network (SOQN) model estimates system performance, optimized through analytical and simulation methods for cost and throughput.

### 13.4.1.3 AI and Machine Learning

AI algorithms analyse historical data to predict demand, optimize inventory levels, and streamline replenishment processes. Zhang et al. (2022) carried out a literature review to understand mobile app usage in logistics systems. The results showed that the usage of mobile apps helps human centricity. They suggested more research to be carried out on the psychosocial aspect. Grosse (2024) examines supportive and substitutive technologies, like augmented reality, analysing their impact on operations and proposing future research for sustainable, human-centred logistics systems.

#### 13.4.1.4 Internet of Things (IoT)

Smart Sensors monitor environmental conditions, equipment status, and inventory levels in real time, providing actionable insights and ensuring optimal operating conditions.

Helm et al. (2024) used new technology Intelligent Video Analysis (IVA) to find the root cause of errors such as incorrect quantity sent to customers, damaged items etc. The study brings out the factors that cause such errors and suggests ways to prevent them. Glock et al. (2021) carried out a literature review to analyse the assistive devices used in warehouse operations by order pickers. They suggested selecting an assistive device depending on the application that can improve the well-being of the picker.

#### 13.4.1.5 Digital Twins

Virtual Modelling creates digital replicas of physical warehouse operations to simulate, analyse, and optimize processes before implementing changes in the real world. Setayesh et al. (2024) evaluate the Warehouse Error Prevention (WEP) tool, consisting of seven yes/no modules to identify pick error sources. Thirty-three participants from 27 organisations across three countries trialled the tool. Survey and interview results indicated high usability and functionality, with the potential for improving order-picking quality.

#### 13.4.1.6 Sustainable Practices

Sustainable systems implement energy-efficient technologies and practices to reduce the carbon footprint of warehouse operations. Baig & Yadegaridehkordi (2024) reviewed 48 articles (2019–2022) highlighting global growth in Industry 5.0's sustainability discourse. Technologies like IoT, AI, and collaborative robots dominate. Four main research themes emerge, with human-centricity prevailing. This study aids in identifying gaps and inspiring further exploration, guiding the effective implementation of Industry 5.0 for sustainability. Zhang and Li (2023) researched Society 5.0 and Industry 5.0, examining their impact on sustainable development in the Chinese industrial sector using normative stakeholder theory. Results show both models positively affect sustainable performance economically, socially, and environmentally. The study offers policy, managerial, and theoretical recommendations for future research.

Circular Economy emphasizes recycling, reusing, and reducing waste to create a more sustainable supply chain. Atif (2023) stated that circular economy (CE) and Industry 4.0 (I4.0) pioneered data-driven, sustainable manufacturing, but lacked a human-centred focus. Industry 5.0 (I5.0) emerges, integrating societal values. Through a systematic literature review, the study reveals I5.0's broader scope, emphasising human-centricity alongside technical innovation for resilient, sustainable smart manufacturing.

#### 13.4.1.7 Enhanced Worker Well-Being

Ergonomics deals with designing workstations and processes to minimize physical strain and enhance worker comfort and safety. Vijayakumar and Sobhani (2023) emphasized that Industry 5.0 advocates for improving the working conditions of the order picker rather than replacing them with an automated system. They proposed a mathematical model based on Pick and Transport Robot (PTR) that optimizes the order picker's productivity, quality, and well-being. De Lombaert et al. (2023) reviewed to analyse how skills, conduct, and perception are integrated into planning models to improve the well-being of the order picker. Modgil et al. (2023) suggested that in the dynamic landscape of Industry 5.0, the demand for skilled supply chain professionals intensifies. Identifying and nurturing talent crucial for organisational strategy is imperative. Through literature review and industry consultations, key competencies are identified.

## 13.5 CONCLUSION

The advancement of warehousing into the realm of Warehousing 5.0 marks a transformative shift in logistics and supply chain management, embodying the core principles of Industry 5.0. This paradigm shift focuses on the seamless integration of human expertise and advanced technologies, creating an environment where human ingenuity and innovation are magnified by intelligent systems and automation.

Central to the philosophy of Warehousing 5.0 is the emphasis on human-centric design. Unlike previous industrial revolutions that predominantly focused on automation and efficiency, Industry 5.0 prioritizes the well-being and empowerment of the workforce. In the context of warehousing, this translates to the deployment of technologies such as cobots, AR, and wearable devices. These technologies assist workers in their tasks, minimize physical strain, and enhance overall job satisfaction. By leveraging these advancements, warehouses can achieve higher productivity while simultaneously improving ergonomic conditions and ensuring the safety of their employees.

The implementation of advanced data analytics and AI further augments the capabilities of Warehousing 5.0. Predictive analytics and AI-driven decision-making tools enable more accurate forecasting, inventory management, and demand planning. These technologies allow warehouses to operate with increased efficiency by optimising stock levels, reducing waste, and ensuring the availability of the right products at the right time. Additionally, the utilisation of real-time data and IoT devices provides valuable insights into warehouse operations, facilitating continuous improvement and rapid response to emerging issues.

Sustainability is another pivotal component of Warehousing 5.0. The integration of green technologies and sustainable practices, such as energy-efficient lighting, renewable energy sources, and eco-friendly packaging solutions, significantly reduces the environmental impact of warehouse operations. Furthermore, the adoption of circular economy principles promotes the reuse and recycling of materials, contributing to more sustainable supply chains. This commitment to sustainability benefits the environment and aligns with the growing consumer demand for environmentally responsible business practices.

The flexibility and adaptability of Warehousing 5.0 are essential to its success in a dynamic market environment. The ability to quickly reconfigure warehouse layouts, processes, and workflows in response to new demands or disruptions is crucial for maintaining operational continuity and meeting customer expectations. This adaptability is supported by modular and scalable technology solutions that can be easily integrated and upgraded as needed, ensuring that warehouses remain agile and responsive to changing conditions.

Moreover, Warehousing 5.0 fosters innovation through the collaborative efforts of humans and technology. By combining human creativity with AI's analytical power and automation's precision, warehouses can develop innovative solutions to complex problems. This synergy drives continuous improvement and operational excellence, setting new standards for efficiency and effectiveness in the industry.

## REFERENCES


Al Shehhi, S., King, N., Amer, S., & Mohammad, J. (2019). Modeling the ergonomics of Goods-to-Man order picking. In *Proceedings of IIE Annual Conference* (pp. 620–625). Institute of Industrial and Systems Engineers (IISE). Orlando, FL.

Atif, S. (2023). Analysing the alignment between circular economy and industry 4.0 nexus with industry 5.0 era: An integrative systematic literature review. *Sustainable Development*, *31*(4), 2155–2175.

Baig, M. I., & Yadegaridehkordi, E. (2024). Industry 5.0 applications for sustainability: A systematic review and future research directions. *Sustainable Development*, *32*(1), 662–681.

Battini, D., Calzavara, M., Persona, A., & Sgarbossa, F. (2017). Additional effort estimation due to ergonomic conditions in order picking systems. *International Journal of Production Research*, *55*(10), 2764–2774.

Battini, D., Glock, C. H., Grosse, E. H., Persona, A., & Sgarbossa, F. (2016). Human energy expenditure in order picking storage assignment: A bi-objective method. *Computers & Industrial Engineering*, *94*, 147–157.

Botti, L., Galizia, F. G., Mora, C., & Zecchi, G. (2020). A thorough investigation on pushing activities in industry: The impact of the variation in the speed of motion and load conditions on initial and sustained forces. *Applied Ergonomics*, *85*, 103080

Calzavara, M., Glock, C. H., Grosse, E. H., Persona, A., & Sgarbossa, F. (2016). Models for an ergonomic evaluation of order picking from different rack layouts. *IFAC-PapersOnLine*, *49*(12), 1715–1720.

Calzavara, M., Glock, C. H., Grosse, E. H., Persona, A., & Sgarbossa, F. (2017). Analysis of economic and ergonomic performance measures of different rack layouts in an order picking warehouse. *Computers & Industrial Engineering*, *111*, 527–536.

Calzavara, M., Glock, C. H., Grosse, E. H., & Sgarbossa, F. (2019). An integrated storage assignment method for manual order picking warehouses considering cost, workload and posture. *International Journal of Production Research*, *57*(8), 2392–2408.

Chan, F. T., & Chan, H. K. (2011). Improving the productivity of order picking of a manual-pick and multi-level rack distribution warehouse through the implementation of class-based storage. *Expert Systems with Applications*, *38*(3), 2686–2700.

Dai, B., Jin, S., Ning, X., & Mirka, G. A. (2010). The effects of horizontal load speed and lifting frequency on lifting technique and biomechanics. *Ergonomics*, *53*(8), 1024–1032.

Daria, B., Martina, C., Alessandro, P., & Fabio, S. (2015). Linking human availability and ergonomics parameters in order-picking systems. *IFAC-PapersOnLine*, *48*(3), 345–350.

De Koster, R. B., Stam, D., & Balk, B. M. (2011). Accidents happen: The influence of safety-specific transformational leadership, safety consciousness, and hazard reducing systems on warehouse accidents. *Journal of Operations Management*, *29*(7–8), 753–765.

De Koster, R., Le-Duc, T., & Roodbergen, K. J. (2007). Design and control of warehouse order picking: A literature review. *European Journal of Operational Research*, *182*(2), 481–501.

De Lombaert, T., Braekers, K., De Koster, R., & Ramaekers, K. (2023). In pursuit of humanised order picking planning: Methodological review, literature classification and input from practice. *International Journal of Production Research*, *61*(10), 3300–3330.

De Vries, J., De Koster, R., & Stam, D. (2016a). Aligning order picking methods, incentive systems, and regulatory focus to increase performance. *Production and Operations Management*, *25*(8), 1363–1376.

De Vries, J., De Koster, R., & Stam, D. (2016b). Safety does not happen by accident: Antecedents to a safer warehouse. *Production and Operations Management*, *25*(8), 1377–1390.

Diefenbach, H., & Glock, C. H. (2019). Ergonomic and economic optimization of layout and item assignment of a U-shaped order picking zone. *Computers & Industrial Engineering*, *138*, 106094.

Elbert, R. M., Franzke, T., Glock, C. H., & Grosse, E. H. (2017). The effects of human behavior on the efficiency of routing policies in order picking: The case of route deviations. *Computers & Industrial Engineering*, *111*, 537–551.

Gladysz, B., Tran, T. A., Romero, D., van Erp, T., Abonyi, J., & Ruppert, T. (2023). Current development on the Operator 4.0 and transition towards the Operator 5.0: A systematic literature review in light of Industry 5.0. *Journal of Manufacturing Systems*, *70*, 160–185.

Glock, C. H., Grosse, E. H., Abedinnia, H., & Emde, S. (2019). An integrated model to improve ergonomic and economic performance in order picking by rotating pallets. *European Journal of Operational Research*, *273*(2), 516–534.

Glock, C. H., Grosse, E. H., Neumann, W. P., & Feldman, A. (2021). Assistive devices for manual materials handling in warehouses: A systematic literature review. *International Journal of Production Research*, *59*(11), 3446–3469.

Gajšek, B., Vujica Herzog, N., Butlewski, M., & Durkic, G. (2017). Research opportunity: Incorporation of human factors in order picking system models. *Sci J Poznan Univ Technol Ser Organ Manag, 72*, 45–61.

Granotto, G. F., Sgarbossa, F., Glock, C. H., & Grosse, E. H. (2019). The effect of worker fatigue on the performance of a bucket brigade order picking system. *IFAC-PapersOnLine*, *52*(13), 2195–2200.

Grosse, E. H., & Glock, C. H. (2013). An experimental investigation of learning effects in order picking systems. *Journal of Manufacturing Technology Management*, *24*(6), 850–872.

Grosse, E. H., Glock, C. H., & Jaber, M. Y. (2013). The effect of worker learning and forgetting on storage reassignment decisions in order picking systems. *Computers & Industrial Engineering*, *66*(4), 653–662.

Grosse, E. H., Glock, C. H., & Neumann, W. P. (2017). Human factors in order picking: A content analysis of the literature. *International Journal of Production Research*, *55*(5), 1260–1276.

Grosse, E. H. (2024). Application of supportive and substitutive technologies in manual warehouse order picking: A content analysis. *International Journal of Production Research*, *62*(3), 685–704.

Gruchmann, T., Mies, A., Neukirchen, T., & Gold, S. (2021). Tensions in sustainable warehousing: Including the blue-collar perspective on automation and ergonomic workplace design. *Journal of Business Economics*, *91*(2), 151–178.

Helm, M., Malikova, A., & Kembro, J. (2024). Rooting out the root causes of order fulfilment errors: A multiple case study. *International Journal of Production Research*, *62*(11), 3853–3871.

Hilmola, O. P., & Tolli, A. (2016). Warehouse layout implications on picking distance: Case of human factor. *World Review of Intermodal Transportation Research*, *6*(1), 43–58.

Hofstra, N., Petkova, B., Dullaert, W., Reniers, G., & De Leeuw, S. (2018). Assessing and facilitating warehouse safety. *Safety Science*, *105*, 134–148.

Horváthová, B., Dulina, Ľ., Čechová, I., Gašo, M., & Bigošová, E. (2019). Data collection for ergonomic evaluation at logistics workplaces using sensor system. *Transportation Research Procedia*, *40*, 1067–1072.

Jiang, M., Leung, K. H., Lyu, Z., & Huang, G. Q. (2020). Picking-replenishment synchronization for robotic forward-reserve warehouses. *Transportation Research Part E: Logistics and Transportation Review*, *144*, 102138

Koreis, J., Loske, D., & Klumpp, M. (2025). Together, we travel: Empirical insights on human-robot collaborative order picking for retail warehousing. *The International Journal of Logistics Management*, *36*(1), 1–20.

Lambert, D., Stock, J. R., & Ellram, L. M. (1998). *Fundamentals of Logistics Management*. McGraw-Hill/Irwin.

Larco, J. A., De Koster, R., Roodbergen, K. J., & Dul, J. (2017). Managing warehouse efficiency and worker discomfort through enhanced storage assignment decisions. *International Journal of Production Research*, *55*(21), 6407–6422.

Le-Duc, T., & De Koster, R. (2005). *Determining number of zones in a pick-and-pack order-picking system* (No. ERS-2005-029-LIS).

Modgil, S., Singh, R. K., & Agrawal, S. (2023). Developing human capabilities for supply chains: An industry 5.0 perspective. *Annals of Operations Research*, *25*(2), 1–31.

Napolitano, M. (2012). 2012 warehouse/DC operations survey: mixed signals. *Logistics Management (Highlands Ranch, Colo.: 2002)*, *51*(11), 48–56.

Pasparakis, A., De Vries, J., & De Koster, R. (2023). Assessing the impact of human–robot collaborative order picking systems on warehouse workers. *International Journal of Production Research*, *61*(22), 7776–7790.

Petersen, C. G. (1997). An evaluation of order picking routing policies. *International Journal of Operations & Production Management, 17*(11), 1098–1111.

Petersen, C. G., Siu, C., & Heiser, D. R. (2005). Improving order picking performance utilizing slotting and golden zone storage. *International Journal of Operations & Production Management, 25*(10), 997–1012.

Reif, R., Günthner, W. A., Schwerdtfeger, B., & Klinker, G. (2010, March). Evaluation of an augmented reality supported picking system under practical conditions. In *Computer Graphics Forum* (Vol. 29, No. 1, pp. 2–12). Oxford, UK: Blackwell Publishing Ltd.

Reif, R., & Walch, D. (2008). Augmented & virtual reality applications in the field of logistics. *The Visual Computer*, *24*, 987–994.

Roodbergen, K. J., Vis, I. F., & Taylor Jr, G. D. (2015). Simultaneous determination of warehouse layout and control policies. *International Journal of Production Research*, *53*(11), 3306–3326.

Setayesh, A., Grosse, E. H., Greig, M. A., Glock, C. H., & Neumann, W. P. (2024). Evaluating usability, functionality, and usefulness of the warehousing error prevention tool. *International Journal of Production Research*, *62*(5), 1633–1647.

Sun, Y., Zhao, N., & Lodewijks, G. (2021). An autonomous vehicle interference-free scheduling approach on bidirectional paths in a robotic mobile fulfillment system. *Expert Systems with Applications*, *178*, 114932.

Teck, S., & Dewil, R. (2022). A bi-level memetic algorithm for the integrated order and vehicle scheduling in a RMFS. *Applied Soft Computing*, *121*, 108770.

Tompa, E., Dolinschi, R., De Oliveira, C., Amick, B. C., & Irvin, E. (2010). A systematic review of workplace ergonomic interventions with economic analyses. *Journal of occupational rehabilitation*, *20*, 220–234.

Vijayakumar, V., & Sobhani, A. (2023). Performance optimisation of pick and transport robot in a picker to parts order picking system: a human-centric approach. *International Journal of Production Research*, *61*(22), 7791–7808.

Zhang, M., Grosse, E. H., & Glock, C. H. (2022). Potential of mobile applications in human-centric production and logistics management. *IFAC-PapersOnLine*, *55*(10), 151–156.

Zhang, Y., & Li, Y. (2023). Society 5.0 versus Industry 5.0: An examination of industrialization models in driving sustainable development from a normative stakeholder theory perspective. *Sustainable Development*, *31*(5), 3786–3795.

Zhao, X., Liu, N., Zhao, S., Wu, J., Zhang, K., & Zhang, R. (2019). Research on the work-rest scheduling in the manual order picking systems to consider human factors. *Journal of Systems Science and Systems Engineering*, *28*, 344–355.

Zou, B., Xu, X., & De Koster, R. (2018). Evaluating battery charging and swapping strategies in a robotic mobile fulfillment system. *European Journal of Operational Research*, *267*(2), 733–753.

# 14 Strategic Orientation for Corporate Success

## *Innovating Human Factors and Ergonomics in Robotics for Industry 4.0*

*Archana Barman, Nilakshi Teron, and Purna Prasad Sharma*

## 14.1 INTRODUCTION

The fourth industrial revolution (4.0) was built based on the technologies of its predecessor i.e. the third manufacturing swirl, resulting in producing data through the deployment of transistors, sensors, and microelectronics. The phrase "Industry 4.0" initially applied at the 2011 Hannover Messe (Hannover Fair) in Germany (Kagermann et al., 2011). With a focus on robotization, interrelatedness and real-time process control, manufacturing industries are digitally transformed with the help of digital technologies like the Internet of Things (IoT), Machine Learning (ML), Artificial Intelligence (AI), Cyber-Physical Systems (CPS), Cloud computing, Additive Manufacturing (AM), Digital twins, Cyber security, and so forth to discourse and guide each other (Mayer & Schlick, 2012; Rojas et al., 2019). Industry 4.0 is a setting in which computerized and physical environments are ultimately interconnected to enable customized production under different constraints (Magd et al., 2022; Khan & Ansari, 2024; Ansari et al., 2024). It establishes a linkage between every component of the production system to design a flexible and dynamic self-managed production system. Industry 4.0 becomes vital for the significant success of the corporate world in this dynamic and competitive world as it aims to enhance productivity by reducing operational costs and enabling quicker decision-making through data analysis (Magd et al., 2022; Khan & Ansari, 2024; Ansari et al., 2024).

## 14.2 REVIEW OF LITERATURE

The digital revolution and its effect on operations are commonly perceived by experts as a mysterious and complex phenomenon. Organizations and procedures have to be fundamentally modified as a result of the industrial change brought about by paradigmatic and technological forces (Matt et al., 2015) as well as human labour (Kadir et al., 2019).

 DOI: 10.1201/9781003532293-16

Human factors (HF) observation is considered important in this context, although HF has been proven to be vital in four of the eight of "I-4.0" developmental priorities (i.e. handling complicated systems, safety and surveillance, work organization and design, and training and executive development) arranged in the seminal "I4.0"-report (Kagermann et al., 2013). Almost every aspect of human life, work and interaction is changing rapidly due to I4.0 and technology advancement. As a result of these, dramatic changes occurring in the passage of human work are assembled and executed. Using mobile devices to improve workflow and support the workers, such as during maintenance or in order picking are noteworthy instances. Collaborative robots help assembly workers, and cashing empowers production process and workers of logistics, and the solutions of cloud-based software improve the performance of the business enterprises that are rapidly developing. These are just a couple of examples of interaction amid humans and I4.0 mechanisms in business' revolution, nevertheless, it also highlights numerous ways of interactions between humans. Human determinants and workplace efficiency are the knowledge, abilities, and skills; both operators and production and technology experts play a key role in ensuring and optimizing fluid and safe workflows (Raja & Muthuswamy, 2022; Wang & Haghighi, 2016). It is quite evident that it is important to have a deeper understanding about the human factors while designing new technologies, production processes and products for better communication between different players. To meet the challenges, industrial design and management processes should incorporate the engineering of HF/E theories and concepts. This necessitates greater understanding of the complexity of human beings and the interfaces between them in work systems, in addition to the output products and production systems (Raja & Muthuswamy, 2022; Wang & Haghighi, 2016).

This chapter discusses various aspects of Ergonomics and human factor in robotics in "Industry 4.0".

The objective of this study is:

**RO1**: To examine whether robotics is assisting industries in achieving the goals of the successor of Industry 3.0 aligning human elements and ergonomics.

**RO2**: To comprehend the applications of robotics in various industries while taking the human factor into consideration.

The purpose is to provide a glimpse of Ergonomics, human factor, "Industry 4.0" and its understanding on Robotics in different sectors as well as to pinpoint challenges. This study also focuses on the role that implementation of robotics is playing to achieve the objectives of Industry 4.0.

## 14.3 METHODOLOGY

To investigate Innovating "Human Factors and Ergonomics" for "Industry 4.0", a need was developed to grasp the concept of the newly developed technology along with human factor and robotics. For research purpose, secondary data source like Scopus-indexed journals, IEEE journals and research papers from Google Scholar have been considered. Boolean operators have been used to merge the

**TABLE 14.1**
**Briefing of the Methodical Literature Analysis**

| Phase | Steps Followed |
|---|---|
| **Phase 1 and Phase 2 (searching, analysis)** | **Digital directories** such as Scopus, IEEE, Google Scholar is used during the time frame of 2015–2022 |
| | **Admittance principles** Papers with title, abstract which is similar to the goal of research have been initially gathered and, after reviewing the whole contents of the article, these had been appraised to consider ultimate insertion in the directory |
| | **Elimination principle:** Articles and conference papers of different languages instead of English language, as well as those which are not related to the theme of the chapter, are eliminated |
| | **Strings of searching** "industry 4.0", "robotics" as well as other combinations with: "Ergonomics", "human factor", "augmented reality", "Big Data Analysis", etc. have been used for searching |
| **Phase 3 Assessment of outcomes** | **Analysis stage**: Iterative gathering of articles within the directory |
| **Phase 4 Reporting of outcomes** | **Synthesis stage:** Applicable contents are extracted from directory and explained |

exploratory keywords like "industry 4.0", "robotics", "ergonomics", and "human factor". Additionally, the application of key concepts that are identified against the initial research expedition, keyword "augmented reality". "Big Data Analysis" are explored. A total number of 45 papers which are published between 2015 and 2022 were collected. After collecting and assessing all the articles that matched the study's objectives in terms of both title and abstract were evaluated to consider for ultimate incorporation in the directory. On the contrary, non- English language articles were excluded. The result section describes the concept of technology and in what way human factors and robotics have a role in innovation. Table 14.1 illustrates the literature review process that has been followed.

## 14.4 RESULTS

### 14.4.1 "Industry 4.0"

The new Industrial Revolution allows affordable, intelligent, structured, personalized, and customized production. Faster processors, more intelligent devices, compact sensors, and more affordable data transmission and storage enable the product and machines to connect and learn from one another.

Nine pillars of the Fourth Industry Revolution are depicted in Figure 14.1.

### 14.4.2 Internet of Things

The IoT are the digital item, particular computer software that enables connectivity and physical objects incorporated with electronic sensors and actuators (Proctor &

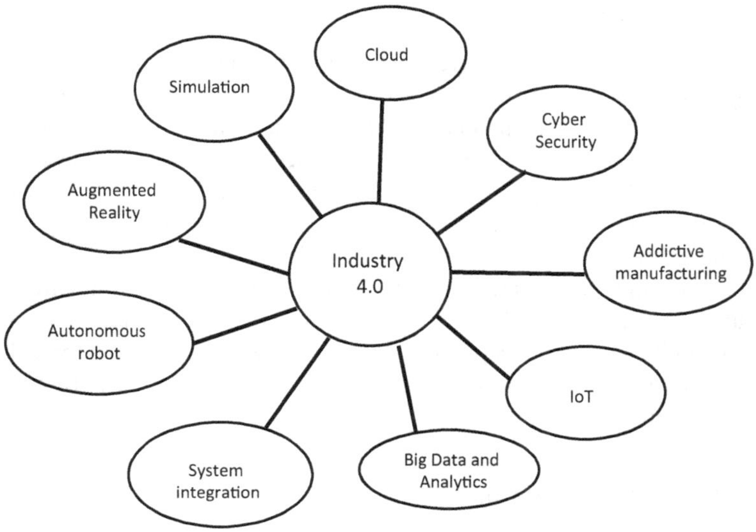


**FIGURE 14.1** Nine pillars of the Fourth Industry Revolution.

Wilkins, 2019; Tan et al., 2019); IoT have a great impact on the day-to-day behavioral life of the people who use it (Zareen & Khan, 2023). Our everyday lives are changing significantly as a result of the growing use of IoT devices and technology. The idea of smart home systems (SHS) and appliances, which include internet-based gadgets, home automation systems, and dependable energy management systems, is one example of an IoT development (Zhou et al., 2017).

### 14.4.3 Cloud and Information Technology

Cloud is a computational service that provides resources over the Internet for various purposes (Wang & Haghighi, 2016; Xu, 2012) that are accessed through cloud computing server. Depending on the need, robots can be considered as a key ingredient of Industry 4.0. For example, robotic arms, a complete assembly line, a rover that looks like a car, or patrol robots with legs or androids. It is evident how important robotics is to make food and beverages, pharmaceuticals, and chemicals (Proctor & Wilkins, 2019).

### 14.4.4 3D Printing (Additive Manufacturing)

It is a manufacturing innovation allowing for the creation of 3D solid structures by using additive or layered development structures. In the past few years, rise in demand and investment for 3D has been increasing steadily (Proctor & Wilkins, 2019). 3D printing is becoming more and more common, mostly in the medical industry. According to the study, 3D components for implants or medical devices account for roughly 11% of the medical industry's revenue (Javaid & Haleem, 2018).

### 14.4.5 Augmented Reality

Augmented reality is an interactive, reality-based display situation that uses virtual images, audio, and other foundations to improve the insight of being in the real world (Bai et al., 2020; Jonathan et al., 2024; Asim et al., 2024). Augmented reality gives access to employees the real-time information to improve workflow and decision-making. For instance, because augmented reality (AR) enables detailed visualization and object animation, it can reduce misconceptions that result from students' incapacity to visualize concepts like chemical bond-making. The ability to visualize concepts and objects at either macro or micro scales that are invisible to the human eye is another benefit of augmented reality. AR allows for the varied display of objects and concepts from various viewing angles, thereby enhancing students' comprehension of the subject matter (Cerqueira & Kirner, 2012; Jonathan et al., 2024; Asim et al., 2024).

### 14.4.6 Autonomous Robots

Autonomous robots or robotics in Industry 4.0 are used in manufacturing which works similarly to human actions (Bai et al., 2020). It enables the manufacturer to have complete control on their factory's setup and data flow (Javaid et al., 2022). Modern examples include self-driving vacuums and cars. Industrial robot arms that work on assembly lines inside factories may also be considered autonomous robots.

### 14.4.7 Big Data and Analytics

This is the way of analyzing huge amount of data when old methods of facts processing and mining are unable to expose the implication and insights that are hidden inside the data (Bai et al., 2020; Khan et al., 2022). Big Data and analytics (e.g. Data mining and machine learning) are seen as primary forces of Industry 4.0, a significant potential source of competition in future. There is a need for Big Data to create the manufacturing unit's digital twins, analytics that allows for enhanced predictive capability by detecting potential production-affected actions before they occur (Frank et al., 2019).

### 14.4.8 Simulation

Simulation mirrors the real-world systems or processes through a computer (Bai et al., 2020) in a virtual world, which involve products, human, and machines, reducing setup times for machines and improving quality. Simulations can be used in the production process to decrease production failures at the startup phase in addition to changing and shortening downtime (Vaidya et al., 2018).

### 14.4.9 System Integration

Integration is one of the major mechanisms in industrial organization (Schuh et al., 2014). In Industry 4.0, three integration dimensions are outlined: end-to-end engineering over the whole lifecycle of product; vertical integration and networked production system; horizontal integration throughout value creation job (Stock & Seliger, 2016).

### 14.4.10 Cyber Security

Cyber security in Industry 4.0 helps as a protective measure against data theft, breach, and attack (Bai et al., 2020) and saves the companies from losing their competitive edge (Lezzi et al., 2018). Protection of sensitive data, network security, and cloud security are the applications of cyber security.

### 14.4.11 Human Factor and Industry 4.0

"The International Ergonomics Association (IEA, 2000)" defined ergonomics as a scientific discipline that deals with comprehending the interaction between humans and other elements of a socio-technical system. "Ergonomics" is the study and practice of applying theory, concepts, data, and design techniques to maximize system performance and human well-being. This is primarily liable for designing, assessing jobs, tasks, settings, and systems to make sure they are appropriate for people's requirements, abilities, and limits. The word "ergonomics" has been inferred from the word "Ergon", meaning work in Greek, "Nomos" means law. Thus, to put it simply, this is the study of work from an etymological perspective. The word has a long history in European culture. Since "Human Factor Engineering" and "Human Factors" are American terminology that both pertain to the same problems, can be used interchangeably and as synonyms. This is proven by "Human Factor Society", instituted in 1957 at Tulsa (Oklahoma), which is commonly referred as "Human Factor and Ergonomics Society" (HFES). An additional phrase used in a similar aspect is "Engineering psychology" (Wickens et al., 2021).

### 14.4.12 Robotics in Industry 4.0

Robotics is an essential element, industrial robots need to be highly productive and highly flexible (Ustundag & Cevikcan, 2018). "Industry 4.0" emphasizes advancement to come up with the shortcomings faced in Industrial 3.0. Like many other fields, robotics also began to start to embrace Industry 4.0- Industrial robots in smart factories focus more and more on tasks like developing new products, identifying and resolving defects, and monitoring, analyzing, and optimizing the manufacturing process. Robots in industrial 4.0 are well equipped with advanced skills, help to work collaboratively with human workforce as well as with other robots in assembly line (Bayram & İnce, 2018).

### 14.4.13 Ergonomics and Human-Robot Collaboration

I-4.0 has significantly transformed the role played by humans in manufacturing process but the part played by humans cannot readily be replaced by advanced technologies. Moreover, the main drawback of the introduction of new technology is that it is making the manufacturing process more complicated. As a result, industry is requiring incredibly skillful and well-trained workforce. The fourth mechanical swirl regenerates a corporeal way in designing new innovations, shifting focus from replacing workers and providing better support and complement humans (ISO, 2011) describes as "scientific discipline concerned with the understanding of interactions among human and other elements of a system, and the profession that applies theory, principles, data and methods to design, for optimization of human well-being and overall system performance". This field emphasizes on designing a best workplace and equipment or products, tools for human use. According to ergonomic concepts, a number of human characteristics, including height, weight, age, hearing, and vision are taken into consideration. Because of this, ergonomics is often known as "human factor engineering". The International Ergonomics Association recognized some areas of specialization in the field of ergonomics based on the unique features and attributes of human being. The key areas of research in ergonomics are divided into areas such as physical, cognitive, and organizational ergonomics.

**Physical ergonomics** studies human anatomy, anthropometry, physiology, and biomechanics since it is concerned with person physical activity. Postures while working, material managing; redundant motions, work-related physical disorders, workplace format, security, and health are among most relevant issues studied in this field. The engineering of consumer goods might gain from the application of physical ergonomics principles, which are frequently used in construction of workplaces and industrial items.

**Cognitive ergonomics** studies how people's minds interact with those of other system components. It addresses a variety of person cognitive functions and mental activities, including perception, memory, motor response, and reasoning. Designing automated, complex, or high-tech systems should be the primary focus of cognitive ergonomics because a badly designed industrial equipment interface of an automated industrial equipment may possibly lead to a serious error in addition to lowering output and quality. Workplace stress, mental strain, human dependability, decision-making, skillful presentation, and human–machine interfaces are among the primary topics which are addressed by this area of study.

**Organizational ergonomics** focuses on optimizing socio-technical systems to maximize efficiency in terms of their structures, procedures, and policies. In order to safeguard employees' dedication and job satisfaction, it creates a completely integrated work system. Thus, organizational ergonomics covers teamwork, community ergonomics, collaborative task, new work prototype, virtual organizations, telework, standard management, transmission, crew resource management, working time design etc.

## 14.5 UNLOCKING THE STRATEGIC INNOVATION POTENTIAL OF ROBOTICS IN INDUSTRY 4.0

This is the age where organizations are automating their many activities, to increase their effectiveness and fostering alliances of human–machine interaction. The business environment is rapidly changing now a days resulting in increasing competition.

It is imperative to make more intelligent judgment at the precise time, which prompts the development of more intelligent mechanisms for taking smart decisions. It has been a decade now that robotics machines are used to carry out specific jobs in the factories, while predetermined tasks has been assigned human-machine collaboration such as examining the standard of a product and to reject the products if it is a defective one. The role of automated machines cannot be ignored in modern industry as it has the capability to complete their tasks smartly focusing on protection, adaptability and collaboration. In the next few years from what it is today, the society will be different due to the extensive usage of robots. AI and Robotics are the main industrial mechanisms and alliances of both in "Industry 4.0" have changed the face of the manufacturing process (Mia & Shuford, 2024). Industrial robots or robots, which are used for industrial applications in comparison to the medical and military fields, observe a slow development in the advancement of artificial intelligence. Intelligent industrial robots require trained specialists and standardized methods for predictive and remedial maintenance (Tantawi et al., 2019). Collaborative robots are used to implement automation to increase production and for the overall growth of economy. Big data, AI, cloud computing, the application and creation of robots in industry are examples of emerging technology. Technological development is not only bringing changes in production but also sales and marketing are also changing. For e.g. drones that Autodesk designs and builds as the capability to deliver the customer more quickly.

Real-time data and robotics is an important element of the industrial revolution which has the market opportunity for intelligent machines such as expert systems, autonomous robots, and digital assistants. As per the latest report by Technavio Research, the growth of Robotics is increasing tremendously. The objectives of I-4.0 for robotics are essentially flourished to smart industry, where items can be navigated through the production process on their own and is able to establish alternatives in event of any hindrance. Internet-based digital technologies are used widely in today's business environment. The strategical innovation technologies such as ICT mechanisms, sensor and its application in robotic innovation is leading augmentation of modern automated machine capable of collaboratively working with the workers in a user-friendly manner. The key component of industrial revolution is nothing but Robotics. Robots are more productive because they use sensors and actuators to generate robust and precise data. The fundamental fact is that robot will not be able to survive until it is included to network system using the newest technologies. Smart factories will be widely used near future. Fact and the evidence suggest that all the factories in future would be fully automatic and robotized. Factories will be converted into extremely productive manufacturing hub with machines at various cycles of production and logistics networks that will continuously share information without a great deal of human interference (Bartodziej, 2017).

In organization nowadays, COBOT, i.e., collaborative robots are used. The ideas behind designing such a robot are to create designs that safely include human factors. The use of sensors and different software in mobile robots can identify their surroundings. Mobile robotics comes in two categories: Autonomous and Non-Autonomous. Most robots are semi-autonomous and are partially independent. But they cannot do anything alone, to perform there is a need to get the command from human expert or external control system.

Autonomous Mobile Robots are mobile robots that are capable of independent study of their surroundings and don't require external guidance. Examples of popular autonomous robots are Pet Robots, GameBot, Delivery robot, and Space Robots, whereas non-autonomous robots require an external guidance system to move and to explore the surroundings. People often refer to these as controllable robots. For instance, a robot that manufactures automobile is programmed to place parts on an assembly line; yet it needs direction or command to know exactly when and where to fix that component in production.

### 14.5.1 Robotics Application in Industry 4.0

The advancement of current manufacturing standard will raise a new lifestyle for coming generations and will raise the economic development all over the globe. In "Industry 4.0", robots and human collaborate using smart sensor, and human-machine interfaces. The best example in manufacturing industries is autonomous robots. In today's world where consumer is considered as king, the primary goal is to offer more values within very less time. Faulty components that are released in the market can lower down trustworthiness which can adversely affect lowering sales of the organization. Automation has usages in the various fields of industries.

#### 14.5.1.1 Manufacturing

It has been observed that manufacturing industry faces various labor-related issues. The main issues are running cost as well as the non-availability of skilled labors. The potential solution to cater to this problem is nothing but automation. For example, in the manufacturing industry, heavy lifting is required, and it is troublesome for workers to lift such things as it is dangerous. This problem can be solved by some automotive robot to relieve the worker. For instance, a robot has the capability to lift a heavy material such as car's interior-finishing materials. According to experts, the machines should be designed to do the tasks that are physically challenging or hazardous for people, such as planting hybrid batteries that are heavier than 100 kilos. A great contribution of Industry 4.0. is "Smart factory." It is a production facility having all the equipment digitally networked into a single system in addition to being automated (Wickens et al., 2021).

#### 14.5.1.2 Autonomous Vehicles

It has been observed that more and more traffic accidents are happening due to the increasing population. The priority is to reduce road accidents. Self-governing vehicles are used by various sectors like manufacturing, mining, etc. Automated guided vehicles (AGVs) are mainly used to deliver the material from the production house to a storehouse. In the era of current industrial reformation, AGVs are shifting their focus across mechanization towards intelligence. Using a different type of technology such as Wi-Fi, laser navigation, an AGV manufacturer has been able to design a robot which is so intelligent that it can elevate and park a car within 120 s.

#### 14.5.1.3 Health Care

Healthcare industry is taking the benefit of robotics to cut down the cost, to increase the operator safety, to enhance the quality in compounding medications. To sustain cutting-edge competition as well as to train the operation to maximize the production, it is an utmost necessary for the healthcare industry to adopt robotics technology. There is a need for error-free robots to carry out tasks like a septic fill finish, where a single mistake or error of human being can ruin an expensive batch of drugs and could cost a lot of money. AURA Robot is constructed, and it got initiated from a traditional robot, in which standard state of devices are still present, but automatic and remote forms are broken up into three other non-selectable states: automatic or remote form, the laser scanner. Robots are accomplishing meticulous incision and making them less costly for their patients. One of the benefits of using robotics is that it can work with hazardous chemicals to humans and can bring greater level of efficiency in operations (Neumann & Village, 2012)

#### 14.5.1.4 Packaging

Computerized systems and augmentation in data stream means that packaging in production can blend, affix key activities such as packaging production needed blueprint, manufacturing, disposition, conservation—all into an exclusive unified approach, instead of relying on separate pieces of automation. Most vital influence of newest robotic platforms will be utilized for transforming and distribution. A computerized application utilized in food industry is Food packaging. For packing different food items robots can be used. Robots can optimize productivity, significantly reducing the time it takes to package a product, even when working alongside human workers. With the help of robots, it becomes easy to carry products, choose and position the food items, load and palletize in an efficient way (Kadir et al., 2019). Robots are designed in such a way that precise amount of food can be filled in packaging modules, whereas this is not at all feasible for a human employee who is allocated for the job.

#### 14.5.1.5 In Mining

The advancement of robotics technology is also adding value to the Mining Industry. To drill in deeper earth robotics is used. It was very difficult as well as dangerous for human to go into deeper earth to get information which has been solved by robotics without any hazard to humans. Nowadays, to inspect the crude oil and gas lines in certain disputable areas, there is a need to use autonomous equipment as it is very hazardous for humans. In fact, to understand the sub-sea conditions, autonomous underwater vehicles have been used by Schlumberger without any assistance from the local team. Remote Operational Vehicles which are used in mining industry are self-powered as well as self-driven. Rio Tinto (Henning et al., 2013), one of the biggest mining firms in the world, makes use of immense quantity of automobiles in Western Australia at its ore mines.

#### 14.5.1.6 Military and Public Safety Industry

Automation has increased the safety standards for publics and military industry is taking advantage of this technology. Drones are used to have the continuous surveillance and to safe the military unit. Robots are also used in military industry to get the support in battlefield. These types are designed in such a way that adequate amount of battle material can be put into and can sense the environment for any danger. As a result of rapid advancement in innovation, the Military sector and Public Safety Industry utilized robots for guard duty, surveying, and battlefield assistance and drones. These robots (drones) are well armed with essential battle material and are competent to free the surrounding from any danger; to ensure the safety of public, remote-controlled drones (Xu, 2012) have been used to offer situational assessments in real time, monitor dangerous situations and to evaluate risk.

#### 14.5.1.7 Supermarket and Malls

Robotics are also used in different supermarkets and malls to increase the efficiency as well as to satisfy the customers. For example, Walmart is taking the advantage of using the robotics technology to place online orders for its products in the event they run out of stock. Moreover, machines are also used to wash floors. The instruments are using sensors to examine individuals (Wang & Haghighi, 2016). Robots can also be used as a shopping assistant assisting buyers to locate items, recover products placed on shelves that they can't reach.

#### 14.5.1.8 Agriculture

In agriculture sector, to increase productivity and to get guidance on the structure of different soil and which soil can be used for which crops, a self-driven robotic tractor is used. The sensors that are used in this type of robot can give information on soil conditions and necessary measures that need to be taken care for the planted crop.

#### 14.5.1.9 3D Printing in the World of Furniture

Ashley furniture makes use of robotics and three-dimensional mechanism to expand furniture creation at low labour merchandise; rather than collecting the many tiny parts necessary to equip robots with adequate tools to make a certain kind of fixtures, Ashley has created those parts in-house in real time and reformed to 3D Printing.

#### 14.5.1.10 Robots in Music

A four-armed robot called 'Shimon' has been constructed by Georgia Institute of Technology and is apt to pay attention to chords. To a greater extent, it can extemporize and play with human musicians in musical shows (Mia & Shuford, 2024). At Georgia Tech's Centre for Melody Technology, Gil Weinberg, the lead researcher, and his team have used Shimon to represent their 12 years of experiments. The main objective was to improve the creative ability of a person with robotics. Shimon developed the ability to master music theory, melodic notes and styles. It is beyond the reach of human ability as robots can add attributes musical performances as well as it can play chord structure.

## 14.6 DISCUSSION AND PROSPECTS

The purpose of this chapter was to describe the innovations that are required in a "smart factory." To that end, systematic literature review was conducted to define the innovation in the technological industry. With the introduction of the "nine pillars" techniques, Industry 4.0 has brought about radiant changes to manufacturing systems, processes, and maintenance operators now require new capabilities. Although the intricacy and dynamism in the process of manufacturing have been increased in this digitalized world, industry can take benefit from technologies like augmented reality (AR) and CPSs that enable real-time feedbacks and more effective training. It seems necessary to have a hands-free technology.

Modern ergonomics is an established field in which experts from several disciplines collaborate to build socio-technical systems with the shared objective of ensuring that they meet human requirements and promote wellness. Since the introduction of Industry 4.0 in 2011, almost every industry has been impacted. In order to facilitate this technological revolution, numerous countries have begun implementing their own policies. Still, we are just at the start of this change. The concepts of Industry 4.0 have been studied in this paper. The industry of the future will benefit from the combined use of robotics and human interaction to help make complex decisions earlier by anticipating risks. Their mutual control over one another is another important characteristic. Whether robotics will eventually replace humans in the work force or eliminate human employment entirely is now the most crucial subject.

## 14.7 CONCLUSION

Industry 4.0, the revolutionary beginning, is portrayed by the incorporation of electronic mechanisms, automation, and material exchange in production. Creative applications such as automatic technology have allowed these robotic systems to exhibit unparalleled levels of flexibility, effectiveness, and self-sufficiency. These systems improve industrial processes' responsiveness, flexibility, and efficiency by utilizing cutting-edge mechanisms. As an outcome of new industrial revolution 4.0, the industrial system's human center is shifted, and production processes are reorganized. Because of these ideal settings, industrial collaborative robots will learn a great deal, prove the value of trustworthy and safe cooperation, and carry out tasks more productively and efficiently. As the importance of collaborative robotics is increasing, industry is trying to convert manual workplaces into collaborative workspaces while also evaluating potential cost, time, and ergonomic improvements. Because of this, man, and robot will work closely together to demonstrate the industry's progress in collaborative robotics and will become increasingly important to evaluate ergonomics in collaborative applications, including both physical and cognitive ergonomics, and new collaborative workplaces should take this into account.

## REFERENCES

Ansari, M. S. A., Khan, S. A., & Bhuyan, U. (2024). Exploring the strategic orientation factors influencing the organizational performance through bibliometric analysis. In Umair, S., Waqas, U., Mrugalska, B., & Al Shamsi, I. R. (Eds.), *Human Perspectives of Industry 4.0 Organizations* (pp. 68–88). CRC Press.

Asim, Z., Vasudevan, A., Khan, S. A., Mahmood, M., Yosof, Y., & Muniyanayaka, D. (2024). Beyond the screen: Ethical dilemmas in the XR frontier. In Kajla T., Kansra P., & Singh N. (Eds.), *Multidisciplinary Applications of Extended Reality for Human Experience* (pp. 15–43). IGI Global. https://doi.org/10.4018/979-8-3693-2432-5.ch002

Bai, C., Dallasega, P., Orzes, G., & Sarkis, J. (2020). Industry 4.0 technologies assessment: A sustainability perspective. *International Journal of Production Economics*, 229, 107776. https://doi.org/10.1016/j.ijpe.2020.107776

Bartodziej, C. J. (2017). *The Concept of Industry 4.0: An Empirical Analysis of Technologies and Applications in Production Logistics*. Springer Fachmedien Wiesbaden GmbH. https://doi.org/10.1007/978-3-658-16502-4

Bayram, B., & İnce, G. (2018). Advances in robotics in the era of industry 4.0. In *Springer* Series in Advanced Manufacturing (pp 187–200), Springer, Cham. https://doi.org/10.1007/978-3-319-57870-5_11

Cerqueira, C. S., & Kirner, C. (2012). Developing educational applications with a non-programming augmented reality authoring tool. In *Proceedings of World Conference on Educational Multimedia, Hypermedia and Telecommunications* (pp. 2816–2825). Association for the Advancement of Computing in Education (AACE), Denver.

Frank, A. G., Dalenogare, L. S., & Ayala, N. F. (2019). Industry 4.0 technologies: Implementation patterns in manufacturing companies. *International Journal of Production Economics*, 210. https://doi.org/10.1016/j.ijpe.2019.01.004

Henning, K., Wolfgang, W., &Johannes, H. (2013). Recommendations for *Implementing* the *Strategic Initiative* INDUSTRIE 4.0 April 2013 *Securing the Future of German Manufacturing Industry Final Report of the INDUSTRIE 4.0 WorkingGroup Forschungsunion*. Forschungsunion, Acatech, April.

IEA (2000). *International Ergonomics Association: Triennial report*. Santa Monica, CA: IEA Press.

International Organization for Standardization (ISO). (2011). *Ergonomics—General Approach, Principles and Concepts (ISO Standard No. 26800)*. https://www.iso.org/standard/42885.html.

Javaid, M., & Haleem, A. (2018) Additive manufacturing applications in medical cases: A literature-based review. *Alexandria Journal of Medicine*, 54, 411–422.

Javaid, M., Haleem, A., Singh, R. P., Suman, R., & Gonzalez, E. S. (2022). Understanding the adoption of Industry 4.0 technologies in improving environmental sustainability. *Sustainable Operations and Computers*, 3. https://doi.org/10.1016/j.susoc.2022.01.008

Jonathan, H., Magd, H., & Khan, S. A. (2024). Artificial intelligence and augmented reality: A business fortune to sustainability in the digital age. In Singh, N., Kansra, P., & Gupta, S. L. (Eds.), *Navigating the Digital Landscape: Understanding Customer Behaviour in the Online World* (pp. 85–105). Emerald Publishing Limited. https://doi.org/10.1108/978-1-83549-272-720241005

Kadir, B. A., Broberg, O., & da Conceição, C. S. (2019). Current research and future perspectives on human factors andergonomics in Industry 4.0. *Computers and Industrial Engineering*, 137. https://doi.org/10.1016/j.cie.2019.106004

Kagermann, H., Lukas, W. D., & Wahlster, W. (2011). Industrie 4.0: Mit dem Internet der Dinge auf dem Weg zur 4. industriellen Revolution. VDI *N*achrichten, 13(1), 2–3.

Kagermann, H., Wahlster, W. and Helbig, J. (2013). Securing the Future of German Manufacturing Industry: Recommendations for Implementing the Strategic Initiative Industrie 4.0. Final Report of the Industrie 4.0 Working Group, Acatech— National Academy of Science and Engineering, 678 p.

Khan, S. A., & Ansari, M. S. A. (2024). Critical factors for relationship between strategic orientation, corporate success, and innovation. In Umair, S., Waqas, U., Mrugalska, B., & Al Shamsi, I. R. (Eds), *Human Perspectives of Industry 4.0 Organizations* (pp. 89–110). CRC Press.

Khan, S. A., Magd, H., & Epoc, F. (2022). Application of data management system in business to business electronic commerce. In Naim, A., & Malik, P. K (Eds.), *Competitive Trends and Technologies in Business Management.* (pp. 109–124). Nova Science Publishers.

Lezzi, M., Lazoi, M., & Corallo, A. (2018). Cybersecurity for Industry 4.0 in the current literature: A reference framework. *Computers in Industry*, 103. https://doi.org/10.1016/j.compind.2018.09.004

Magd, H., Jonathan, H., Khan, S. A., & El Geddawy, M. (2022). Artificial intelligence—The driving force of industry 4.0. In Chatterjee, J. M., Garg, H., & Thakur, R. N. (Eds.), A Roadmap for Enabling Industry 4.0 by Artificial Intelligence (pp. 1–15). Wiley. https://doi.org/10.1002/9781119905141.ch1

Matt, C., Hess, T., & Benlian, A. (2015). Digital Transformation Strategies. In *Business & Information Systems Engineering*. (pp. 339–343). Springer. https://doi.org/10.1007/s12599-015-0401-5

Mayer, M. P., & Schlick, C. C. (2012). Improving operator's conformity with expectations in a cognitively automated assembly cell using human heuristics. In Trzcielinski, S. & Karwowski, W. (Eds), *Advances in Ergonomics in Manufacturing*. CRC Press. https://doi.org/10.1201/b12322

Mia, M. R., & Shuford, J. (2024). Exploring the synergy of artificial intelligence and robotics in industry 4.0 applications. *Journal of Artificial Intelligence General Science (JAIGS)*, 1(1). https://doi.org/10.60087/jaigs.v1i1.31

Neumann, W. P., & Village, J. (2012). Ergonomics action research II: a framework for integrating HF into work system design. *Ergonomics*, *55*(10), 1140–1156. https://doi.org/10.1080/00140139.2012.706714

Proctor, M., &Wilkins, J. (2019). 4.0 Sight—*Digital Industry Around the World.* Technical report, EU Automation. https://www.4sightbook.com

Proctor, M., & Wilkins, J. (2019). *4.0 Sight—digital industry around the world.* Technical report, EU Automation. http://www.4sightbook.com

Raja Santhi, A., & Muthuswamy, P. (2022). Pandemic, war, natural calamities, and sustainability: Industry 4.0 technologies to overcome traditional and contemporary supply chain challenges. *Logistics*, 6(4). https://doi.org/10.3390/logistics6040081

Rojas, R. A., Garcia, M. A. R., Wehrle, E., & Vidoni, R. (2019). Avariational approach to minimum-jerk trajectories for psychological safety in collaborative assembly stations. *IEEE Robotics and Automation Letters*, 4(2). https://doi.org/10.1109/LRA.2019.2893018

Schuh, G., Potente, T., Wesch-Potente, C., Weber, A. R., & Prote, J. P. (2014). Collaboration mechanisms to increase productivity in the context of industrie 4.0. *Procedia Cirp*, 19, 51–56.

Stock, T., & Seliger, G. (2016). Opportunities of *sustainable manufacturing in industry* 4.0. *Procedia CIRP*, 40. https://doi.org/10.1016/j.procir.2016.01.129

Tan, Q., Tong, Y., Wu, S., & Li, D. (2019). Anthropocentric approach for smart assembly: Integration and collaboration. *Journal of Robotics*, 2019(1), 3146782.

Tantawi, K. H., Sokolov, A., & Tantawi, O. (2019). Advances in industrial robotics: From industry 3.0 automation to industry 4.0 collaboration. In *TIMES-ICON2019-2019 4th Technology Innovation Management and Engineering Science International Conference*. https://doi.org/10.1109/TIMES-iCON47539.2019.9024658

Ustundag, A., & Cevikcan, E. (2018). Industry *4.0: Managing the digital transformation.* Springer Series in Advance Manufacturing. Springer Cham. https://doi.org/10.1007/978-3-319-57870-5

Vaidya, S., Ambad, P., & Bhosle, S. (2018). Industry 4.0-A glimpse. *Procedia Manufacturing*, 20. https://doi.org/10.1016/j.promfg.2018.02.034

Wang, L., & Haghighi, A. (2016). Combined strength of holons, agents and function blocksin-cyber-physical systems. *Journal of Manufacturing Systems*, 40. https://doi.org/10.1016/j.jmsy.2016.05.002

Wickens, C. D., Helton, W. S., Hollands, J. G., & Banbury, S. (2021). *Engineering Psychology and Human Performance*. Routledge. https://doi.org/10.4324/9781003177616

Xu, X. (2012). From cloud computing to cloud manufacturing. *Robotics and Computer-Integrated Manufacturing*, 28(1). https://doi.org/10.1016/j.rcim.2011.07.002

Zareen, S. & Khan, S. A. (2023). Exploring dependence of human resource management (HRM) on internet of things (IoT) and digital marketing in the digital era. In Naim A., & Devi V. (Eds.), *Global Applications of the Internet of Things in Digital Marketing* (pp. 51–66). IGI Global. https://doi.org/10.4018/978-1-6684-8166-0.ch004

Zhou J., Cap Z., Dong X., & Vasilakos A. V. (2017). Security and privacy for cloud-based IoT: Challenges. *IEEE Communications Magazine*, 55(1):26–33. https://doi.org/10.1109/MCOM.2017.1600363CM

# 15 Implication of the Dark Side of Artificial Intelligence on Corporate Success from the perspective of Human Factors and Ergonomics

*Preeti Mehra and Pooja Kansra*

## 15.1 INTRODUCTION

### 15.1.1 Background of Artificial Intelligence

Modern technology of the 21st century is transforming many facets of human life, one of which is artificial intelligence (AI). AI is essentially the creation of computer systems that are able to carry out tasks that normally require human intelligence, like learning, reasoning, problem-solving, and making decisions (Russell & Norvig, 2021). The origins of AI can be traced to innovators such as Alan Turing, who proposed the Turing Test in the mid-1900s as a way to evaluate a machine's potential for intelligent behavior (Turing, 1950). Since then, significant advancements in computing power, the inventiveness of algorithms, and the accessibility of vast amounts of data have propelled the development of AI.

The goal of machine learning is to educate computers on how to solve issues by utilizing sample data or existing expertise. Machine learning powers innovative technology like speech recognition, self-driving cars, and translation apps (Alpaydin, 2016). Deep learning is a kind of machine learning that enables computers to learn from experience and understand the world by employing a hierarchy of concepts. Because computers learn from experience, a human-computer operator does not need to explicitly communicate all the knowledge that the computer needs (Goodfellow et al., 2016). Computers can learn from experience and gain an understanding of the world through deep learning, a sort of machine learning that employs a hierarchy of concepts. Since computers pick up knowledge via experience, human computer operators do not need to explicitly teach them everything (Jurafsky & Martin, 2019). Computer vision is the study and application of teaching computers to comprehend

DOI: 10.1201/9781003532293-17

and analyze visual data. Applications such as autonomous driving, object identification, and image recognition are made possible by computer vision (Jurafsky & Martin, 2019).

In the modern world, AI has become commonplace. Its technology is used in many different sectors of the economy, such as healthcare, banking, transportation, education, and entertainment. AI-powered diagnostic tools in healthcare could evaluate imaging data, identify anomalies, and assist physicians in making previously unheard-of-accurate disease diagnoses (Esteva et al., 2019). The creation, distribution, and consumption of financial goods and services have all benefited from the tremendous growth that financial technology, or fintech, has experienced in recent years. AI algorithms are used in algorithmic trading, fraud detection, risk assessment, and strategy optimization. They also aid in loss minimization (Huang et al., 2019). AI-driven tutoring systems have the potential to enhance educational outcomes by customizing learning experiences and meeting each student's unique demands. With an impact size of 0.76, intelligent tutoring systems are nearly as successful as human tutoring (VanLehn, 2011).

### 15.1.2 Importance of AI Integration

The potential benefits of incorporating AI into several facets of society are enormous, including increased productivity, innovation, and competitive advantage. First off, AI technologies have the potential to greatly increase productivity and efficiency in a variety of industries. AI technologies can expedite decision-making, optimize operations, and cut costs by automating repetitive jobs, streamlining workflows, and extracting insights from massive datasets. Humans will benefit from amazing personal technology, state-of-the-art infrastructure, and nearly limitless access to the cultural artifacts that give our life significance when digital technologies completely take off (Brynjolfsson and Mcafee, 2017). With the constant advancement of technology, businesses can now operate more efficiently from start to end and take advantage of predictive maintenance options in addition to preventative maintenance to reduce downtime, increase safety, and increase income (Wang et al., 2016).

Furthermore, AI encourages creativity by making it possible to create original answers to challenging issues. Through the utilization of sophisticated algorithms and computational methods, scientists and inventors can address problems that were previously thought to be unsolvable. These deep neural networks are trained using an unusual blend of supervised learning from human expert games and reinforcement learning from self-play games (Silver et al., 2016). AI techniques used in drug discovery, for example, can assess chemical structures, predict drug interactions, and identify promising candidates for new treatments. These capacities will improve patient outcomes and speed up the medication development process. By developing more efficient solar panels, wind turbines, and energy storage devices with the help of AI-powered optimization algorithms, the renewable energy sector may contribute to a sustainable and environmentally friendly future. AI is able to evaluate the past, improve the present, and predict the future. AI will therefore be able to handle the majority of these issues. AI, despite its complexity, increases energy intelligence by minimizing error and aiming for higher accuracy (Chen et al., 2016).

Organizations that use AI technologies also benefit from a competitive edge in the quickly changing market environment of today. Businesses may make data-driven decisions, spot new business possibilities, and outperform rivals by utilizing AI to monitor market trends, consumer behavior, and rival strategy. The impact of user expectations on digital disruption is good but insignificant (Bughin et al., 2017). For example, AI-driven recommender systems can improve customer engagement, personalize product recommendations, and increase retail sales. In terms of cybersecurity, AI-powered threat detection systems are able to quickly detect and neutralize cyberthreats, protecting digital assets of businesses and upholding customer confidence (Shakarian et al., 2016).

### 15.1.3 Relevance to Human Factors and Ergonomics

The integration of AI technology has significant implications for optimizing human-system interactions, improving user experience, and enhancing safety and performance in the context of Human Factors and Ergonomics (HF&E). The multidisciplinary discipline of HF&E is concerned with creating environments, systems, and products that are safe, effective, and simple for human operators to operate (Wickens et al., 2021). HF&E specialists work to produce user-centric designs that minimize errors, reduce cognitive burden, and optimize usability by understanding human capabilities, limitations, and preferences (Norman & Draper, 1986).

Predictive analytics driven by AI may forecast user demands, modify system behavior in real time, and offer tailored recommendations, all of which increase user engagement and happiness (Herlocker et al., 2000). AI-driven natural language processing technologies in human-computer interaction allow for more conversational and intuitive user interfaces, increasing computer systems' accessibility and usability for people of different backgrounds and skill levels (Jurafsky & Martin, 2019).

Additionally, AI technologies are essential to the development of human-centered automation systems, which combine human and machine cooperation to safely and successfully complete complicated jobs. Our model's major evaluative criteria for automation design are the effects on human performance that specific automation kinds and levels have (Parasuraman et al., 2000). Human operators can concentrate on high-level tasks that call for human judgment and creativity by utilizing AI algorithms for intelligent decision support and adaptive automation, while AI systems take care of repetitive or routine work on their own (Rouse & Morris, 1986). This cooperative approach to automation lowers workload and weariness, improves system performance overall, and lessens the possibility of human error. However, automation of the task's implementation phase improves human operators' performance the most—but only in typical operational environments (Endsley & Kaber, 1999).

To ensure the proper integration of AI technologies in HF&E, we will examine these obstacles and ethical considerations in detail in the sections that follow. We will also offer ways for overcoming these issues.

### 15.1.4 Purpose and Scope of the Chapter

This chapter's main goal is to give readers a thorough understanding of the difficulties and moral dilemmas raised by the use of AI in HF&E. By utilizing interdisciplinary viewpoints from the domains of ethics, HF&E, and AI, we hope to dispel prevalent misconceptions, clarify the negative aspects of AI, and offer useful advice to HF&E professionals as they navigate the challenges of integrating AI. This chapter covers a broad range of subjects such as algorithm bias, privacy issues, transparency, and autonomy.

## 15.2 UNDERSTANDING THE DARK SIDE OF AI

The "dark side" of AI encompasses a wide range of moral dilemmas, prejudices, invasions of privacy, and perhaps detrimental effects on human welfare. This section explores the various facets of AI's dark side in further detail, highlighting the moral conundrums and societal ramifications that arise when AI is used in HF&E contexts.

### 15.2.1 Definition of the Dark Side of AI

Fundamentally, the term "dark side of AI" alludes to the unexpected outcomes and moral dilemmas that arise while using AI systems. These effects could include algorithmic biases, invasions of privacy, incentives that aren't aligned, greed, geopolitics, job displacement, and a decline in human autonomy and dignity, among other things (Floridi et al., 2018). In order to ensure responsible and ethical AI deployment, it is critical to comprehend and mitigate the negative aspects of AI as these technologies become more and more interwoven into daily life.

### 15.2.2 Ethical Dilemmas in AI Applications

Algorithmic prejudice is a major ethical conundrum arising from AI applications in HF&E. When AI systems behave discriminatorily and produce unjust results for particular people or groups, this is known as algorithmic bias. Vulnerable groups may become less involved in society as a result of an overreliance on data mining (Barocas & Selbst, 2016). Biased algorithms, for instance, that favor some demographic groups over others during the employment process may contribute to systemic inequality. To maintain fairness and equity in AI-driven decision-making processes, algorithmic bias must be addressed with careful attention to data collection, model construction, and continuous monitoring.

The application of AI-driven surveillance technologies, which raise serious privacy issues, presents another moral issue. AI surveillance techniques, like facial recognition and behavioral tracking tools, have the potential to violate people's right to privacy and undermine public confidence in institutions (Dwork et al., 2012). Furthermore, the widespread use of surveillance AI raises concerns about how to strike a balance between security and individual liberty by potentially suppressing free speech and social cohesiveness.

### 15.2.3 Real-World Examples of the Dark Side of AI

The detrimental consequences of AI systems on people and society are demonstrated by a number of real-world cases. In order to predict criminal activity and future crime, police agencies are using software more and more to detect potential offenders. This raises questions about the possibility that police-recorded datasets contain systemic bias that targets particular populations, as discussed by Lum and Isaac (2016). In a similar vein, AI-powered social media algorithms have come under fire for spreading false information (Floridi et al., 2018).

## 15.3 DEBUNKING MISCONCEPTIONS

Misconceptions about AI abound, particularly in the realm of HF&E. This section aims to identify and debunk common myths and misconceptions surrounding AI, providing a nuanced understanding of its capabilities and limitations in HF&E contexts. These illustrations highlight how crucial it is to take into account the societal ramifications of AI deployment as well as the possible harm it could do to populations that are at risk.

In conclusion, realizing the negative aspects of AI is crucial to identifying and reducing its ethical ramifications in HF&E settings. Organizations may work to implement AI technology responsibly and ethically by eliminating algorithmic bias, defending privacy rights, and closely analyzing real-world examples. But reaching this objective calls for interdisciplinary cooperation, stakeholder involvement, and a dedication to maintaining human-centered design principles in the creation of AI systems.

### 15.3.1 Identification of Common Myths and Misconceptions

In today's conversation, AI is a relevant issue that is frequently accompanied by many myths and misconceptions. These false beliefs can originate from a number of places, such as headlines, media depictions, and social media culture. This section will address some of the most common myths and misconceptions about AI, look at how they came to be, and offer fact-based corrections.

#### 15.3.1.1 Myth 1: AI Will Completely Replace Human Workers

The idea that AI will eventually result in widespread unemployment due to robots taking over human jobs is one of the most common fallacies regarding the technology. This misconception about the potential and constraints of AI technologies frequently gives rise to this fallacy. Although it is possible for AI to automate some jobs that have historically been done by humans, it is false to believe that AI will replace human labor entirely. According to Brynjolfsson and McAfee (2017), the enhanced "bounty" that technology offers would merely shift demand to various forms of employment, as it always has, dispelling the Luddite worry that clever machines must necessarily lower the aggregate demand for labor jobs.

AI systems are best at doing certain, well-defined tasks that are simple to automate and codify. But they frequently find it difficult to complete tasks that call for human abilities like creativity, emotional intelligence, and complicated decision-making. Because of this, it is doubtful that AI would completely automate many tasks requiring these peculiarly human skills. AI technologies, on the other hand, are more likely to be a benefit to human laborers, relieving them of menial duties and freeing them up to concentrate on more strategic, higher level work. Bughin et al. (2017) emphasize the necessity of quickening the digital transition, particularly for companies that are early adopters, as AI provides certain immediate advantages.

Furthermore, historical data indicates that, notwithstanding the obsolescence of some tasks, technological developments—including automation—have traditionally resulted in the emergence of new work opportunities. For example, completely new businesses like software development, cybersecurity, and digital marketing emerged as a result of the widespread use of computers and the internet in the late 20th century. Similarly to this, it is anticipated that the development of AI would lead to a need for new positions requiring human abilities like interpersonal communication, critical thinking, and problem-solving (Brynjolfsson & McAfee, 2017).

#### 15.3.1.2 Myth 2: AI Systems Are Completely Unbiased and Objective

The idea that AI systems make decisions in a totally impartial and objective manner is another widespread misconception regarding AI. In actuality, AI systems are frequently trained on data that replicates societal prejudices and inequities, which causes these prejudices to be reinforced in AI-driven decision-making processes (Barocas & Selbst, 2016).

The method that algorithmic bias appears can vary based on the type of data that is utilized to train AI models. For instance, a hiring algorithm may learn to reproduce biases against specific demographic groups, like women or minorities, by favoring candidates from privileged backgrounds if the historical data used to train the algorithm contains such biases (O'Neil, 2016). O'Neil calls on modelers to take more responsibility for their algorithms and on policymakers to regulate their use. Ultimately, though, we must learn more about the models that guide our lives. Similarly to this, facial recognition algorithms may find it difficult to correctly identify the faces of people with darker skin tones if they are predominantly trained on data that consists of photographs of people with lighter complexion. This could result in biased results in applications like surveillance and law enforcement.

If commercial organizations want to create facial analysis algorithms that are truly fair, transparent, and accountable, they must give immediate attention to the significant discrepancies in the accuracy of classifying darker females, lighter females, darker males, and lighter males in gender categorization systems (Buolamwini & Gebru, 2018). In order to mitigate algorithmic bias, mitigation measures must be carefully considered at every level of the AI development lifecycle, from data pretreatment and collecting to model training and evaluation. Before deploying AI systems in real-world situations, this entails making sure datasets are representative and diverse, putting fairness-aware algorithms into place, and carrying out exhaustive audits and impact assessments to find and reduce biases (Barocas & Selbst, 2016).

#### 15.3.1.3 Myth 3: AI Poses Existential Threats to Humanity

The widespread belief that AI presents existential risks to humanity gives rise to visions of superintelligent computers taking over the planet and eradicating or subjugating people. An AI capable of self-improvement could trigger an explosion of intelligence leading (perhaps quickly) to a superintelligence. Significantly, enhanced abilities might be possessed by such a superintelligence, particularly in the areas of economic production, social manipulation, strategy, and hacking (Bostrom, 2014).

It is true that AI technologies have the potential to disrupt society and raise ethical concerns, particularly as they become more advanced and autonomous. However, the idea of a rogue AI spontaneously developing malevolent intentions and launching a global takeover is largely speculative and unlikely to occur in the near future. To shape the future of AI, one must have an open mind, a solid understanding of fundamental computer science concepts, and a strong desire to learn about the latest advancements (Russell & Norvig, 2021).

Rather, more pressing and concrete problems like privacy, security, responsibility, and economic injustice are the main worries related to AI (Floridi et al., 2018). These worries are a result of the increasing pervasiveness of AI technology in many facets of daily life, from social media algorithms and monitoring to autonomous weapon systems and predictive policing tools. Fears about AI's potential effects are diverting researchers' attention from the actual risks associated with deployed systems (Crawford & Calo, 2016).

Adopting a proactive approach to AI governance that prioritizes accountability, transparency, and ethical supervision is crucial to allaying these worries. This entails creating strong legal frameworks, defining best practices and standards for the creation and application of AI, and encouraging interdisciplinary cooperation among stakeholders in the public sector, business sector, academic institutions, and civil society (Floridi et al., 2018).

#### 15.3.1.4 Myth 4: AI Technologies Are Inherently Secure and Cannot Be Exploited for Malicious Purposes

It's a common misperception that AI technologies are safe by nature and cannot be abused by bad people. Through methods like automated incident response, threat prediction, and anomaly detection, AI can improve cybersecurity defenses. However, it also creates new attack surfaces and weaknesses that adversaries can exploit. Despite its strength, the internet is not secure. If we do nothing now, the risks will increase as "smart" devices become more common (Schneier, 2018).

Furthermore, as AI technologies become increasingly linked and autonomous, they could be the target of cyberattacks meant to compromise vital infrastructure or steal confidential data (Cimpanu, 2018). Autonomous vehicles may be susceptible to cyberattacks that alter sensor data or control systems, resulting in collisions or accidents (Miller & Valasek, 2015).

AI security issues call for a comprehensive strategy that incorporates technological advancements, governmental regulations, and global collaboration. In order to foresee and prevent new dangers, this entails creating strong cybersecurity procedures and standards for AI systems, including secure-by-design concepts in AI development processes, and encouraging cooperation between researchers, industry practitioners, and legislators (Schneier, 2018).

In conclusion, promoting a more knowledgeable and nuanced understanding of AI's capabilities, limitations, and ramifications requires the identification and dispelling of prevalent myths and misunderstandings about the technology. We can make sure that AI technologies are created, implemented, and managed in a responsible and ethical manner, optimizing their advantages for society while limiting potential risks and negatives, by busting these myths and encouraging evidence-based discourse.

### 15.3.2 Addressing Job Displacement Concerns

While AI may be able to automate some work, it also opens up new avenues for employment creation and creativity (Brynjolfsson & McAfee, 2017). AI may free up humans to concentrate on higher value jobs that call for creativity, problem-solving, and interpersonal skills by enhancing human talents. Initiatives for retraining and reskilling employees can also assist them in adjusting to the shifting nature of the labor market and assuming new responsibilities in AI-driven industries.

#### 15.3.2.1 Mitigating Algorithmic Bias

The first step in reducing the influence of biases in AI systems is acknowledging their existence. Bias in AI models can be found and addressed with the use of methods like bias detection algorithms and fairness-aware machine learning (Barocas & Selbst, 2016). Furthermore, by including a variety of viewpoints and experiences in the design process, encouraging diversity and inclusivity in AI development teams can help reduce the danger of prejudice.

#### 15.3.2.2 Fostering Informed Discussions about AI Risks

It is necessary to promote educated dialogues that take into account the advantages and disadvantages of using AI in order to dispel the misconception that it poses existential concerns (Bostrom, 2014). Organizations may foster trust and confidence in AI technologies while reducing unwarranted worries and misconceptions by involving stakeholders in dialogue and encouraging transparency about AI capabilities and limitations.

#### 15.3.2.3 Enhancing AI Security Measures

In order to dispel the myth that AI technologies are inherently secure, strong cybersecurity protections must be put in place at every stage of the AI lifecycle (Goodfellow et al., 2015). This includes putting access controls and authentication procedures in place, protecting data transfer and storage, and routinely patching and updating AI systems to fix new security flaws.

In conclusion, dispelling myths around AI is crucial to promoting a fair awareness of both its advantages and disadvantages in HF&E settings. Organizations may minimize dangers and optimize the benefits of AI technologies by dispelling common misconceptions and encouraging educated dialogues.

## 15.4 ETHICAL CONSIDERATIONS IN AI DESIGN

This includes putting access controls and authentication procedures in place, protecting data transfer and storage, and routinely patching and updating AI systems to fix new security flaws. When designing, developing, and implementing AI systems, ethical issues are crucial. In order to guarantee that AI technologies prioritize human well-being, safety, and fairness, ethical challenges pertaining to AI integration into various HF&E components must be addressed. The ethical frameworks, guidelines, and tactics for encouraging moral AI creation in HF&E settings are examined in this section.

### 15.4.1 Ethical Principles for AI Design

AI design and development procedures are guided by ethical standards, which guarantee that AI technologies are consistent with social norms and moral ideals. The following fundamental ethical precepts apply to AI design in HF&E contexts:

#### 15.4.1.1 Beneficence

AI systems should aim to maximize benefits and minimize harm to individuals and society (Floridi et al., 2018). Respecting the principle of beneficence requires designing AI systems with an eye toward improving social welfare, safety, and human well-being.

#### 15.4.1.2 Non-Maleficence

AI systems ought to refrain from hurting people or having unfavorable effects on stakeholders (Brynjolfsson & McAfee, 2014).

#### 15.4.1.3 Autonomy

AI systems ought to respect and maintain people's capacity for autonomy, agency, and decision-making (Floridi et al., 2018). Respecting autonomy in AI deployment requires designing AI solutions that empower users, encourage informed consent, and allow for user transparency.

#### 15.4.1.4 Justice

In both their development and use, AI systems ought to prioritize impartiality, justice, and equity (Barocas & Selbst, 2016). Promoting justice in AI design involves addressing algorithmic prejudice, guaranteeing equitable access to AI-driven services.

#### 15.4.1.5 Transparency

AI systems ought to be accountable for their choices and acts, transparent, and explicable (Dwork et al., 2012). It is crucial to give consumers knowledge about how AI algorithms work, how they make decisions, and whether they have any biases. This will help to build responsible and trustworthy AI systems.

#### 15.4.1.6 Privacy

AI systems should respect individuals' privacy rights and uphold data protection principles (Floridi et al., 2018). Safeguarding privacy in AI design necessitates putting strong data protection safeguards in place, getting informed consent for data collection and processing, and minimizing data retention.

#### 15.4.1.7 Responsibility

The ethical and societal effects of AI systems should be the responsibility of AI practitioners and developers (Brynjolfsson & McAfee, 2014). This entails carrying out in-depth ethical analyses, involving stakeholders in the design phase, and resolving any ethical issues that may arise during the AI lifespan.

These ethical principles serve as guiding principles for AI design and development, helping AI practitioners navigate complex ethical challenges and dilemmas in HF&E contexts.

### 15.4.2 Frameworks for Ethical AI Development

To make ethical concepts in AI design and development more practically applicable, a number of frameworks and recommendations have been put out.

A set of principles and guidelines for AI practitioners and developers has been produced by the IEEE, with an emphasis on societal effect assessment, accountability, and transparency (IEEE, 2019). These guidelines provide a comprehensive framework for addressing ethical concerns in AI design and deployment. Many leading tech companies, including Google, Microsoft, and IBM, have adopted ethical AI principles and guidelines to govern their AI development practices (Google AI Principles, 2022; Microsoft AI Principles, 2022; IBM AI Ethics Principles, 2022). These principles emphasize transparency, fairness, inclusivity, and accountability in AI deployment.

Seven essential conditions for the creation of ethical AI have been outlined in the European Commission's ethics guidelines for reliable AI (European Commission, 2019). The aforementioned prerequisites encompass human agency and supervision, technological resilience and security, privacy and data management, openness, diversity, impartiality, and non-discrimination and ecological welfare and responsibility.

These frameworks assist organizations include ethical issues in their AI design and deployment processes by offering an organized approach to ethical AI development.

## 15.5 MAXIMIZING THE BENEFITS OF AI

AI presents enormous potential for improving accessibility, safety, and productivity, among other elements of HF&E. This section discusses best practices for incorporating AI technologies into human-centered design processes and showcases successful AI implementations from a variety of industries.

### 15.5.1 Enhancing Productivity with AI

By streamlining workflows, automating repetitive tasks, and producing insightful data, AI technologies present chances to boost productivity. AI-driven tools and systems in HF&E can optimize workflows and enhance task performance. AI-enabled virtual assistants, for instance, can help users manage complicated activities, make appointments, and retrieve pertinent information all of which increase productivity and efficiency.

### 15.5.2 Improving Safety and Risk Management

AI technologies are essential for enhancing risk management and safety in a number of industries, such as manufacturing, transportation, and healthcare. AI-driven decision support systems and diagnostic tools can help medical personnel diagnose conditions, forecast patient outcomes, and suggest individualized treatment regimens. In order to fully comprehend their patients, doctors will find machine learning to be an invaluable tool. Its function will only increase along with the complexity of medical technologies and patient circumstances, challenging clinical medicine to keep up with this growth (Obermeyer & Emanuel, 2016). AI systems can improve patient outcomes and diagnostic accuracy by using machine learning algorithms to evaluate medical imaging data. By doing so, they can identify abnormalities and help radiologists understand complex pictures (Esteva et al., 2019).

AI-powered traffic management and autonomous vehicle systems have the potential to improve road safety and lower accident rates in the transportation sector. With time, this new technology could significantly alter the way people travel in the United States by lowering crash rates, easing traffic, increasing fuel efficiency, reducing the need for parking, and providing mobility to those who are unable to drive (Fagnant & Kockelman, 2015). In order to save travel time and fuel consumption, AI algorithms can improve route planning, forecast traffic congestion, and evaluate real-time traffic data. Additionally, AI-powered DSS can identify possible threats, warn users of hazardous conditions, and help prevent crashes, all of which contribute to increased road safety.

AI has the ability to improve safety procedures, spot possible dangers, and reduce the likelihood of workplace mishaps and injuries in manufacturing and industrial environments. Predictive analytics systems driven by AI, for instance, can examine past safety data, spot dangerous behavior trends, and proactively put preventive measures in place to lower the risk of accidents. The AI application in industries, what Lee et al. (2018) called Industrial AI, can produce useful insights and indicators to identify flaws in industrial processes and make predictions to support automated actions and decision-making (Khan et al., 2023). Organizations can cut down on the likelihood of occupational hazards and establish safer work conditions by combining AI-driven safety monitoring systems with current safety regulations.

### 15.5.3 Enhancing Accessibility and Inclusivity

AI has the capacity to improve accessibility by offering tailored support and accommodations to people with impairments. AI-driven assistive technologies in HF&E can enable people with cognitive or physical limitations to get beyond obstacles to engagement and participation. AI-driven speech recognition and natural language processing systems, for instance, can help people with speech impairments communicate more efficiently and get information. The process of creating digital systems and services that are accessible to people with disabilities, such as those who have visual, auditory, motor, or cognitive impairments, is known as digital accessibility (Chemnad & Othman, 2024). People with disabilities could benefit from increased accessibility and an overall improvement in their quality of life if AI is implemented. In a similar vein, AI-powered motion tracking and gesture recognition technologies can help people with mobility problems operate electronics and engage with their surroundings more freely (Patel et al., 2018).

Furthermore, AI-powered tailored recommendations and adaptive interfaces can improve the user experience for people with a range of requirements and preferences. Recommendation engines driven by AI have the ability to recommend tailored information and dynamic user interfaces depending on user input and behavior, improving accessibility and user experience. Developers can now include a range of AI functionalities in user-facing systems thanks to advancements in AI (Amershi et al., 2019).

### 15.5.4 Showcase of Successful AI Applications

Several real-world examples illustrate the successful integration of AI technologies into HF&E contexts, showcasing the diverse benefits and applications of AI across different industries:

#### 15.5.4.1 Healthcare

With the help of AI, IBM's Watson Health platform analyzes scientific literature, patient data, and medical records to help medical practitioners diagnose conditions, choose the best course of treatment, and forecast patient outcomes. AI offers chances to help lower human mistake, support medical personnel and professionals, and offer round-the-clock patient care. AI has the potential to be used much more in analyzing medical photos, X-rays, and scans; diagnosing medical issues; and developing treatment plans as these tools continue to advance (IBM Watson Health, 2022).

#### 15.5.4.2 Transportation

Automated lane changes, adaptive cruise control, and lane-keeping assistance are just a few of the semi-autonomous driving features made possible by Tesla's Autopilot system, which makes use of AI-driven computer vision and machine learning algorithms (Tesla, 2022).

#### 15.5.4.3 Manufacturing

In industrial manufacturing environments, Siemens' Industrial AI platform uses AI algorithms to detect equipment breakdowns, streamline production processes, and enhance product quality. For Siemens, the goal of integrating generative AI into predictive maintenance is to provide manufacturers with real benefits in addition to technological advancements. It boosts efficiency, encourages sustainability, and quickens the organization's digital transformation by making maintenance decisions faster and simpler (Siemens, 2022).

#### 15.5.4.4 Retail

Amazon's recommendation engine makes use of AI algorithms to examine user preferences, past purchases, and browsing patterns in order to provide tailored product recommendations and improve customers' shopping experiences (Amazon, 2022).

These examples highlight the diverse applications and benefits of AI technologies in enhancing productivity, safety, and accessibility across different domains.

### 15.5.5 Best Practices for Integrating AI into HF&E

Design principles, user requirements, and ethical implications must all be carefully considered when integrating AI technologies into HF&E. Organizations can optimize the advantages of AI while maintaining ethical deployment and human-centered design by implementing many best practices:

#### 15.5.5.1 User-Centered Design

When designing and developing AI-driven systems, give users' requirements and preferences first priority. This will guarantee that user interfaces are user-friendly, easily accessible, and sensitive to user input.

#### 15.5.5.2 Ethical Considerations

Throughout the AI lifecycle, address ethical issues and potential hazards related to the deployment of AI, such as algorithmic bias, privacy violations, and user permission. Organizations need to deal with moral issues like human dignity, privacy, accountability, justice, and openness.

#### 15.5.5.3 Interdisciplinary Collaboration

Encourage cooperation between domain experts, human factors specialists, and AI developers to guarantee that AI technologies are customized to satisfy the unique demands of stakeholders and end users. In addition to designing and implementing AI, human factors and ergonomics researchers researchers should dynamically evaluate how AI technologies affect patient outcomes, workflow, and interaction (Asan & Choudhury, 2021).

#### 15.5.5.4 Continuous Evaluation and Improvement

By following these best practices, organizations can harness the full potential of AI technologies to enhance productivity, safety, and accessibility while ensuring that HF&E principles are integrated into the design and deployment of AI-driven systems.

## 15.6 FUTURE DIRECTIONS AND RECOMMENDATIONS

AI is rapidly evolving and is expected to play an increasingly significant role in HF&E. As we look towards the future, it's essential to understand the emerging trends, challenges, and opportunities in AI integration within HF&E contexts. This section delves into the potential future directions of AI in HF&E and offers recommendations for ensuring its responsible and effective integration.

### 15.6.1 Emerging Trends in AI Research

To make sure that AI technologies continue to be efficient, dependable, and sensitive to shifting user demands and preferences, implement methods for ongoing review and development of AI-driven systems.

#### 15.6.1.1 Explainable AI

There is an increasing demand for interpretability and transparency as AI systems get more complicated. The fact that AI-based systems frequently lack transparency is a major barrier to their deployment, despite their historically significant breakthroughs. Strong forecasts are made possible by these systems' black-box design, which makes them impossible to explain (Adadi & Berrada, 2018). AI algorithms can be made more trustworthy and accountable by using techniques like interactive visualization tools and model interpretability methodologies, which let users know how the algorithms arrive at their findings.

#### 15.6.1.2 Ethical AI Design

The role of ethical considerations in AI research and development is growing. Future research will concentrate on incorporating moral values like justice, responsibility, and openness into the development and application of AI systems (Floridi et al., 2018). Addressing algorithmic biases, safeguarding user privacy, and making sure AI systems put the well-being of people and society first are all part of ethical AI design.

#### 15.6.1.3 Human-AI Collaboration

AI-human cooperation has the potential to revolutionize the creation and application of AI technology. Novel interaction paradigms and interfaces that enable smooth human-AI system collaboration will be the subject of future study. AI and humans working together in the workplace of the future will foster creativity and increase output. AI is more than simply a tool; it enhances skill and facilitates decision-making, which disrupts entire sectors. Accepting this collaboration is essential since it promotes ethical issues, skill development, and educational investment. It will also assist in creating a framework that will support the application of AI and be supported by the law (Quarks, 2024). Organizations can use AI skills in conjunction with human experience to address difficult HF&E challenges by combining human and machine learning.

#### 15.6.1.4 Responsible AI Governance

The demand for appropriate AI governance structures is growing as AI technologies become more widely used. The development of standards, rules, and regulatory frameworks will be the main focus of future efforts to guarantee the moral and responsible application of AI technologies. In order to succeed in the future, humans will need to stay up with robots and devise plans and guidelines that businesses and individuals may utilize to combine digital processing power and human creativity (Brynjolfsson et al., 2020). In order to handle the dangers and new problems that come with the use of AI, governments, business stakeholders, and civil society organizations must work together to practice responsible AI governance.

### 15.6.2 Future Directions in AI Applications

In HF&E, developments in AI algorithms, data analytics, and human-centered design principles will influence the direction of AI applications in the future. AI applications are expected to increase and innovate in a few key areas.

#### 15.6.2.1 Healthcare

AI will keep changing the way that healthcare is delivered, diagnosed, and treated. In order to enhance patient outcomes and save healthcare costs, future research will concentrate on creating AI-driven diagnostic tools, predictive models, and tailored therapy suggestions. In all tasks, the deep convolutional neural networks outperform all tested specialists, proving that AI is able to classify skin cancer at a level of proficiency that is comparable to dermatologists (Esteva et al., 2019). Virtual assistants and telehealth platforms driven by AI will also improve access to healthcare treatments, especially in underprivileged areas.

#### 15.6.2.2 Transportation

Autonomous vehicles and intelligent transportation systems will have a significant impact on transportation in the future. Future research will focus on enhancing AI-driven navigation, vehicle automation, and traffic management to raise the efficiency, safety, and sustainability of transportation networks. Furthermore, AI technology will enable the creation of mobility-as-a-service and smart city solutions that support inclusivity, accessibility, and environmental sustainability (Fagnant & Kockelman) (Floridi & Cowls, 2019).

#### 15.6.2.3 Manufacturing

AI technology will be used in Industry 4.0 and smart manufacturing initiatives to streamline production processes, improve product quality, and increase worker safety. Future studies will concentrate on creating digital twins, collaborative robots, and AI-driven predictive maintenance systems that allow for real-time industrial operation monitoring, analysis, and optimization (Khan et al., 2023). AI-powered supply chain management and logistics systems will also improve the supply chain's adaptability and resistance to external shocks.

#### 15.6.2.4 Retail

By facilitating predictive analytics, streamlining inventory control, and personalizing consumer experiences, AI technology will revolutionize the retail sector. In order to improve consumer engagement and satisfaction, future research will concentrate on creating AI-driven recommendation systems, virtual shopping assistants, and customer sentiment analysis tools (Amazon, 2022). Moreover, businesses will be able to adjust to shifting consumer tastes and market dynamics more skillfully thanks to AI-powered demand forecasting and pricing optimization algorithms.

### 15.6.3 Recommendations for AI Integration in HF&E

Several suggestions can be made in order to optimize the advantages of AI technology in HF&E while minimizing potential dangers and difficulties.

#### 15.6.3.1 Interdisciplinary Collaboration

To guarantee that AI technologies are developed and implemented responsibly, ethically, and inclusively, foster interdisciplinary collaboration among ethicists, domain experts, human aspects professionals, and AI researchers. It has recently been acknowledged that the targeted success in manufacturing can only be achieved through the integrated actions of the human, technology, and organization due to the clear limitations of technology-oriented methods (Salvendy, 2012).

#### 15.6.3.2 Ethical AI Design Principles

When designing and developing AI systems, take into account ethical AI design principles like responsibility, transparency, justice, and privacy (Floridi et al., 2018). When implementing AI, give human welfare, safety, and society welfare a top priority. This will help to guarantee that AI follows ethics and moral standards.

#### 15.6.3.3 User-Centered Design

When developing AI, use a user-centered design stance, emphasizing comprehension of user requirements, preferences, and behaviors. Use feedback loops, usability testing, and participatory design techniques to involve end users in the design process and make sure AI systems fulfill their needs and expectations. Computational fairness tools are useful, but must be researched and developed in and with the messy contexts that will shape their deployment, rather than just for imagined situations (Veale & Binns, 2017).

#### 15.6.3.4 Continuous Evaluation and Improvement

Implement mechanisms for continuous evaluation and improvement of AI systems, including performance monitoring, user feedback, and iterative design cycles (Norman & Nielsen, 2010). Regularly assess the ethical and societal implications of AI deployment, and iterate on AI systems' design and functionality to address emerging challenges and opportunities.

#### 15.6.3.5 Responsible AI Governance

Advocate for responsible AI governance mechanisms, including regulatory frameworks, standards, and policies that promote ethical AI deployment and mitigate potential risks (Brynjolfsson et al., 2020). Engage with policymakers, industry stakeholders, and civil society organizations to shape AI governance policies and ensure that AI technologies are deployed in a manner that benefits society as a whole.

By heeding these suggestions, companies can successfully negotiate the challenging terrain of AI integration in HF&E and capitalize on the revolutionary potential of AI technology while preserving public safety, welfare, and well-being.

## 15.7 CONCLUSION

There are several dilemmas with integrating AI into different facets of society. We have covered a wide range of topics in this conversation, including ethical conundrums and technical obstacles, and we have also suggested solutions for integrating AI. As we come to the end of this investigation, it is clear that stakeholders from a variety of disciplines must work together to implement AI technologies effectively.

First and foremost, it's critical to understand that integrating AI is a customized process. Different AI applications and industries provide different problems that call for customized solutions. Every industry, including healthcare, banking, transportation, and education, has its own set of technological needs, legal frameworks, and ethical issues. As a result, a sophisticated strategy is needed in different situations.

Second, tackling the ethical issues related to AI integration necessitates a dedication to openness, equity, and responsibility. From data gathering and model training to deployment and evaluation, ethical AI design principles must be given top priority by AI developers and practitioners. In order to do this, it must be made sure that AI systems are accountable to humans, transparent, and devoid of algorithmic biases.

Moreover, overcoming the challenges of AI integration requires cooperation and participation with a variety of stakeholders. This includes active involvement with affected communities, advocacy groups, and disenfranchised people, as well as interdisciplinary collaboration involving AI researchers, ethicists, policymakers, and domain specialists. Stakeholders can obtain important insights into the moral ramifications and societal effects of AI deployment through inclusive discourse and involvement, which will help them make better informed decisions and use AI responsibly.

Regulatory frameworks and standards also have a significant impact on the ethical and legal context of AI integration. Governments, international organizations, and industry associations must work together to establish robust regulatory frameworks that safeguard individual rights and the general welfare of society while promoting the moral application of AI. To do this, specific guidelines for algorithmic accountability and transparency, data security and privacy, and auditing and supervisory processes must be established.

In addition to legislative initiatives, education, and awareness, efforts are essential for promoting AI literacy and empowering stakeholders to make informed decisions about AI technologies. This entails teaching algorithmic bias mitigation, ethical AI design, and responsible data practices to AI engineers, decision-makers, and end users. Improving stakeholders' comprehension of the moral ramifications and issues surrounding the integration of AI can assist them in making more morally and responsibly informed choices.

The proper integration of AI technology ultimately requires a comprehensive and collaborative approach that achieves a balance between technological innovation and ethical considerations, social values, and human well-being. By addressing the ethical issues and challenges associated with AI integration head-on, stakeholders can take advantage of the transformative potential of AI technologies while making sure they serve the greater good and contribute to a more equitable and sustainable future.

## REFERENCES

Adadi, A., & Berrada, M. (2018). Peeking inside the black-box: A survey on explainable artificial intelligence (XAI). *IEEE Access*, *6*, 52138–52160.

Alpaydin, E. (2016). *Introduction to Machine Learning*. MIT Press.

Amazon Personalize. (2022). *Elevate the Customer Experience with ML-Powered Personalization*. Retrieved from https://aws.amazon.com/personalize/

Amershi, S., Weld, S., Vorvoreanu, M., Fourney, A., Nushi, B., Collisson, P., & Weld, D. S. (2019). Guidelines for Human-AI Interaction. In *Proceedings of the 2019 CHI Conference on Human Factors in Computing Systems* (pp. 1–14). Glasgow, UK.

Asan, O., & Choudhury, A. (2021). Research trends in artificial intelligence applications in human factors health care: Mapping review. *JMIR Human Factors, 8*, 2. Retrieved from https://doi.org/10.2196/28236

Barocas, S., & Selbst, A. D. (2016). Big data's disparate impact. *California Law Review, 104*(3), 671–732.

Bostrom, N. (2014). *Superintelligence: Paths, Dangers, Strategies*. Oxford University Press.

Brynjolfsson, E., & McAfee, A. (2017). *The Second Machine Age: Work, Progress, and Prosperity in a Time of Brilliant Technologies*. WW Norton & Company.

Brynjolfsson, E., & Mcafee, A. (2017). The business of artificial intelligence. *Harvard Business Review*, 7, 3–11. https://starlab-alliance.com/wp-content/uploads/2017/09/The-Business-of-Artificial-Intelligence.pdf

Brynjolfsson, E., Mitchell, T., & Rock, D. (2020). What can machines learn, and what does it mean for occupations and the economy? *AEA Papers and Proceedings, 110*, 326–330.

Bughin, J., Hazan, E., Ramaswamy, S., Allas, T., Dahlstrom, P., Henke, N., & Trench, M. (2017). Artificial intelligence: The next digital frontier? *Open Journal of Business and Management*, *9*, 4.

Buolamwini, J., & Gebru, T. (2018). Gender shades: Intersectional accuracy disparities in commercial gender classification. *Proceedings of Machine Learning Research, 81*, 77–91. https://proceedings.mlr.press/v81/buolamwini18a/buolamwini18a.pdf

Chemnad K, & Othman A. (2024, February 16) Digital accessibility in the era of artificial intelligence-bibliometric analysis and systematic review. *Front Artificial Intelligence*, *7*, 1349668. https://doi.org/10.3389/frai.2024.1349668

Chen, J., Zhang, C., Xue, Y., & Li, H. (2016). Artificial intelligence techniques for renewable energy systems. In Bashir, A.K., Fortino, G., Khanna, A., & Gupta, D. (Eds) *Proceedings of International Conference on Computing and Communication Networks*. Springer. https://doi.org/10.1007/978-981-19-0604-6_25

Cimpanu, C. (2018). Hacker infects 100K routers in latest botnet attack aimed at sending email spam. *Zdnet*. Retrieved from https://www.zdnet.com/article/iot-botnet-infects-100000-routers-to-send-hotmail-outlook-and-yahoo-spam/#google_vignette

Crawford, K., & Calo, R. (2016). There is a blind spot in AI research. *Nature*, *538,* 7625, 311–313.

Dwork, C., Hardt, M., Pitassi, T., Reingold, O., & Zemel, R. S. (2012). Fairness through awareness. In *Proceedings of the 3rd Innovations in Theoretical Computer Science Conference* (pp. 214–226). Massachusetts, USA.

Endsley, M. R., & Kaber, D. B. (1999). Level of automation effects on performance, situation awareness and workload in a dynamic control task. *Ergonomics*, *42*(3) 462–492.

Esteva, A., Kuprel, B., Novoa, R. A., Ko, J., Swetter, S. M., Blau, H. M., & Thrun, S. (2019). Dermatologist-level classification of skin cancer with deep neural networks. *Nature, 542*(7639) 115–118.

European Commission. (2019). Ethics guidelines for trustworthy AI. Retrieved from https://digital-strategy.ec.europa.eu/en/library/ethics-guidelines-trustworthy-ai. Accessed on August 12, 2023.

Fagnant, D., & Kockelman, K. (2015). Preparing a nation for autonomous vehicles: Opportunities, barriers and policy recommendations, *Transportation Research Part A: Policy and Practice, 77,* 167–181.

Floridi, L., & Cowls, J. (2019). A unified framework of five principles for AI in society. *Harvard Data Science Review*, *1*, 1.

Floridi, L., Cowls, J., Beltrametti, M., Chatila, R., Chazerand, P., Dignum, V., & Schafer, B. (2018). AI4People - An ethical framework for a good AI society: Opportunities, risks, principles, and recommendations. *Minds and Machines*, *28*(4)689–707.

Goodfellow, I., Bengio, Y., & Courville, A. (2016). *Deep Learning* (Vol. 1). MIT Press.

Goodfellow, I., Shlens, J., & Szegedy, C. (2015). Explaining and harnessing adversarial examples. In *International Conference on Learning Representations*. New Orleans.

Google AI Principles. (2022). AI Principles. Retrieved from https://ai.google/responsibility/principles/. Accessed on Jan 19, 2023.

Herlocker, J. L., Konstan, J. A., & Riedl, J. (2000). Explaining collaborative filtering recommendations. In *Proceedings of the 2000 ACM Conference on Computer Supported Cooperative Work* (pp. 241–250). ACM. https://doi.org/10.1145/358916.358995

Huang, T., Li, D., Zhang, J., & Zhu, Z. (2019). Research and development on AI in finance: A survey. *IEEE Access*, *7*, 7833–7850.

IBM AI Ethics Principles. (2022). *The Benefits of AI in Healthcare*. IBM Education. Retrieved from https://www.ibm.com/think/insights/ai-healthcare-benefits

IBM Watson Health. (2022). Healthcare technology solutions and services. Retrieved from https://www.ibm.com/industries/healthcare. Accessed on August 11, 2023.

Jurafsky, D., & Martin, J. H. (2019). *Speech and Language Processing* (Vol. 3). Pearson.

Khan, A. N., Jabeen, F., Mehmood, K., Soomro, M. A., & Bresciani, S. (2023). Paving the way for technological innovation through adoption of artificial intelligence in conservative industries, *Journal of Business Research*, *165*, 3–10.

Lee, J., Davari, H., & Singh, J. (2018). Industrial Artificial Intelligence for industry 4.0-based manufacturing systems. *Manufacturing Letters*, *18*, 20–23.

Lum, K., & Isaac, W. (2016). To predict and serve? *Significance*, *13*(5), 14–19.

Microsoft AI Principles. (2022). Responsible AI at Microsoft. Retrieved from https://www.microsoft.com/en-us/ai/responsible-ai#:~:text=Microsoft's%20responsible%20AI%20principles%20include,ingation%20based%20on%20personal%20characteristics. Accessed on July 14, 2023.

Miller, C., & Valasek, C. (2015). *Remote Exploitation of an Unaltered Passenger Vehicle*. Black Hat USA.

Norman, D. A., & Draper, S. W. (1986). *User Centered System Design: New Perspectives on Human-Computer Interaction.* Lawrence Erlbaum Associates.

Norman, D., & Jakob, N. (2010). Gestural interfaces: A step backward in usability, design and usability. *Interactions*, *9*(11), 46–49.

O'Neil, C. (2016). *Weapons of Math Destruction:* How Big *Data Increases Inequality and Threatens Democracy*. Broadway Books.

Obermeyer Z., & Emanuel E. J. (2016). Predicting the future – Big data, machine learning, and clinical medicine. *New England Journal of Medicine, 375*(13), 1216–1219. https://doi.org/10.1056/NEJMp1606181. PMID: 27682033; PMCID: PMC5070532.

Parasuraman, R., Sheridan, T. B., & Wickens, C. D. (2000). A model for types and levels of human interaction with automation. *IEEE Transactions on Systems, Man, and Cybernetics-Part A: Systems and Humans*, *30*(3), 286–297.

Patel, R., Dhakad, J.B., Desai, K., Gupta, T., & Correia, S. (2018). Hand gesture recognition system using convolutional neural networks. In *2018 4th International Conference on Computing Communication and Automation (ICCCA)* (pp. 1–6). Greater Noida, India.

Quarks. (2024). *The Future of Human-AI Collaboration, Quarks*. Retrieved from https://www.linkedin.com/pulse/future-human-ai-collaboration-quarks-technosoft-pvt-ltd--qfq5f/ on March 19, 2024.

Rouse, W. B., & Morris, N. M. (1986). On looking into 'The Black Box': Prospects and limits in the search for mental models. *Psychological Bulletin, 100*(3), 349.

Russell, S., & Norvig, P. (2021). *Artificial Intelligence: A Modern Approach* (4th ed.). Pearson.

Salvendy, G. (2012). *Handbook of Human Factors and Ergonomics*. John Wiley & Sons, Inc. https://doi.org/10.1002/9781118131350.

Schneier, B. (2018). *Click Here to Kill Everybody: Security and Survival in a Hyper-Connected World.* WW Norton & Company.

Shakarian, G, I. Simari, Moores, G., Paulo, D., Parsons,. S., Falappa, M., & Aleali. (2016). A belief revision in structure probabilistic argumentation: Model and application to cyber security. *AMAI, 78*(3–4), 259–301.

Siemens (2022). *Generative Artificial Intelligence Takes Siemens' Predictive Maintenance Solution to the Next Level*. Retrieved from https://press.siemens.com/global/en/pressrelease/generative-artificial-intelligence-takes-siemens-predictive-maintenance-solution-next

Silver, D., Huang, A., Maddison, C. J., Guez, A., Sifre, L., Van Den Driessche, G., & Dieleman, S. (2016). Mastering the game of go with deep neural networks and tree search. *Nature, 529*, 7587, 484–489.

Tesla. (2022). *Autopilot and Full Self-Driving Capability*. Retrieved from https://www.tesla.com/support/autopilot

Turing, A. M. (1950). Computing machinery and intelligence. *Mind, 59*(236), 433–460.

VanLehn, K. (2011). The relative effectiveness of human tutoring, intelligent tutoring systems, and other tutoring systems. *Educational Psychologist, 46*(4), 197–221.

Veale, M., O. & Binns, R. (2017). Mitigating discrimination without collecting sensitive data. *Big Data & Society*, *4*, 2.

Wang, D., Cui, P., Zhu, W., & Deng, H. (2016). Pdwatcher: Predictive maintenance of mechanical equipment with multi-sensor time series data. *MDPI, 29,* 18 available at https://www.mdpi.com/1424-8220/19/18/3884

Wickens, C. D., Helton, W. S., Hollands, J. G., & Banbury, S. (2021). *Engineering Psychology and Human Performance*. Routledge. https://doi.org/10.4324/9781003177616

# Section III

## HF&E in Action

### Strategies for Employee Well-Being and Corporate Success

# 16 Application of Tailored Ergonomic Intervention Strategy for Reducing Work-Related Stress

*Julie Vardhan, Mona Ratnesh, and Prabhat Kumar Tripathi*

## 16.1 INTRODUCTION

Human stress is a complicated reaction that affects the mental health of employees substantially and is a key contributor to chronic illnesses, reduced productivity, and cost to organisation (Hassard et al., 2014). It is an inevitable phenomenon of contemporary organisational life. Workplace stress is imminent that significantly affects employees across various industries, thereby leading to adverse effects on both individual's well-being and organisational performance. According to International Labour Organization (ILO), workplace stress is a consequence of discrepancy between employee's capabilities, resources, and demand of the job (ILO, 2016). Over the decades, there has been a notable surge in stress level especially at the workplace. Prolonged stress has a significant effect on both body and the mind. These include physical health conditions like cardiovascular disease and musculoskeletal illnesses, as well as mental health conditions like anxiety and depression (Leka et.al., 2003). A study by Anderson and Pulich (2001) revealed that stress is an inexorable part of any organisation irrespective of its size and have a detrimental impact on employees' performance. Further research by Salama et al. (2022) supported this finding and examined that job burnout acts as a mediator between work stress and turnover intention. Additionally, the study revealed that job burnout, which in turn influences turnover intention, is significantly impacted by work stress. Eventually these workplace stress results in musculoskeletal discomfort (Hämmig, 2020). Several studies have established the association between mentally stressful work and musculoskeletal discomfort (Freimann et al., 2016; Hämmig, 2020).

The modern workplace has undergone a remarkable transformation, characterised by an ever-increasing demand for productivity, technological advancements, and a relentless pursuit of efficiency. However, this drive towards enhanced productivity has often come at the expense of employee well-being, leading to a concerning rise in work-related stress (Soylu & Campbell, 2012). The contemporary workplace has grown to be a haven for many types of stress, which can have detrimental effects on

DOI: 10.1201/9781003532293-19

employees' physical and mental health. Musculoskeletal discomforts are one such effect that can have a big influence on a person's productivity and general quality of life (Korhan & Memon, 2019; Haghshenas et al., 2018).

Ergonomics plays a crucial role in mitigating work-related stress and enhancing the overall well-being of employees. By designing workstations and work environments that are compatible with the potentiality and capacity of employees, ergonomic interventions can substantially deflate the physical strain and mental fatigue. This can lead to a more efficient and healthier workforce (Afroz & Haque, 2021). The field of ergonomics involves the application of pertinent "scientific knowledge that is relevant to a situation through methods which are appropriate to both the body of knowledge and the situation." It involves a sequence of suggested adjustments to workplace policies, procedures, individual work practices and standard operating procedures (Rothmore et al., 2015).

The fundamental principle of ergonomics is to optimise the interaction between employees and their work environment. This entails job creation, designing of workstations, controls, displays, lighting, ventilation, handling of tools and equipment that are aligned with the physical capacities and limitations of employees (Koirala & Nepal, 2022). Using a qualitative study, this study aimed to examine the effectiveness of tailored ergonomic intervention on reducing work-related stress. A semi-structured interview with IT professionals who are known to work long hours on their system was taken to understand their perception on ergonomic interventions, their beneficial effects and factors that influence intervention strategies in their organisations. This chapter is organised with an overview of Stress and its effects in the next section followed by Ergonomic Principles and their relevance to mitigating stress. The following section is regarding the methodology adopted with details on the data collection and interview procedure. The next section analyses the results of the study followed by Discussion and Conclusion. The findings of this study suggested that the various issues and facilitators for ergonomic interventions in organisations could be classified into four major themes as explored through the interactions- reduction of stress, musculoskeletal discomfort, organisational culture, availability of resources. Moreover, the study highlights the importance of the role of organisation culture, leadership and training and development that supports employee holistic well-being.

## 16.2 REVIEW OF LITERATURE

### 16.2.1 Understanding the Work-Related Stress and Its Effects

Work-related stress is a predominant issue that has a prominent effect on both individuals and organisations. Due to its detrimental effects on employees' emotions, psychological states, and physical health worldwide, work-related stress (WRS) is one of the significant contributors to the global burden of illness. The intricate psychological concept of workplace stress must be understood within the parlance of its parent concept, stress.

Stress has been defined as "the change in one's physical or mental state in response to situations (stressors) that pose challenge or threat" (Krantz et al., 1985; Colligan & Higgins, 2006). However, work-related stress is a sense of work-related difficulty, irritation and anguish (Grossi, Keil, & Vito, 1996). It is a consequence of negative effects in the workplace which results in deviation from normal functioning as absenteeism, turnover intent, or burnout. In his study, Pickering (2001) also emphasises how stress can have a significant impact on employees, including higher rates of illness, absenteeism, and staff turnover. Furthermore, stress that is work related is often cost high to organisation due to low efficiency and health issues (Hassard et al., 2018; García-Tudela et al., 2022). Hence, stress at work has a detrimental impact on employees and organisational efficiencies.

### 16.2.2 Ergonomic Principles to Mitigate Work-Related Stress

The study of ergonomics which examines the interaction between humans and their work environment, has relatively vital in tackling the issues of stress at work. By implementing ergonomic concepts, organisations can design customise workstations and modify tasks and workflows as per the physical and cognitive requirement of their employees, hence mitigating the consequences of stress on both individual and organisational (Feuerstein et al., 2004; Harmsen et al., 2019). The basic theory of ergonomics is to create a work environment that revolves around the individual rather than compelling employees to adapt to suboptimal conditions. Ergonomics has its roots in ancient civilisations when people realised how important it was to modify jobs, settings, and instruments to better fit the physical and mental capacities of users (Nag & Nag, 2019). Ergonomics comes from the Greek words 'Ergo', which means work, and 'Nomos', which means health. The Polish scientist Wojciech Jastrzębowski first used the term "ergonomics" in the 19th century, defining it as the "science of work" (Charytonowicz, 2009).

The core principles of ergonomics incorporate physical, cognitive, and organisational factors, all of which play a crucial role in addressing work-related stress. Physical ergonomics emphasises on the anatomical, anthropometric, physiological, and biomechanical aspects of the human body, along with how these aspects are linked with physical activity (Hendrick, 2000). This involves customisation of workstations, handling of equipment and tools to mitigate the discomfort of musculoskeletal disorders, such as back pain, injuries, joint problems and repetitive strain. Intervention of physical ergonomics in organisations can personalise the work environments that promote proper posture, minimise the repetitive strain on the body, and support the overall physical well-being of employees (Segar & Rahman, 2019).

Cognitive ergonomics delves into the cognitive processes involved in work, like memory, logical reasoning, perception and motor reactions. It aims to compose the information and control systems, along with tasks organisation to coordinate the cognitive capabilities and limitations of the human mind. With respect to work-related stress, Cognitive ergonomics can help reduce factors that lead to higher cognitive demands and mental fatigue, such as task complexity, information overload, and poorly designed user interfaces (Hernadewita et al., 2020).

However, organisational ergonomics accentuate on the organisational structures, procedures, and policies in order to optimise the social-technical system (Kata & Prasad, 2020). This embraces the consideration of team dynamics, work schedules, channels of communication, and management practices. Proper implication of organisational ergonomics concepts can contribute to the development of holistic work environment that promotes quality of work life, employee well-being, reduces strain at work and enhances overall organisational performance and employees' satisfaction (Jordan & Escobales, 2015).

### 16.2.3 Tailored Ergonomic Intervention Strategies

Ergonomic interventions have long been recognised as an effective means of mitigating the adverse effects of work-related stress (McSweeney et al., 2002). To address this pressing issue, a tailored ergonomic intervention approach presents a promising solution. In this study, a few cases of tailored ergonomic interventions have been reviewed to assess the efficacy of these interventions in reducing stress and increasing the well-being of the employees at workplace.

Rasotto et al. (2015) examined the efficacy of a tailored exercise program in reducing symptoms of work-related musculoskeletal disorders in the upper limbs of metalworkers. In this randomised controlled trial, 143 metalworkers were divided into two groups: one received a tailored exercise program, while the other received a controlled program. The results showed that the tailored exercise program was more effective in reducing pain and improving function in the upper limbs compared to the controlled program.

A similar study (So et al., 2019) examined the efficacy of two distinct interventions for shoulder and neck pain associated with the workplace. 101 people with work related neck and shoulder pain (WRNSP) diagnoses participated in the study and were split into two groups. While the ergo motor group received a 12-week programme that combined specialised motor control training with ergonomic advice, the control group underwent 12 weeks of traditional physiotherapy. While there were notable improvements in both groups' functional outcomes and pain levels, the ergo motor group reported noticeably higher improvements in perceived exertion in relation to job demands and workstyle, compared to the control group. The study suggests combining motor control training with ergonomic advice may improve occupational health outcomes for individuals.

Visser et al. (2018) conducted a process evaluation of two participatory ergonomics (PE) intervention strategies in construction companies. The study compared a more participative bottom-up approach with the conventional top-down method. The results indicate that participatory approaches which encourage worker ownership and involvement in the intervention process show greater efficacy in identifying workable solutions and effective implementation. Participatory ergonomics therefore within companies is beneficial as it fosters a culture of safety and continuous improvement in addition to addressing ergonomic risks.

Earlier studies have also mentioned promising outcomes of various ergonomics training techniques (Robertson et al., 2008, 2009). The workers' understanding of MSDs increased because of educational interventions, which also altered their posture when using computers in the neck, shoulders, and hands (Marcoux et al., 2000). In contrast to individuals who did not receive training, those who did receive education programmes, such as participatory training and traditional training (lectures and handouts), reported less pain and discomfort and a positive perception of psychosocial work stress (Bohr, 2002; Kivimäki & Kawachi, 2015).

The effect of ergonomic interventions on psychosocial factors and musculoskeletal symptoms in office workers was investigated by Choobineh et al. (2011). The study used ergonomic modifications, including chairs with adjustable desks, ergonomic chairs, and appropriate workstation layouts along with training programs on ergonomic principles and practices. The findings reveal significant reductions in musculoskeletal symptoms and improvements in psychosocial factors including job satisfaction and perceived stress levels. Furthermore, Choobineh et al. (2021) reported a significant enhancement in participants' ergonomics knowledge and observed that participatory ergonomics measures amplified workforce motivation to accept ergonomic practices. This results in better ergonomics awareness and a strong participatory culture. The study concluded that such a comprehensive intervention effectively enhances employee wellness and productivity.

With approximately 60% of the working population in developing nations employed in various informal or unorganised sector jobs, reffered in Gangopadhyay and Dev (2014) and conducted a study among Indian workers in the informal sector for work related musculoskeletal disorders (WMSDs) (Lindegard, 2014). Based on the analysis of data from six distinct case studies, the study concluded that workers in the agricultural sector are required to exert the greatest amount of physical effort with the least amount of safety. Although several interventions have been planned, these are either poorly designed or not well maintained. However, the study also identified that ergonomic interventions have a significant positive impact on the unorganised or informal sector workers in India, particularly goldsmiths, carpenters, and workers making sand cores.

Effective interventions are required given the rise in the number of people engaged in sedentary occupations and the mounting evidence of the significance of sedentary behaviour as a major lifestyle risk factor. Straker et al. (2013) sourced data from 15 different call centres located throughout Sweden and proposed sit-stand desks could be a helpful solution, especially in office settings. Additionally, if ergonomics awareness is increased and supported by the work organisation it could help to further reduce sedentary behaviour (Kristianti et al., 2020).

The review highlights that regardless of the kind of industries, workplace stress and MSD are extensively prevalent across workplaces globally. Stakeholders across these industries are concerned because these occupational disorders usually take time to manifest, diminishing working life, having an impact on employees' wellness and reducing productivity for the organisations.

The reviewed studies also show that a variety of risk factors contribute to workplace stress and MSDs and controlling these factors effectively becomes more difficult, particularly in informal sectors. Further, the risk factors might also change especially in manufacturing, mining or service industries like call centres. Different types of ergonomic interventions, either alone or in combination have been shown to not only mitigate physical health risks but also enhance psychosocial well-being, demonstrating the multifaceted benefits of ergonomics in improving workplace safety, health, and productivity.

## 16.3 METHODOLOGY

With an objective to gain an in-depth understanding of the benefits and barriers in implementing ergonomic interventions in the IT sector, semi-structured interviews were conducted. The interview was held among those working groups who were in the core profile of programming, or data analysis.

### 16.3.1 Participants

This group of respondents were selected based on a snowball technique where they were contacted by the researchers and further invited for a telephonic interview. Each interaction was for a mean duration of 12 min and 2–3 interviews were held by each researcher in a day. The participants were informed about the purpose of the study prior to the interview and their consent was taken. The location of the interview was not considered to be of importance to the results as most of the questions were not contextually dependent. Moreover, as the respondents were majorly from the IT sector, and based on the general information it was known that the employees had to work long hours in a particular posture. Interviews were conducted in the last fortnight of January 2024. A total of 27 participants consented and participated in the interview.

### 16.3.2 Data Analysis

All the interviews were transcribed verbatim. To minimise researcher's bias, each researcher interchanged the responses and extracted the information. To illustrate the meaning of the perceived benefits and the impending organisational barriers, phrases that were considered representative of these were reported.

The discussions and extraction of data were re-evaluated by the researchers to arrive at consensus on the themes generated. Four major themes came across through the extraction process: reduction of stress, musculoskeletal discomfort, organisational culture, and availability of resources.

## 16.4 RESULTS

Table 16.1 presents the explanations of the perceived benefits and barriers in implementing ergonomic interventions at the organisational level. These appeared as keywords during the extraction process of the interviews.

Ergonomic interventions can mitigate the adverse effects of work-related stress (McSweeney et al., 2002). Tailored exercise program has a significant impact on reducing symptoms of work-related stress (Rasotto et al., 2015). These have been supported by many respondent "At work, I recently had a customised ergonomic intervention. After evaluating my workstation, they made a number of changes. They also changed the height of my workstation and monitor to relieve pressure on my neck and shoulders and also customised my chair to better support my back. Additionally, they suggested some stretches and exercises that I can perform all day long". Ergonomic interventions have had a positive impact in reducing worker's stress level has been

**TABLE 16.1**
**Major Themes**

| Sl. No | Themes | Keywords |
|---|---|---|
| 1 | Reduction of stress | • Relaxed work atmosphere, reduced pressure, better work-life balance |
| 2 | Musculoskeletal discomfort | • Productivity, self-efficacy |
| 3 | Organisational culture | • Leadership, co-workers, work pressure, deadlines, tailored ergonomic intervention |
| 4 | Availability of resources | • Facilities according to hierarchy, protocols |

supported by many respondents "I feel positive regarding tailored made ergonomic intervention. It is evident that a one-size-fits-all strategy is ineffective, therefore having something customised to meet my unique requirements was essential. Having a remedy that doesn't merely treat the symptoms is a relief because it gets to the source of the issue".

### 16.4.1 Musculoskeletal Discomfort

Musculoskeletal discomforts are one such effect that can have a big influence on a person's productivity and general quality of life (Korhan & Memon, 2019; Haghshenas et al., 2018). Musculoskeletal disorders (MSDs) are a wide range of illnesses affecting the muscles, tendons, ligaments, joints, and nerves, frequently linked to poor ergonomics, prolonged static postures, and repetitive strain (Shin & Park, 2019; Subramanya et. al., 2021; Sen et al., 2020). Many respondents reported constant ache in shoulder and lower back which impact productivity and efficiency of employees (Tsang et al., 2019). Statement of respondent states that "I have been sitting too long for my work, sometimes beyond my limits, which cause strain in my shoulder. That constant pain is frustrating and sometimes makes me feel irritable or down". Another respondent states that "It's constantly in the back of my mind, and the ache never goes away, especially in my shoulder. It is difficult to concentrate on work or other tasks. I'm scared about how long this will persist and whether it will become worse, which is why I'm tight and nervous. Searching finding a comfortable position and knowing if my actions are making it worse is frustrating. Worry and discomfort together just make everything I do more stressful."

The respondents mentioned a number of musculoskeletal discomforts, as mentioned in Table 16.2 and the frequency of each being mentioned during the interview process shows that majority had shoulder aches, eye stress and headache during their work hours.

### 16.4.2 Organisational Culture

Organisational culture refers to the extent to which the organisation was able to align the ergonomic intervention strategies with the organisation goals.

**TABLE 16.2**
**Musculoskeletal Discomforts**

| Sl No. | Musculoskeletal Discomforts | Frequency |
|---|---|---|
| 1 | Shoulder aches | 8 |
| 2 | Eye stress and headache | 9 |
| 3 | Back pain | 6 |
| 4 | Muscles strain | 5 |

**TABLE 16.3**
**Organisational Culture**

| Sl No. | Organisational Culture Aspects | Frequency |
|---|---|---|
| 1 | Work environment | 9 |
| 2 | Leadership styles | 9 |
| 3 | Learning and development | 4 |
| 4 | Communication pattern | 7 |
| 5 | Co-workers behaviour | 7 |

It was found that work environment that emphasises employee well-being is considered to be of high importance among the respondents coupled with the leadership styles. Another aspect that came out during the discussions was the openness of communication and behaviour of the co-workers (Table 16.3). While almost all the respondents agreed to the fact that work environment and organisational culture including leadership and co-workers' behaviour have an effect on the productivity of employees, some of the respondents had a differing view. For instance, one of the respondents mentioned, 'these interventions are not considered a priority by several of my co-workers and they are not ready to voice their concerns…'. Another respondent's view was, 'we are not encouraged by our leadership to come up with such ideas on ergonomic interventions… it is considered as making up excuses for low performance.' Few respondents also mentioned the role of learning and development in making the employees aware of various intervention strategies that could be helpful in increasing their productivity.

### 16.4.3 Availability of Resources

Most of the respondents reported that while ergonomic interventions would definitely be good, yet the major constraint is that the organisation would not have adequate financial resources to be able to change the infrastructure (Table 16.4).

The respondents mentioned only these two organisational constraints; however, they also mentioned that these resource constraints are related to the organisational culture issues. As one respondent reported, 'The management, I would say is aware of the benefits of these ergonomic interventions. Yet, we are often told regarding the

**TABLE 16.4**
**Organizational Resources**

| Sl. No | Organisational Resources | Frequency |
|---|---|---|
| 1 | Financial resources | 11 |
| 2 | Infrastructural resources | 5 |

financial constraints and may be by next quarter, it would be sanctioned for us.' Or another respondent who mentioned, 'recently we have had new systems and chairs which are more comfortable to work as we work long hours, and this would have been allotted earlier, but there was a resource constraint. Moreover, once the budget is allocated, it is sanctioned according to the hierarchical levels!'

## 16.5 DISCUSSION

A systematic review of various ergonomic interventions by Heidarimoghadam et al. (2022) offers an evaluation of their effectiveness. These interventions range from engineering controls, such as redesigning workstations and tools, to administrative controls, including training and organisational changes. The review emphasises how ergonomic interventions can raise worker comfort levels, lower the prevalence of musculoskeletal disorders (MSDs) and increase productivity in general. The study highlights the significance of adopting a comprehensive strategy that incorporates various forms of interventions customised to specific workplace circumstances and needs.

Similar to the studies reviewed, the findings of this study suggested that the various issues and facilitators for ergonomic interventions in organisations could be classified into four major themes as explored through the interactions- reduction of stress, musculoskeletal discomfort, organisational culture, availability of resources. These issues were frequently mentioned by the respondents during the interview process. In an earlier study van den Berg et al. (2008) demonstrated a strong correlation between the work performance of white-collar employees in the commercial services industry and psychosocial work factors like self-development, teamwork, and stress management. High stress levels among employees may increase the likelihood of compensatory movements or bad postures that further strain the musculoskeletal system (Alyan et al., 2021; Palliser et al., 2005; Seppala, 2001). Although Bhanderi et al. (2008) mention that musculoskeletal disorders can arise due to a variety of factors, including individual characteristics, workstation design, and work organisation, in addition to the direct and indirect effects of workplace stress. Organisational culture, leadership and co-workers behaviours play a crucial role in the implementation of ergonomic intervention strategies along with the availability of resources. Although not explicitly mentioned in previous studies, several of the earlier studies (Visser et al., 2018; Straker et al., 2013) imply the role of organisational factors and resources in planning and implementing ergonomic interventions.

## 16.6 CONCLUSION

The findings of this study suggested that the various issues and facilitators for ergonomic interventions in organisations could be classified into four major themes as explored through the interactions- reduction of stress, musculoskeletal discomfort, organisational culture, and availability of resources. This chapter highlights the potential for tailored ergonomic interventions to address work-related stress and improve overall employee well-being. Based on the various intervention strategies adopted, it may be suggested that the possibility of involving all the concerned stakeholders should be done at the workplace to implement the ergonomic initiatives. Several studies (Feyen et al., 2000; Peteri, 2017; Hassard et al., 2014) through decades suggest that ergonomic interventions have had a positive impact in reducing worker's stress levels across various industries. Similar to these studies, the findings of the study offer an insight into the benefits and barriers in implementing the intervention strategies that would be useful for organisations in planning and understanding the unforeseen factors that have an effect on implementation practices. Tailored interventions are required to suit the specific industry requirements and the job-specific roles in the various industries which could either be through workplace design, structure and layout of the workplace infrastructure coupled with the employee's awareness and training needs. Moreover, the study highlights the importance of the role of organisational culture that supports employee holistic development and well-being. The results may contribute towards the improvement of the ergonomic intervention strategies thereby resulting in increased productivity and well-being of the employees.

**Conflict of Interest:** All authors declare no conflict of interest.

## REFERENCES

Afroz, S., & Haque, M. I. (2021). Ergonomics in the workplace for a better quality of work life. In *Ergonomics for Improved Productivity: Proceedings of HWWE 2017* (pp. 503–511). Springer Singapore.

Alyan, E., Saad, N. M., Kamel, N., & Rahman, M. A. (2021). Workplace design-related stress effects on prefrontal cortex connectivity and neurovascular coupling. *Applied Ergonomics*, *96*, 103497.

Anderson, P., & Pulich, M. (2001). Managing workplace stress in a dynamic environment. *The Health Care Manager*, *19*(3), 1–10.

Bhanderi, D., Choudhary, S. K., Parmar, L., & Doshi, V. (2008). A study of occurrence of musculoskeletal discomfort in computer operators. *Indian Journal of Community Medicine*, *33*(1), 65–66.

Bohr, P. C. (2002). Office ergonomics education: A comparison of traditional and participatory methods. *Work*, *19*(2), 185–191.

Charytonowicz, J. (2009). Evolutionary changes in the traditional ergonomics. In *Universal Access in Human-Computer Interaction. Addressing Diversity: 5th International Conference*, UAHCI 2009, Held as Part of HCI International 2009, San Diego, CA, USA, July 19-24, 2009. Proceedings, Part I 5 (pp. 450–459). Springer Berlin Heidelberg.

Choobineh, A., Motamedzade, M., Kazemi, M., Moghimbeigi, A., & Pahlavian, A. H. (2011). The impact of ergonomics intervention on psychosocial factors and musculoskeletal symptoms among office workers. *International Journal of Industrial Ergonomics*, *41*(6), 671–676.

Choobineh, A., Shakerian, M., Faraji, M., Modaresifar, H., Kiani, J., Hatami, M., ... & Kamali, G. (2021). A multilayered ergonomic intervention program on reducing musculoskeletal disorders in an industrial complex: A dynamic participatory approach. *International Journal of Industrial Ergonomics*, *86*, 103221.

Colligan, T. W., & Higgins, E. M. (2006). Workplace stress: Etiology and consequences. *Journal of Workplace Behavioral Health*, *21*(2), 89–97.

Feuerstein, M., Nicholas, R. A., Huang, G. D., Dimberg, L., Ali, D., & Rogers, H. (2004). Job stress management and ergonomic intervention for work-related upper extremity symptoms. Applied *Ergonomics*, *35*(6), 565–574.

Feyen, R., Liu, Y., Chaffin, D., Jimmerson, G., & Joseph, B. (2000). Computer-aided ergonomics: A case study of incorporating ergonomics analyses into workplace design. *Applied Ergonomics*, *31*(3), 291–300.

Freimann, T., Pääsuke, M., & Merisalu, E. (2016). Work-related psychosocial factors and mental health problems associated with musculoskeletal pain in nurses: A cross-sectional study. *Pain Research and Management*, 2016(1), 9361016.

Gangopadhyay, S., & Dev, S. (2014). Design and evaluation of ergonomic interventions for the prevention of musculoskeletal disorders in India. *Annals of Occupational and Environmental Medicine*, *26*, 1–6. https://doi.org/10.1186/2052-4374-26-18

García-Tudela, Á., Simonelli-Muñoz, A. J., Rivera-Caravaca, J. M., Fortea, M. I., Simón-Sánchez, L., González-Moro, M. T. R., González-Moro, J. M. R., Jiménez-Rodríguez, D., & Gallego-Gómez, J. I. (2022). Stress in emergency healthcare professionals: The stress factors and manifestations scale. *International Journal of Environmental Research and Public Health*, *19*(7), 4342.

Grossi, E. L., Keil, T. J., & Vito, G. F. (1996). Surviving "the joint": Mitigating factors of correctional officer stress. *Journal of Crime and Justice*, *19*(2), 103–120.

Haghshenas, B., Habibi, E., Haji Esmaeil Hajar, F., Ghanbary Sartang, A., van Wijk, L., & Khakkar, S. (2018). The association between musculoskeletal disorders with mental workload and occupational fatigue in the office staff of a communication service company in Tehran, Iran, in 2017. *Journal of Occupational Health and Epidemiology*, *7*(1), 20–29.

Hämmig, O. (2020). Work-and stress-related musculoskeletal and sleep disorders among health professionals: A cross-sectional study in a hospital setting in Switzerland. *BMC Musculoskeletal Disorders*, *21*, 1–11.

Harmsen, R., Helms-Lorenz, M., Maulana, R., van Veen, K., & van Veldhoven, M. (2019). Measuring general and specific stress causes and stress responses among beginning secondary school teachers in the Netherlands. *International Journal of Research & Method in Education*, *42*(1), 91–108.

Hassard, J., Teoh, K., Cox, T., Cosmar, M., Gründler, R., Flemming, D., Cosemans, B., & van den Broek, K. (2014). *Calculating the cost of work-related stress and psychosocial risks* (Technical Report). Publications Office of the European Union.

Hassard, J., Teoh, K. R., Visockaite, G., Dewe, P., & Cox, T. (2018). The cost of work-related stress to society: A systematic review. *Journal of Occupational Health Psychology*, *23*(1), 1.

Heidarimoghadam, R., Mohammadfam, I., Babamiri, M., Soltanian, A. R., Khotanlou, H., & Sohrabi, M. S. (2022). What do the different ergonomic interventions accomplish in the workplace? A systematic review. *International Journal of Occupational Safety and Ergonomics*, *28*(1), 600–624.

Hendrick, H. W. (2000). The technology of ergonomics. *Theoretical Issues in Ergonomics Science*, *1*(1), 22–33.

Hernadewita, H., Hendra, H., Kristianti, R., Asih, I. A., Satria, D., & Yuliani, E. N. S. (2020). *Similarity B32: The ergonomic Factor Application for Improvement of Performance Office Staff.*

International Labour Organization. (2016). *Workplace Stress. A Collective Challenge*. Turin: International Training Centre of the ILO.

Jordan, S., & Escobales, M. (2015). VAMC Orlando's ergonomic program. *Procedia Manufacturing*, *3*, 4823–4827.

Kata, A., & Prasad, D. V. S. (2020). A study on ergonomics for employee wellness. *International Journal of Innovative Research in Computer Science &* Technology (IJIRCST), *8*(4), 315–318.

Kivimäki, M., & Kawachi, I. (2015). Work stress as a risk factor for cardiovascular disease. *Current Cardiology Reports*, *17*, 1–9.

Koirala, R., & Nepal, A. (2022). A literature review on ergonomics, ergonomics pactices, and employee performance. *Management*, *4*(2), 273–88.

Korhan, O., & Memon, A. A. (2019). Introductory chapter: Work-related musculoskeletal disorders. In *Work-Related Musculoskeletal Disorders*. IntechOpen.

Krantz, D. S., Grunberg, N. E., & Baum, A. (1985). Health psychology. *Annual Review of Psychology*, *36*(1), 349–383.

Kristianti, R., Asih, I., Dhimas, S., & Yuliani, E. N. S. (2020, December). The ergonomic factor application for improvement of performance office staff. In IOP Conference Series: *Materials Science and Engineering* (Vol. 909, No. 1, p. 012087). IOP Publishing.

Leka, S., Griffiths, A., Cox, T., & World Health Organization. (2003). *Work Organisation and Stress: Systematic Problem Approaches for Employers, Managers and Trade Union Representatives*. World Health Organization.

Marcoux, B. C., Krause, V., & Nieuwenhuijsen, E. R. (2000). Effectiveness of an educational intervention to increase knowledge and reduce use of risky behaviors associated with cumulative trauma in office workers. *Work*, *14*(2), 127–135.

McSweeney, K. P., Craig, B. N., Congleton, J. J., & Miller, D. (2002). Ergonomic program effectiveness: Ergonomic and medical intervention. *International Journal of Occupational Safety and Ergonomics*, *8*(4), 433–449.

Nag, P. K. (2019). Fundamentals of office ergonomics. In A. Chakrabarti (Series Ed.), *Office buildings: Health, safety and environment* (pp. 129–146). Springer Nature Singapore. https://doi.org/10.1007/978-981-13-2577-9_5

Palliser, C. R., Firth, H. M., Feyer, A. M., & Paulin, S. M. (2005). Musculoskeletal discomfort and work-related stress in New Zealand dentists. *Work & Stress*, *19*(4), 351–359.

Peteri, V. (2017). Bad enough ergonomics: A case study of an office chair. *Sage Open*, *7*(1), 2158244016685135.

Pickering, T. (2001). Job stress, control, and chronic disease: Moving to the next level of evidence. *Psychosomatic Medicine*, *63*(5), 734–736.

Rasotto, C., Bergamin, M., Simonetti, A., Maso, S., Bartolucci, G. B., Ermolao, A., & Zaccaria, M. (2015). Tailored exercise program reduces symptoms of upper limb work-related musculoskeletal disorders in a group of metalworkers: A randomized controlled trial. *Manual Therapy*, *20*(1), 56–62.

Robertson, M., Amick III, B. C., DeRango, K., Rooney, T., Bazzani, L., Harrist, R., & Moore, A. (2009). The effects of an office ergonomics training and chair intervention on worker knowledge, behavior and musculoskeletal risk. *Applied Ergonomics*, *40*(1), 124–135.

Robertson, M. M., Huang, Y. H., O'Neill, M. J., & Schleifer, L. M. (2008). Flexible workspace design and ergonomics training: Impacts on the psychosocial work environment, musculoskeletal health, and work effectiveness among knowledge workers. *Applied Ergonomics*, *39*(4), 482–494.

Rothmore, P., Aylward, P., & Karnon, J. (2015). The implementation of ergonomics advice and the stage of change approach. *Applied Ergonomics*, *51*, 370–376.

Salama, W., Abdou, A. H., Mohamed, S. A. K., & Shehata, H. S. (2022). Impact of work stress and job burnout on turnover intentions among hotel employees. *International Journal of Environmental Research and Public Health*, *19*(15), 9724.

Segar, G., & Rahman, M. N. A. (2019, August). Musculoskeletal symptoms and ergonomic risk assessment among production operators at manufacturing industries: A review. In *IOP Conference Series: Materials Science and Engineering* (Vol. 607, No. 1, p. 012009). IOP Publishing.

Sen, A., Sanjog, J., & Karmakar, S. (2020). A comprehensive review of work-related musculoskeletal disorders in the mining sector and scope for ergonomics design interventions. *IISE Transactions on Occupational Ergonomics and Human Factors*, *8*(3), 113–131.

Seppala, P. (2001). Experience of stress, musculoskeletal discomfort, and eyestrain in computer-based office work: A study in municipal workplaces. *International Journal of Human-Computer Interaction*, *13*(3), 279–304.

Shin, W., & Park, M. (2019). Ergonomic interventions for prevention of work-related musculoskeletal disorders in a small manufacturing assembly line. *International Journal of Occupational Safety and Ergonomics*, *25*(1), 110–122.

So, B. C., Szeto, G. P., Lau, R. W., Dai, J., & Tsang, S. M. (2019). Effects of ergomotor intervention on improving occupational health in workers with work-related neck-shoulder pain. *International Journal of Environmental Research and Public Health*, *16*(24), 5005.

Soylu, A., & Campbell, S. S. (2012). Physical and emotional stresses of technology on employees in the workplace. *Journal of Employment Counseling*, *49*(3), 130–139.

Straker, L., Abbott, R. A., Heiden, M., Mathiassen, S. E., & Toomingas, A. (2013). Sit–stand desks in call centres: Associations of use and ergonomics awareness with sedentary behavior. *Applied Ergonomics*, *44*(4), 517–522.

Subramanya, K. N., Rajeswara Rao, K. V. S., & Shobha, N. S. (2021). Analysis of working postures leading to musculoskeletal disorders among employees in garment manufacturing units—A case study. In *Ergonomics for Improved Productivity: Proceedings of HWWE 2017* (pp. 251–262). Springer Singapore.

Tsang, S. M., So, B. C., Lau, R. W., Dai, J., & Szeto, G. P. (2019). Comparing the effectiveness of integrating ergonomics and motor control to conventional treatment for pain and functional recovery of work-related neck–shoulder pain: A randomized trial. *European Journal of Pain*, *23*(6), 1141–1152.

van den Berg, T. I., Alavinia, S. M., Bredt, F. J., Lindeboom, D., Elders, L. A., & Burdorf, A. (2008). The influence of psychosocial factors at work and life style on health and work ability among professional workers. *International Archives of Occupational and Environmental Health*, *81*, 1029–1036.

Visser, S., van der Molen, H. F., Sluiter, J. K., & Frings-Dresen, M. H. (2018). The process evaluation of two alternative participatory ergonomics intervention strategies for construction companies. *Ergonomics*, *61*(9), 1156–1172.

# 17 Optimizing Employee Well-being and Performance
## *The Role of Human Factors and Ergonomics for Corporate Success*

*Swati Sisodia and Syed Hameedur Rahman Zaini*

### 17.1 INTRODUCTION

The dynamics of workforce, including the origins of stress and pressure, may differ significantly between industries due to this inherent diversity. In comparison to other sectors, the IT industry is characterized by an exceptionally high level of anxiety. A survey conducted by Management Today magazine in 2003 determined that the workplace environment is an essential element in the paradigm shift toward innovative work practices and technologies that boost productivity and creativity. The process entails comprehending these interactions and using appropriate approaches and theories, which may include utilizing data to improve employee welfare and maximize business achievement. Prior research (Aryee et al., 2012; Pousette et al., 2008) has examined the effects of human factors and ergonomics (HFE) on the well-being of employees, but these studies were conducted independently. Nevertheless, there is a dearth of studies regarding the most efficient methods for cultivating an environment that concurrently encourages achievement and overall welfare. An effective method to demonstrate both of these principles is by promoting a systems approach, which is often utilized in the field of ergonomics. The main aim of introducing ergonomic interventions is to cultivate employee well-being and achieve corporate success. These aims are achieved through the use of ergonomic interventions (Araya León et al., 2023).

Ergonomics, a widely recognized field within Human Factors/Ergonomics (HF/E), seeks to develop systems that facilitate optimal cognitive functioning and human welfare during task execution, as defined by the International Ergonomics Association (Mowrer, 1939). This discipline investigates human cognitive abilities (including but not limited to retention, perception, focus, decision-making, and problem-solving), their interactions with specific components of a working

 DOI: 10.1201/9781003532293-20

system (including but not limited to devices, controls, guidelines, and interfaces), performance-influencing factors (including but not limited to mental workload, stress, and distractions), and performance outcomes (including but not limited to task execution time or error rate). Ergonomics consists of three basic areas: physical, cognitive, and organizational. Physical ergonomics is concerned with how the human body responds to physical and physiological labour requirements. It directly affects physical well-being by minimizing the risk of injury and improving comfort. Cognitive ergonomics lays emphasis on the interaction between mental developments, such as insight, reminiscence, cognitive, and motor reaction, and the elements of a system. This interaction has an impact on psychological well-being. Multiple perspectives exist on the multi-faceted concept of employee well-being (McMahan & Estes, 2011; Zheng et al., 2015). Its scope is broad in this chapter, including variables such as physical and mental health of an employee (Hoffmeister et al., 2015). Furthermore, mental well-being is measured by psychological, social and emotional aspects of individuals (Reznik et al., 2022). The concept of "psychological well-being" is commonly related to both an individual's emotional state and professional performance. The research conducted in the past highlights that, to achieve work-life balance, improve satisfaction, and uphold motivation so as to enhance productivity, there is a need to manage anxiety and emotions at workplace (Ren et al., 2023). All these will help in cultivating a robust work culture and will build an inclusive environment. The literature discloses that having access to nutritious food and having probabilities for physical activity at the workplace helps in obtaining physical well-being of employees (Varghese & Raman, 2016). Consequently, this reduces mental stress, anxiety, and turnover. When employees experience good mental and physical well-being and work in an atmosphere enhanced through ergonomics, they are more actively involved, efficient, and committed, which in turn helps in creating a decrease in absenteeism, a decline in healthcare expenses, and the cultivation of a more favourable workplace culture (Adiga, 2023). Ultimately, placing employee well-being and ergonomics as a top priority contributes to the long-term profitability of a company by cultivating a healthier and more engaged staff (Bolis et al., 2023).

Organizational ergonomics aims to increase emotional well-being by enhancing sociotechnical systems, including policies, procedures, and team dynamics. This can be achieved by providing a supportive work environment and promoting positive social interactions. Together, these areas of knowledge contribute significantly to the overall health and productivity of the workforce, highlighting the interconnectedness of physical, psychological, and emotional well-being in the workplace (Karwowski, 2005).

Ergonomic interventions at the workplace are a major factor involved in increasing productivity and are significantly related to motivated and committed employees. Conversely, low standards result in employee dissatisfaction and increase in turnover, ultimately affecting the organizational performance and productivity. The implementation of an ergonomic design (such as lighting, setting arrangement, furniture, air, and space) may assure the continued productivity of personnel. There is a higher probability that indoor workers will encounter mental stress and fatigue, resulting in a decline in their efficiency and workmanship.

Furthermore, the extended computer use that is mandatory for professionals working in IT industry raises the susceptibility of these individuals to developing musculoskeletal and ocular disorders. Musculoskeletal and visual discomfort experienced by workers is significantly influenced by ergonomics and computer use. Numerous environmental factors, including design, location, computer usage hours, and illumination, can contribute to occupational visual and physical pain.

Modern businesses are becoming more aware of the significant influence that employee well-being and workplace design have on the profitability of a company. In a competitive market, organizations aim for excellence. HFE is a critical contributor towards enhancing the performance and productivity of both employees and employers. This chapter aims to uncover the complex relationships between HFE, employee's well-being and corporate success. Moreover, this chapter will help in identifying the role of workplace design is attaining this relationship. Further, it will explore how a sustainable environment can encourage both efficiency in operation and the contentment of employees by maintaining work-life balance.

Section 17.1, briefly describes the variables and the need of the study, Sections 17.2 and 17.3 include the extensive literature review and theoretical framework. The research methodology is depicted in Section 17.4. In Section 17.5, the in-depth discussion on the proposed relationships is described. Lastly, Section 17.6 highlights the implications of the study along with some future directions for research.

## 17.2 LITERATURE REVIEW

The restructuring of business organizations along with the implementation of cutting-edge technology and the introduction of workplace enhancement strategies have contributed towards making them competitive in the global marketplace. As an illustration, different industries have engaged in the adoption of ergonomics systems that give importance to the avoidance of injuries and accidents at the workplace (Lima and Coelho, 2018; Zare et al., 2020). The relationship between ergonomics and workplace design is required for promoting the employee well-being, productivity, and safety of workers in varied organizations. Adiga (2023) has concluded that by identifying the basic guidelines of ergonomics and using them for the design of workspaces, tools, and tasks, organizations can create environments that promote employee health, heighten job satisfaction, and lead to overall corporate success. Also, Khattak (2021) has proposed that rethinking job designs with a focus on ergonomic factors can result in a deeper reconsideration of elements that empower and improve organizational performance. Both operational performance as well as employee well-being can be enhanced through ergonomics interventions (Hoffmeister et al., 2015). Research by Faez et al. (2021) collates prior studies to summarize that organizations and their employees are benefitted by applying ergonomics principles, which work to improve workers health and productivity on the job. It has been found from the literature that physical, cognitive, and organizational ergonomics are among the most frequently cited HFE specializations (Karwowski & Zhang, 2021). This categorization of the HFE field into three specialized domains: physical, cognitive, and organizational was first offered by the International Ergonomics Association (IEA) in 2018. Other studies (eg. Bernard et al., 2020; Velasco et al., 2022) also explored the field from a similar domain-wise categorization. Physical ergonomics concentrates on elements, interactions, and activities pertaining to the body. "Cognitive ergonomics delves into human perceptions and mental processes while organizational ergonomics aims to optimize the aspects emanating from the organizational settings of the system within which human workers operate (Kadir et al., 2019)".

HFE has been identified as a systems discipline (Wilson, 2014). From a systems perspective, HFE focuses on how humans interact with their environment (Dul et al., 2012). The seemingly complex environment can be sub-divided into the "physical environment ('things'), the organizational environment (how activities are organized and controlled), and the social environment (other people, culture) for the purpose of simplification (Moray, 2000; Wilson, 2000; Carayon, 2006)". The discipline of HFE is accepted as a scientific field that aims to understand the relationship between humans and the different components of a system. It has been found to integrate theoretical aspects, data, and methodologies to improve employee well-being and performance (Kadir & Broberg, 2020; Read et al., 2022). The primary aim of the HFE discipline is to optimize the effectiveness and efficiency of tasks and various activities to achieve goals related to well-being and performance, such as safety, fatigue and stress reduction, and improved quality of life (Khattak, 2021; Naeini, 2020).

## 17.3 THEORETICAL FRAMEWORK

The Job Demands-Resources (JD-R) model provides a comprehensive framework for comprehending the complex connections between employee well-being, ergonomics, organizational success, and job satisfaction (Schaufeli & Bakker, 2004). Job needs and job resources are the two main components of any profession, according to this concept. Physical, social, or organizational components of a job that necessitate ongoing mental or physical exertion and are linked to specific physiological and psychological expenses are known as job demands. Conversely, workplace resources are those parts of the job—whether they be physical, mental, social, or organizational—that help workers reach their goals, do their jobs better, and save money (Fletcher & Jones, 1996). In this model, ergonomics plays a key role since it influences a job's requirements and available resources. Poor ergonomic procedures can intensify job demands by necessitating greater and prolonged physical or mental exertion from employees, potentially resulting in physical strain and psychological stress.

Moreover, the JD-R model proposed by Bakker and Demerouti (2007) is used to analyse the requirements and assets associated with mental strain in IT workers. According to the concept, an imbalance between demands and resources is the underlying cause of professional distress and illness. The notion is founded upon two processes: the first is hypothesized to result in health decline as a consequence of substandard job design, and the second is considered to enhance work engagement and performance through the provision of job resources and motivation (Bakker & Demerouti, 2007). Several employee and organizational outcomes, including intentions to quit a company, commitment, burnout, and absenteeism, have been the subject of research utilizing the JD-R model. This framework offers a lens through which this complex relationship will be studied.

## 17.4 RESEARCH QUESTIONS

Following are the research questions that the current study aims to address:

**RQ1**: How does the relationship between human factors & ergonomics (HFE) and employee well-being contribute to fostering healthier work-life balance?
**RQ2**: How does the relationship between HFE and employee well-being impact workplace design?
**RQ3**: How does the relationship between HFE and employee well-being contribute to sustainable efforts?
**RQ4**: How does the relationship between human factors & ergonomics (HFE) and employee well-being impact the long-term success and competitiveness of the organizations?

## 17.5 METHOD

Arksey and O'Malley (2005) provided the methodological foundation for scoping reviews, and this study followed their recommendations. The findings were reported using the PRISMA criteria. A scoping analysis was done to analyse literature from December 2000 to January 2024, focusing on the relationship between human factors, ergonomics, the employee well-being and their overall impact on corporate success.

A literature review's stated goal is to identify knowledge gaps, provide ideas for new areas of study, and synthesize existing research. One sort of conclusive study is descriptive research. It makes an effort to provide a systematic depiction of a scenario, occurrence, or issue; it also shows the traits of people or groups and the degree to which they are related to the variables under investigation. The creation of precise forecasts is aided by it. Finding, reviewing, and organizing research articles in a structured and organized manner is what it is all about (Peters et al., 2015).

Following an examination of widely recognized prestigious data sources including the likes of Google Scholar, Science Direct, Inderscience, Taylor & Francis, Elsevier, Emerald, and Springer, and using a combination of keywords such as HFE; Employee well-being; corporate success; sustainability; Work-life balance; Workplace design, relevant research articles were sought in these databases. Articles published no later than December 2000 were considered for inclusion in this study.

The initial search resulted in a total of 128 hits, after which these research papers were filtered in two steps. In the first step of filtration, the manual scanning of the titles of all the research papers was conducted so as to remove the irrelevant ones. This resulted in the segregation of research papers, Whitepapers, E-papers, conference papers, book chapters, proceedings papers and opinion articles, and the elimination of duplicates. The second step of the filtration involved reading the abstract and full text of the research papers, to ascertain its relevance. By applying this process of filtration, it was possible to remove papers that didn't talk about Ergonomics & human factors and well-being as the key concept. After this step, only

48 peer-reviewed papers remained which were found to be closely related with the concepts of the ergonomics, well-being, and corporate success. The current chapter serves as a review that helps to identify the relationship between ergonomics, well-being, and corporate success.

Figure 17.1 Representation of the PRISMA flow diagram.

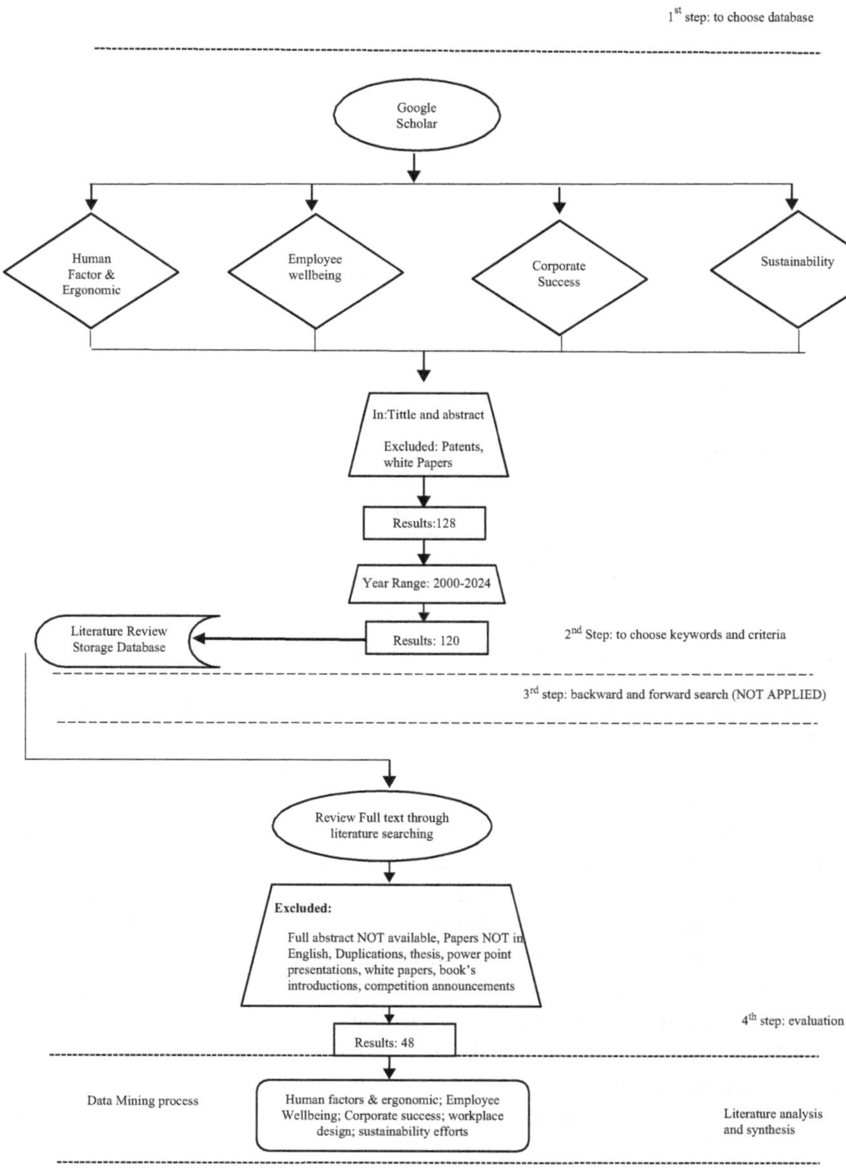


**FIGURE 17.1** Flow diagram for conducting a literature review.

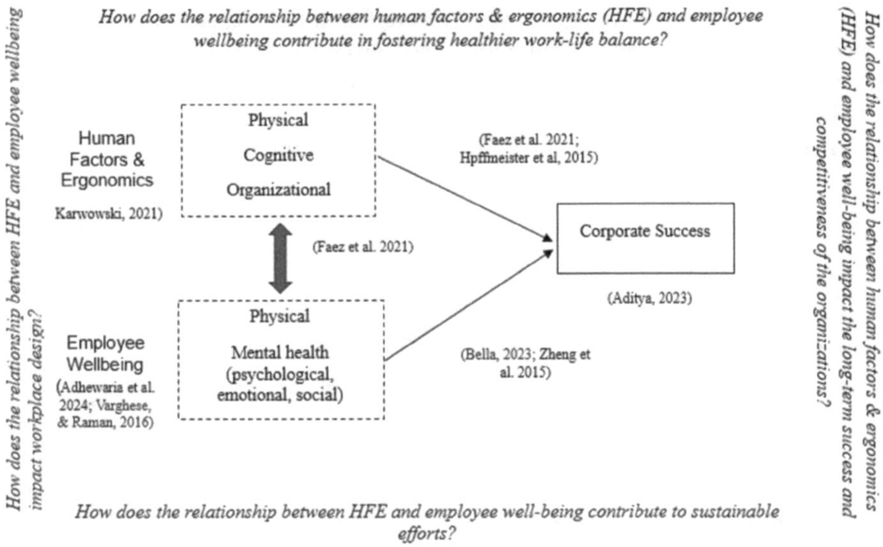


**FIGURE 17.2** Research framework.

## 17.6 DISCUSSION

Based on the review conducted, the different relationships between HFE, employee, well-being and corporate success were identified as depicted in Figure 17.2. These relationships can be examined individually as given in the discussions that follow:

### 17.6.1 Employee Well-being and Work-Life Balance

The study of HFE is essential for creating a healthy workplace environment, as it directly affects the physical, cognitive, and organizational aspects of employee well-being. Ergonomic principles aim to minimize strain and discomfort by designing workstations and instruments in a way that reduces the risk of musculoskeletal illnesses and promotes physical wellness. When it comes to thinking and understanding, factors like the amount of work, how decisions are made, and how difficult a task is can affect mental stress and cognitive tiredness. Organizational factors include elements such as communication, leadership styles, and work design, which can either facilitate or impede employee satisfaction and productivity.

Performance (e.g., productivity, efficiency, effectiveness, quality, innovation, flexibility, security and stability, reliability, and sustainable development) and well-being (e.g., health and safety, satisfaction, education, and personal development) are two interrelated system outcomes that can be achieved by providing healthier work environment. Specialists in HFE manage the practical and ethical trade-offs inside systems, balancing these, and other results (e.g. Wilson et al., 2009). There is a two-way street between performance and well-being; one may affect the other. This will help in attaining a healthy work-life balance.

It is impossible to exaggerate the importance of ergonomics in affecting work-life balance. The integration of ergonomic concepts into the workplace is crucial in creating a balanced relationship between personal well-being and professional obligations. The emotional and physical well-being of workers may be enhanced via the use of ergonomic workplace practices. Organizations and people alike may gain a lot by emphasizing ergonomics, and not only in terms of physical comfort. Human resources experts have taken the lead in fostering a positive work atmosphere after realizing the correlation between employees' mental health and their engagement, productivity, and contentment in their jobs.

There are two main categories of ergonomic interventions: engineering & administrative. Changes, redesigns, and replacements of machinery and tools are all part of engineering progress. Reducing the amount of heavy lifting and increasing the amount of light lifting are some administrative changes that may help employees. Other suggestions include making adjustments to the work schedule, offering recuperation time, and rotating workers through various positions occasionally to exercise different muscle groups. The physical well-being of an employee can be improved by incorporating such interventions (Yan et al., 2020). The responsibility for creating and enforcing policies that give attention to mental health and well-being lies with HR departments (Bella, 2023). Flexible work arrangements, paid time-offs, reconfiguring workstations, and installing ergonomic furniture can help in reducing stress of the workers. HR department may foster supportive and inclusive work environment by incorporating mental health considerations into policies and procedures (Marhaendra, 2023). All this will help in maintaining mental health, thereby, promoting healthy work-life balance. Moreover, it will aid in reducing stress and burnout, leading to enhanced performance and productivity.

### 17.6.2 Workplace Design

Several health issues such as muscular-skeletal disorders, lower back pain, stress, and burnout caused by an ergonomically undesirable setup. In order to prevent injuries, it is important to consider the prerequisite for ergonomic furniture and equipment in the office space. One of the best aids to implement ergonomics intervention at workplace is increased efficiency and productivity.

Moreover, employers are legally required to adhere to local occupational safety and health requirements. Establishing an ergonomic and secure work environment also promotes employee well-being. An organization's dedication to cultivating a secure and conducive work atmosphere improves human productivity and employee welfare, ultimately leading to business prosperity.

The majority of studies investigating the impact of ergonomic factors on workplace design have mostly examined industrial and production-oriented work settings. When contemplating workplace design, it is emphasized that attention should be given not just to physical conditions but also to psychosocial situations. A deficient psychosocial work environment has been consistently demonstrated as a causative element in musculoskeletal problems (Rubenowitz, 1997).

A well-designed workplace should give priority to ergonomic concepts that accommodate a wide range of users. It is important to incorporate these principles starting from the early stages of design in order to reduce the likelihood of musculoskeletal injuries and promote a conducive work environment. Ergonomic principles should be implemented not only in work environments but also in non-work situations. Engaging an ergonomist in the assessment and choice of furniture, with an emphasis on user requirements, can expedite procurement choices and avert the necessity for expensive replacements or the need to provide compensation to injured personnel resulting from non-ergonomic designs (Chim, 2019). It is imperative to highlight that a workplace designed with ergonomics in mind should improve employee satisfaction and well-being. This involves taking into account three degrees of comfort: physical, functional, and psychological. In order to accomplish this, it is crucial to include staff members at every stage of the design process, as their direct experiences can offer valuable feedback throughout the entire process.

### 17.6.3 Sustainability

Given the challenges posed by sustainable development adoption, its vision presents a wide scope for HFE involvement. While attention to sustainability may vary from public and political vantage points, it is important to understand that sustainability should not be considered merely as a passing trend but as a critical requirement for the future of humanity (Irfan & Qadeer, 2020).

Sustainability parameters not limited to health risks, climate change, biodiversity loss, global justice, etc., call for radical changes across disciplines, including HFE. Different methods and skills can be adopted from the HFE discipline, which are valuable for contributing towards the creation of more sustainable development paths (Fischer et al., 2021). Three key outcomes showcase HFE's role as a key contributing discipline for sustainability adoption and practice:

1. Its systems approach, substantiated by contemporary work system concepts.
2. The way it can fall in place with multidisciplinary approaches, highlighted by the contributions to global concerns and comprehensive design applications.
3. Its focus is on an ethical adoption of an overall systems performance, which considers global target groups and, instead of seeking short-term economic gains, prioritizes long-term benefits.

In order to leverage these potentials completely, traditional HFE concepts and models must make way for concepts that are more adaptable considering future challenges. In this light, HFE's focus needs expansion so as to include various phases, target groups, and sustainability dimensions across product and service lifecycles. HFE must adapt to address the complexity of contemporary and future challenges by embracing new perspectives and approaches that encompass broader social, environmental, and economic considerations.

### 17.6.4 Corporate Success

Physical, psychological, emotional, and social dimensions have been used in constructs to examine the linkages between human resource practices, ergonomics, and employee well-being. Past research shows that the relationships can contribute to long-term competitiveness in organizations. Studies indicate that the integration of ergonomics principles within human resources frameworks can contribute towards lowering the occurrences of musculoskeletal problems and workplace injuries, thereby uplifting physical well-being significantly. For instance, Robertson et al. (2021) concluded that ergonomics integration in workplace conditions and training practices resulted in fewer risk factors and improved employee comfort, resulting in enhanced physical health outcomes.

Hendrick (2008) has empirically found that a conducive work environment that promotes autonomy, job satisfaction, and lower stress levels leads to overall psychological well-being which indicates that it is impacted positively through the integration of ergonomic considerations in HR strategies. Oude Hengel et al. (2012) have shown a similar positive association through the findings of a meta-analysis using dimensions of psychological outcomes like employee engagement, job satisfaction, and perceived organizational support. However, some studies (eg. Reznik et al., 2022) opine that the relationships may not always be so straightforward as there can be factors that may lead to stress and anxiety at work. Organizations need to support the emotional needs of employees so that the threat of burnout may be countered. This will lead to sustained engagement and productivity.

Research also finds social support to be a critical determinant towards promoting employees' well-being and organizational competitiveness. Through their study which established the importance of HR initiatives towards networking and social collaboration, (Chevalier et al., 2019) demonstrated that social support within the workplace significantly predicted employee engagement and performance. Through the integration of ergonomic principles into HR practices, organizations can create environments that facilitate social interaction, trust, and teamwork, thereby nurturing a cohesive workforce capable of driving long-term innovation and adaptability (Wodajeneh et al., 2024). Past research contributes vital evidence that the relationship between HR, ergonomics, and employee well-being is vital to achieving sustained competitiveness in organizations, signifying the importance of strategic investments in human capital to encourage holistic employee flourishing (Bolis et al., 2023).

## 17.7 IMPLICATIONS

Practically speaking, this research implies that organizations might greatly benefit by giving priority to ergonomic measures that aim to improve employee well-being. Investment in ergonomic assessments, training, feedback, and workplace modifications by organizations can lower employees' physical strain, reduce the risk of injuries, and improve overall employee comfort. As a result, this brings about increased employee morale, job satisfaction, and retention rates, all leading to the success of the company.

Furthermore, the theoretical underpinnings of this research emphasize the need for a more robust comprehension of the interaction between human variables, organizational dynamics, and employee outcomes. The addition of the JD-R Model as a theoretical framework provides useful insights into the influence of ergonomic issues on job demands and resources. This, in turn, affects employee engagement, performance, and welfare. By utilizing these theoretical insights, policymakers and organizational leaders may make well-informed decisions to maximize workplace circumstances and foster beneficial results for both people and the organization as a whole.

## 17.8 CONCLUSION

This review paper emphasizes the crucial significance of HFE in influencing both employee well-being and corporate success. This research provides a thorough examination of HFE, covering physical, cognitive, and organizational ergonomics. It aims to clarify the complex connection between ergonomic interventions, employee wellness, and business performance. The results emphasize the significance of incorporating ergonomic concepts into workplace design, policies, practices, and HR initiatives to cultivate a healthier and more efficient workforce.

This research highlights the significance of implementing a comprehensive strategy to workplace design and management that gives priority to the well-being of employees by including ergonomic concepts and practices. A healthy, engaged and resilient staff may lead to sustained performance and competitiveness in the long term for companies when they acknowledge the interdependence between human factors, ergonomics, and organizational success.

## 17.9 FUTURE DIRECTION

In the future, researchers could focus on quantitative methods to examine the effects of ergonomic changes on certain organizational measures, such as productivity, absenteeism rates, and healthcare expenditures. Future studies can offer empirical evidence to back their claims by utilizing robust research procedures and statistical analyses to measure the impact of implementing ergonomic techniques on enhancing organizational performance. Furthermore, further investigation might explore the significance of emerging technologies, such as artificial intelligence and wearable devices, in enhancing ergonomic treatments and fostering employee well-being in the changing workplace environment. By exploring these paths, academics can further enhance our comprehension of how HFE principles might be utilized to establish healthier, more sustainable, and ultimately more prosperous organizations.

## REFERENCES

Adiga, U. (2023). Enhancing occupational health and ergonomics for optimal workplace well-being: A review. *International Journal of Chemical and Biochemical Sciences*, *24*(4), 157–164.

Araya León, M. J., Guasch, R., Estévez, A. T., & Peña, J. (2023). Interaction between the interior built environment and the human being. An integrative review in relation to perception, health, and well-being. *Theoretical Issues in Ergonomics Science*, *24*(6), 698–728.

Arksey, H., & O'Malley, L. (2005). Scoping studies: Towards a methodological framework. *International Journal of Social Research Methodology*, *8*(1), 19–32.

Aryee, S., Walumbwa, F. O., Seidu, E. Y., & Otaye, L. E. (2012). Impact of high-performance work systems on individual-and branch-level performance: Test of a multilevel model of intermediate linkages. *Journal of Applied Psychology*, *97*(2), 287.

Bakker, A. B., & Demerouti, E. (2007). The job demands-resources model: State of the art. *Journal of Managerial Psychology*, *22*(3), 309–328.

Bella, K. M. J. (2023). The power of employee well-being: A catalyst for organizational success. *International Journal of Scientific Research in Modern Science and Technology*, *2*(4), 20–26.

Bernard, F., Zare, M., Sagot, J. C., & Paquin, R. (2020). Using digital and physical simulation to focus on human factors and ergonomics in aviation maintainability. *Human Factors*, *62*(1), 37–54.

Bolis, I., Sigahi, T. F. A. C., Thatcher, A., Saltorato, P., & Morioka, S. N. (2023). Contribution of ergonomics and human factors to sustainable development: A systematic literature review. *Ergonomics*, *66*(3), 303–321.

Carayon, P. (2006). Human factors of complex sociotechnical systems. *Applied Ergonomics*, *37*(4), 525–535.

Chevalier, S., Fouquereau, E., Bénichoux, F., & Colombat, P. (2019). Beyond working conditions, psychosocial predictors of job satisfaction, and work engagement among French dentists and dental assistants. *Journal of Applied Biobehavioral Research*, *24*(1), e12152.

Chim, J. M. (2019). 6Ws in ergonomics workplace design. In *Proceedings of the 20th Congress of the International Ergonomics Association (IEA 2018) Volume VII: Ergonomics in Design, Design for All, Activity Theories for Work Analysis and Design, Affective Design 20* (pp. 1282–1286). Springer International Publishing.

Dul, J., Bruder, R., Buckle, P., Carayon, P., Falzon, P., Marras, W. S., ... & Van der Doelen, B. (2012). A strategy for human factors/ergonomics: Developing the discipline and profession. *Ergonomics*, *55*(4), 377–395.

Faez, E., Zakerian, S. A., Azam, K., Hancock, K., & Rosecrance, J. (2021). An assessment of ergonomics climate and its association with self-reported pain, organizational performance and employee well-being. *International Journal of Environmental Research and Public Health*, *18*(5), 2610.

Fischer, K., Thatcher, A., & Zink, K. J. (2021). Human factors and ergonomics for sustainability. In Salvendy, G. & Karwowski, W. (Eds.), *Handbook of Human Factors and Ergonomics* (pp. 1512–1527). John Wiley & Sons.

Fletcher, B., & Jones, F. (1996). Job control and health. In Schabracq, M. J., Winnubst, J. A. M., & Cooper, C. L. (Eds.), *Handbook of Work and Health Psychology* (pp. 121–142). John Wiley & Sons.

Hendrick, H. W. (2008). Applying ergonomics to systems: Some documented "lessons learned". *Applied Ergonomics*, *39*(4), 418–426.

Hoffmeister, K., Gibbons, A., Schwatka, N., & Rosecrance, J. (2015). Ergonomics climate assessment: A measure of operational performance and employee well-being. *Applied Ergonomics*, *50*, 160–169.

Irfan, D. S., & Qadeer, F. (2020). Employers' investments in job crafting for sustainable employability in pandemic situation due to COVID-19: A lens of job demands-resources theory. *Journal of Business & Economics*, *12*(2). 124–140.

Kadir, B. A., & Broberg, O. (2020). Human well-being and system performance in the transition to industry 4.0. *International Journal of Industrial Ergonomics*, *76*, 102936.

Kadir, B. A., Broberg, O., & da Conceicao, C. S. (2019). Current research and future perspectives on human factors and ergonomics in industry 4.0. *Computers & Industrial Engineering*, *137*, 106004.

Karwowski, W., & Zhang, W. (2021). The discipline of human factors and ergonomics. In *Handbook of Human Factors and Ergonomics* (pp. 1–37).

Khattak, S. A. (2021). Role of ergonomics in re-designing job design in call centres. *International Journal of Occupational Safety and Ergonomics*, *27*(3), 784–793.

Lima, T. M., & Coelho, D. A. (2018). Ergonomic and psychosocial factors and musculoskeletal complaints in public sector administration–a joint monitoring approach with analysis of association. *International Journal of Industrial Ergonomics*, *66*, 85–94.

Marhaendra, T. B. P. (2023). Ergonomics strategy to create work-life balance. *International Journal of Social Science Research and Review*, *6*(7), 575–585.

McMahan, E. A., & Estes, D. 2011. Measuring lay conceptions of well-being: The beliefs about well-being scale. *Journal of Happiness Studies 12*(2), 267–287.

Moray, N. (2000). Culture, politics and ergonomics. *Ergonomics*, *43*(7), 858–868.

Mowrer, O. H. (1939). A stimulus-response analysis of anxiety and its role as a reinforcing agent. *Psychological Review*, *46*(6), 553.

Naeini, H. S. (2020). Ergonomics on the context of sustainability: A new approach on quality of life. *International Journal of Architectural Engineering & Urban Planning*, *30*(2), 265–276.

Oude Hengel, K. M., Blatter, B. M., Joling, C. I., Van der Beek, A. J., & Bongers, P. M. (2012). Effectiveness of an intervention at construction worksites on work engagement, social support, physical workload, and need for recovery: Results from a cluster randomized controlled trial. *BMC Public Health*, *12*(Dec), 1–10.

Peters, M. D. J., Godfrey, C. M., Khalil, H., McInerney, P., Parker, D., & Soares, C. B. (2015). Guidance for conducting systematic scoping reviews. *International Journal of Evidence Based Healthcare*, 13(3), 141–146.

Pousette, M. T. A., Mikkelsen, P. K. K. L., Lappalainen, J., Tharaldsen, J., & Tómasson, K. (2008). *A Nordic questionnaire for assessing safety climate (NOSACQ)*. Working on Safety.

Read, G. J. M., Madeira-Revell, K. M. A., Parnell, K. J., Lockton, D., & Salmon, P. M. (2022). Using human factors and ergonomics methods to challenge the status quo: Designing for gender equitable research outcomes. *Applied Ergonomics*, *99*, 103634.

Ren, B., Zhou, Q., & Chen, J. (2023). Research on risk assessment model of subjective and objective integration in manual handling work. *Human Factors and Ergonomics in Manufacturing & Service Industries*, *33*(6), 464–475.

Reznik, J., Hungerford, C., Kornhaber, R., & Cleary, M. (2022). Home-based work and ergonomics: Physical and psychosocial considerations. *Issues in Mental Health Nursing*, *43*(10), 975–979.

Robertson, M. M., Tubbs, D., Henning, R. A., Nobrega, S., Calvo, A., & Murphy, L. A. (2021). Assessment of organizational readiness for participatory occupational safety, health and well-being programs. *Work*, *69*(4), 1317–1342.

Rubenowitz, S. (1997). Survey and intervention of ergonomic problems at the workplace. *International Journal of Industrial Ergonomics*, *19*(4), 271–275.

Schaufeli, W. B., & Bakker, A. B. (2004). Job demands, job resources, and their relationship with burnout and engagement: A multi-sample study. *Journal of Organizational Behavior: The International Journal of Industrial, Occupational and Organizational Psychology and Behavior*, *25*(3), 293–315.

Varghese, P., & Raman, R. (2016). Employee and corporate wellbeing: Evidence from the Indian corporate sector. *NHRD Network Journal*, *9*(4), 82–87.

Velasco, L. S. F., Revilla, P. E. R., Rodríguez, L. V. R., Santa Hincapié, M. P., Saavedra-Robinson, L. A., & Jiménez, J. F. (2022). A human-centred workstation in industry 4.0 for balancing the industrial productivity and human well-being. *International Journal of Industrial Ergonomics*, *91*, 103355.

Wilson, J. R. (2000). Fundamentals of ergonomics in theory and practice. *Applied Ergonomics*, *31*(6), 557–567.

Wilson, J. R. (2014). Fundamentals of systems ergonomics/human factors. *Applied Ergonomics*, *45*(1), 5–13.

Wilson, G. F., Estepp, J., & Davis, I. (2009, October). A comparison of performance and psychophysiological classification of complex task performance. In *Proceedings of the Human Factors and Ergonomics Society Annual Meeting* (Vol. 53, No. 3, pp. 141–145). Sage CA: Los Angeles, CA: SAGE Publications.

Wodajeneh, S. N., Azene, D. K., Berhan, E., & Sileyew, K. J. (2024). Impacts of ergonomic risk factors on the well-being and innovation capability of employees in the manufacturing industry. *International Journal of Occupational Safety and Ergonomics*, *30*(2), 412–424.

Yan, R., Basheer, M. F., Irfan, M., & Rana, T. N. (2020). Role of psychological factors in employee well-being and employee performance: An empirical evidence from Pakistan. *Revista Argentina de Clínica Psicológica*, *29*(5), 638.

Zare, M., Black, N., Sagot, J. C., Hunault, G., & Roquelaure, Y. (2020). Ergonomics interventions to reduce musculoskeletal risk factors in a truck manufacturing plant. *International Journal of Industrial Ergonomics*, *75*, 102896.

Zheng, X., Zhu, W., Zhao, H., & Zhang, C. H. I. (2015). Employee well-being in organizations: Theoretical model, scale development, and cross-cultural validation. *Journal of Organizational Behavior*, *36*(5), 621–644.

# 18 Enhancing Employee Well-Being and Operational Efficiency in the Oil Sector

## *A Holistic Examination of Ergonomics, Corporate Practices, and Human Factors Integration*

*Henry Jonathan,*
*Mohammad Sultan Ahmad Ansari,*
*and Raneem Al Ruhaili*

## 18.1 INTRODUCTION

In any workplace, workers should always be given top priority, and organizations are continually looking for methods to restructure workspaces and office equipment to support safety, healthier routines, and safer conditions. Therefore, ergonomics plays a crucial role in organizations as it focuses on optimizing the interaction between individuals and their workstation environment. It is a study of the fit between workers and their demands, work conditions, and the environment in which they work, and it focuses on enhancing the cognitive and physical aspects. It is a comprehensive term that describes the process of setting up a workspace, systems, and equipment in a way that makes it simple for employees to utilize them. The main goal of ergonomics is to decrease the possibility of workplace accidents and to ease any discomfort to improve workplace comfort. It aims to ensure that the environment, tasks, equipment, and information fit each other. Furthermore, to create a work environment that supports the employee's productivity, comfort, and well-being, which will result in improved satisfaction and job performance (UNC, 2023).

Nowadays, ergonomics has gained significant importance in high-risk environments such as the Oil and Gas Industry due to the increase in physically demanding work and the hazardous environment in the industry. It contributed to the employees'

 DOI: 10.1201/9781003532293-21

safety, well-being, quality, and productivity in the Oil and Gas industry. As we know, the workers in the Oil and Gas industry are working on offshore rigs and platforms, and they are exposed to many ergonomic injury risks, for example, pushing and pulling heavy loads, lowering, performing repetitive tasks, bending, and working in awkward body postures. (United States Department of Labor, n.d.) If ergonomics is not considered in the Oil and Gas Industry, employees will be injured, which will affect productivity and production. Ergonomic procedures and processes such as Ergonomic meetings and ergonomic job assessments are necessary to prevent employees' injuries in drilling operations phases and in the Oil and Gas industry's daily work tasks (Roth, 2006).

In Oman, many rules and regulations give ergonomics importance. For example, according to workshops and industrial safety specifications for offices and laboratories done by PDO, many requirements must be applied in the construction and design of workshops and offices. First, avoid sitting, working, or squatting at floor level in any area of work because it will cause medium- or long-term ergonomic-related health effects such as musculoskeletal injuries. Thus, providing workbenches for the employees is necessary. Second, all working surfaces and workbenches must be fit for working task purposes and constructed of materials of sound design so that the task can be completed safely without putting undue strain on the worker (Petroleum Development Oman L.L.C., 2020).

Nine to five employees are susceptible to ergonomic disorders. It can be avoided by taking some steps, such as encouraging the employees to alternate motions and postures, avoiding eye strain, maintaining neutral joint positions, using adjustable tools, and redesigning equipment. Ergonomic safety always considers the employee's physical demands, and it covers everything from the tools and methods the worker uses to do their work to whether those requirements are within their purview. Musculoskeletal disorders account for (30%) of all cases of days absent from work reported to (OSHA, n.d), according to the United States Bureau of Labor Statistics. MSDs cost organizations billions of dollars in lost productivity and time. Moreover, musculoskeletal conditions caused 37.5% of workplace accidents in the United States in 2019, which resulted in missed days at work. It serves as a sobering reminder of the prevalence of musculoskeletal problems in the workplace environment, emphasizing the necessity for companies to give ergonomic safety priority to safeguard their workers from harm and injuries. The risk of musculoskeletal diseases (MSDs) at work affects between (11%) and (30%) of construction employees, and (43.9%) of injuries related to ergonomics suffered by workers in the manufacturing sector occurred in 2016. Moreover, by utilizing an ergonomic keyboard, office employees can decrease their risk of musculoskeletal problems by (25%) (Stanch et al., 2023).

This study will help to conduct an ergonomic assessment and evaluation of the risk of ergonomic hazards for Petrogas (PGR) corporate office employees. To manage the ergonomic hazards that the employees face in their daily tasks working in the office. Moreover, to implement, control, and develop an ergonomic assessment for the workers that will decrease the hazards related to ergonomics in the office. In addition, to raise awareness of safe ergonomic work practices among the workforce.

### 18.1.1 Study Objectives

The purpose of the study is to understand the ergonomic assessment and evaluation of risk to safeguard employees at corporate offices from ergonomic hazards that will harm the musculoskeletal system. Because the workplace conditions in the oil and gas sector are typically very stressful, involving the handling of many manual work procedures, they tend to lead to ergonomic work-related hazards, hence the aim is to understand the impact, effect, and significance of the work activities on health. This study is undertaken to promote health and workplace safety. Workers must be aware of the common ergonomic hazards that can cause serious work-related injuries at the office. The following objectives are framed to accomplish the aims of the study.

**RO1**: Investigate and address the potential and existing ergonomic hazards at the corporate office.
**RO2**: Evaluate the risk and implement risk rating strategies.
**RO3**: Identify and assess the ergonomic factors and conditions in the office that can cause health concerns and injuries.
**RO4**: Propose a strategic measure that would enable them to reduce ergonomic hazards among the employees and recommend safer management practices related to ergonomic hazards.

## 18.2 LITERATURE REVIEW

According to Eyayo (2014), most of the hazards from working in the oil and gas sector are mechanical and ergonomic-related hazards from the survey conducted among oil and gas industry workers. Mechanical and ergonomics health hazards were the highest percentage (78.8%) of the identified health hazards compared with other types of health hazards. The study helped to increase awareness of the occupational health hazards that the workers in refineries face regularly. Moreover, it aimed to improve the organization's occupational health and safety management system by including the health risk assessment of the hazards, and its evaluation to mitigate the impact of the hazards. Hernan and Paola (2012) report the incidence of MSDs in the oil and gas sector increased by 67%. Studies show that the prevalence of MSDs in the shoulders, neck, wrist/hand, and knee parts of the body is particularly high among workers. A Colombian oilfield services firm is addressing MSDs by creating and putting into practice a plan to detect ergonomic hazards through many processes. First, using two distinct job assessment methods MAC and REBA. Then locating engineering, administrative, and behavioral controls as well as whole-job risk concerns. After completing the steps, ergonomic evaluation, a workplace intervention is implemented, which may involve designing mechanical aids, rearranging jobs, and making other adjustments to the workspace.

The research framework for the identification of possible ergonomic risk factors and mitigation solutions for Malaysian oil and gas drilling sectors is briefly highlighted in this study's research (Asad, et al., 2019). One of the main factors contributing to life-altering injuries and chronic disabilities among oil and gas drilling crews is an ergonomic hazard. The goal of this proposed study framework is to offer

Malaysian businesses the best possible hazard-control methods to address the ergonomic risk factors that are most likely to be connected to oil and gas drilling activities both onshore and offshore. Similarly, this suggested study will provide safety and health officials with prevention tactics and strategies to create efficient safety training and instructional designs based on the perspectives of ergonomic safety specialists.

Chim (2014), intended to create and develop a FITS model office ergonomics program, and its goal is to offer a methodical approach to managing the potential risk of musculoskeletal problems in office workers. It mentioned that a good office ergonomics program can provide beneficial outcomes in terms of lowering the incidence of musculoskeletal injuries, increasing productivity, and raising worker well-being and job satisfaction. The FITS Model, which is proposed in this work as a comprehensive office ergonomics program, considers four key components of an efficient program: stretching exercises and rest breaks, furniture evaluation and selection, individual workstation assessment, and training and education. Karhu et al. (1977) focused on explaining OWAS, which is an Ovako Working Posture Analyzing System that is a practical method that aims to evaluate and identify hazardous working postures. The method consists of an observational technique for evaluating working postures, and the second is to set criteria that aim to redesign working places to decrease ergonomic hazards. The evaluations of experienced employees and ergonomics specialists created the starting point for the criterion. They consider things like health and safety, but the major focus is on the discomfort brought on by the working positions. Koehl and Childre (2009) highlighted the importance of occupational health among nurses to enhance their ergonomic assessment and evaluation tasks. Many companies have a significant proportion of office-based workers. It is often thought that these workers operate in a safe environment. However, several hazards at work (such as ergonomics, inadequate indoor lighting, trip and fall risks, and poor indoor) may negatively impact employees' well-being. Occupational health nurses seek to minimize or eliminate these risks, as well as job-related and non-work-related injuries and cases of short- and long-term incapacity and use wellness programs to strengthen workers.

Madhwani and Nag (2017) aimed to assess the effectiveness of knowledge, attitude, and practice interventions on office ergonomics, an innovative strategy for preventing musculoskeletal pain (MSD) in corporate offices that can influence workers' behavior. Creating awareness among the workers of office ergonomics has become a top priority because of the increasing use of computers and hand-held devices globally. It aimed to prevent MSD caused by using this equipment, emphasis must be placed on maintaining proper work postures, ergonomically arranging workstations, maximizing chair functions, and doing desk stretch. This will encourage safe work habits in workplaces. Funminiyi et al. (2014) introduced an assessment among the academic and non-academic staff of a university was done to examine ergonomic hazards. They mentioned in the study the impact of ergonomic hazards on staff productivity, and the results showed that university employees are exposed to ergonomic risks while working in the office, and many of them are unaware of the health issues that ergonomic hazards can cause. Chaiklieng et al. (2010) emphasized "Work Ergonomic Hazards for Musculoskeletal Pain among University Office Workers." He aimed to investigate workplace ergonomic risks for university office workers who suffer from musculoskeletal pain. Musculoskeletal pain was 69.0%

prevalent within one month. The symptoms began while employees were at work, and prolonged sitting in the same position was the most stated reason. Moreover, muscular pain was substantially correlated with seat height. The results imply that ergonomic workstations should be adjusted according to the needs of each worker, and that standard requirements-compliant modifications in working circumstances should also be considered. Lima and Coelho (2011) examined ergonomic risk factors in 12 newly constructed office workstations. Based on data on pain complaints, a checklist was created to help in the process of spotting ergonomically poor circumstances. To calculate the risk of musculoskeletal disorders (MSDs), the Rapid Upper Limb Assessment (RULA) was chosen due to its applicability and suitability for the subject under investigation. An occupational health and safety technician used this technique to complete the most important task they had seen at work. Criticality was determined by observations and the methodical use of a checklist designed specifically for the office environment. To enhance working conditions and give guidance for all office settings, change recommendations were made. Quemelo et al. (2015) studied to evaluate the risk, prevalence, and symptoms of MSD among the employees of a Brazilian dairy company who work in offices. Thus, the results show that 74% (37 workers) had MSD symptoms. 70% (26 workers) suffered from upper limb pain, and 53% suffered from lower back pain that causes mild to moderate disability. The back and upper limbs were most often infected, and MSD symptoms were extremely common among the workers.

Sonne et al. (2012) report that over the past 20 years, there has been a significant rise in working with computers. In 2000, 60% of employees had computer use requirements as part of their jobs, and 80% of those employees said they used computers every day. For all those working with computers, it was estimated that MSDs were present between (10% and 62%). The Rapid Office Strain Assessment (ROSA) was created to determine the risks and dangers of computer utilization and provide a change level based on worker complaints of discomfort reports. ROSA is an effective and efficient tool for determining the use of computers' risk factors associated with discomfort. Mohammadipour et al. (2018) intended to determine the possibility that office employees at Kerman University of Medical Sciences would suffer from MSDs and ergonomic hazards. Two direct observations using ROSA and RULA methods were used. The findings show that the lower back (72.4%) and neck (55.2%) had the greatest prevalence rates of MSDs. According to the findings of the postural evaluation, (68.8%) of those who participated need more investigation to improve their posture, and (27.6%) need to change their posture immediately. Most office workers were categorized as medium-risk (55.2%) or high-risk (27.6%), according to the workstation investigation. According to the findings, organizations must also implement ergonomic training and workshops, and workstation designs must be altered.

According to research published by Samaei et al. (2015), his study is intended to identify the ergonomic risk factors impacting these illnesses and the prevalence of musculoskeletal conditions in a population of office employees. Due to the nature of office workers' jobs, they have reported a high frequency of muscular-skeletal issues. The findings show that in the last 12 months, 60.4% of workers reported having symptoms of (MSDs). The likelihood of developing MSDs has increased by (17.2%). According to the findings, the ROSA assessment is a useful tool for categorizing and identifying factors that impact the rate of MSDs among office employees. It was observed that implementing safety measures in hazardous workplaces, lowering the amount of time spent on computers, and engaging in regular sports activities can all help to lower the prevalence of MSDs in the research group (office employees).

The purpose of the study "Computer use increases the risk of MSD among newspaper office workers" which was done by Ortiz-Hernández et al. (2003) was to determine the frequency of musculoskeletal conditions among employees working at a newspaper in Mexico City. Moreover, examines the connection between musculoskeletal conditions and PC utilization and related ergonomic factors, and investigates the moderating role that psychological factors play in the relationship between ergonomic factors and musculoskeletal conditions. The findings reveal that workers at the newspaper who used computers performed editing work, and took uncomfortable postures were more likely to develop MSD.

## 18.3 METHODOLOGY

### 18.3.1 Study Area

The study deals with ergonomic assessment and evaluation of risks from ergonomic hazards in employees of some principal oil and gas companies. Both corporate office and field site offices are included in the study to assess the office workplace arrangements that contribute to ergonomic risks.

### 18.3.2 Data Collection Methods

Information was collected by combining visits and administering survey questionnaires. Periodical visits were made to both corporate offices and field site offices for the target oil and gas companies to achieve the objectives of the study. In addition, survey questionnaires are used to collect information from the employees during the visit. The questionnaire was prepared consisting of a total of 19 questions in six main sections to gather information about the objectives. In a few instances, interviews with some employees not placed in office work are also conducted to meet the requirements of the study.

Furthermore, secondary data sources are also referred to for gathering information on management practices, organization policies, and regulatory requirements. Sources include documents from oil and gas companies, websites, scientific studies published by researchers on ergonomic risks in the oil and gas sector, etc. are reviewed. Moreover, certain procedures followed by the oil and gas companies in ergonomic risk management are also referred to during the site visits.

## 18.4 RESULTS AND DISCUSSION

### 18.4.1 Demographic Results

The survey was distributed to the employees both at the corporate office and the field site office. From the survey conducted in the corporate office, we have received 257 responses in total, out of them 52.6% of the respondents are female, and 47.4%

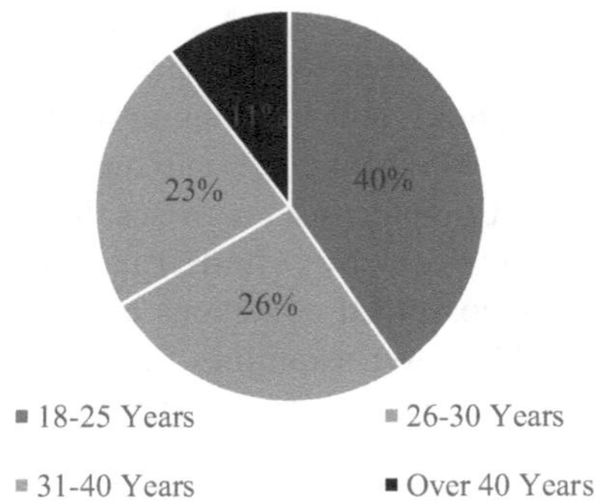


**FIGURE 18.1** Graph showing the age distribution of participants in the corporate office survey.

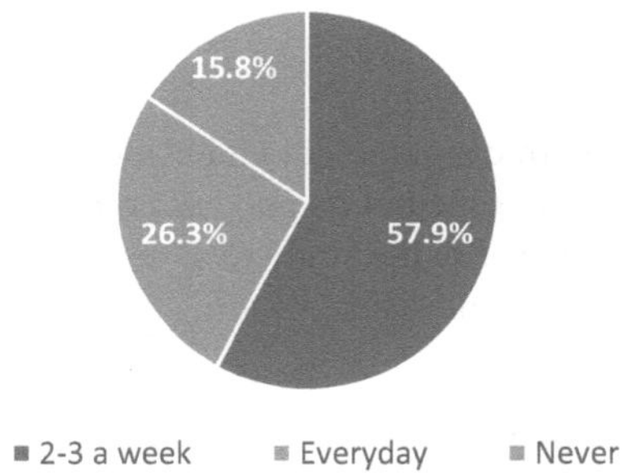


**FIGURE 18.2** Graph showing suffering from discomfort.

of them are male respondents in the corporate office. From the number of respondents who participated in the survey, most of the participants are in the age group of 18–25 years (40%) with the least number of participants in the age group over 40 years (11%) (Figure 18.1).

### 18.4.2 Potential and Existing Ergonomic Hazards at the Corporate Office and Field Site of Oil and Gas Companies

Analysis of potential and existing hazards at the corporate office shows that the majority, (57.9%) of the employees, were observed to be suffering from discomfort for 2–3 weeks, while around 26.3% of employees were facing discomfort every day, whereas only a few employees (15.8%) who are working in the corporate office have not experienced any ergonomic discomfort (Figure 18.2). Moreover, these frequent discomforts are known to affect the ability and interfere with the work, sometimes slightly (40.4%) and more severely (38.6%) among a smaller number of employees (Figure 18.3).

Investigation of employees mostly confined to desk work jobs revealed pain in wrists, and upper arms when operating keyboard and mouse which constitutes 61.4%

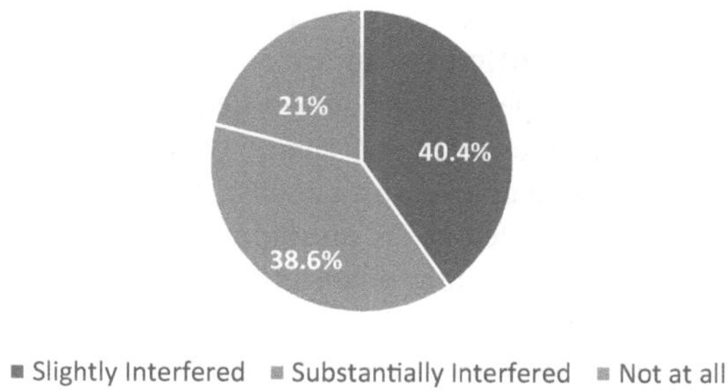


**FIGURE 18.3** Graph showing interferences with the work.

of them. However, 38.6% of employees have reported not experiencing such pain in the wrists and upper arms, as they were following precautions while working with desktop computers.

Observations to field site employees, in most of the oil and gas companies, oil well drillers manually lift the heavy hoists to place them into the drilling hole. Working for 12-h shifts, causing repetitive, twisting, bending and standing postures can significantly lead to MSD, neck pain, and back pain. While it is a fact that oil well drillers work in harsher environments and face tough tasks, ergonomic injuries are also part of their work exposures.

### 18.4.3 Risk Appraisal and Risk Rating Strategies

Risk appraisal forms an important part of oil and gas companies especially to assess the risk from different hazards and risks that workers are exposed to both at the office environment and field site. To counter these risks oil and gas companies have adopted risk management strategies such as hazards and effects management process (HEMP). The risk appraisal tool enables the oil and gas companies to hazard identification more systematically in the field operations and analyze the consequences or effects of the risks that arise from drilling and processing. HEMP facilitates the assessment and quantification of the risks by assigning ratings to risks depending on the impact and severity of ergonomic injuries to workers. Oil and gas companies have started to develop and adopt mechanisms for risk rating strategies from risk matrix, risk scoring, risk categorization, qualitative assessment, dynamic risk assessment, etc. that are enabling them to prioritize the resources and manage ergonomic-related issues.

### 18.4.4 Recognition and Gauging the Ergonomic Factors and Conditions Causing Health Concerns and Injuries

On examination of work-related disorders, it was observed that 72% of the employees in the corporate office were known to suffer from Musculo-skeletal, repetitive strain, and chronic back and neck pain while 28% of them have not reported suffering from any of the above work-related disorders. Around 58% of the employees reported they have well-designed chairs to support their lower back, while the rest feel the chairs do not provide much relief to the lower back. In addition, adjusting the seat height of chairs shows a significant difference in knee pain, where 47.4% of employees in the corporate office have experienced comfort and relief with proper seat height adjustments. Survey on the adjustability of chairs, 68.4% of them are using adjustable chairs that can assist the employees in increasing, or decreasing the height and flexible back support rest. This feature is an important factor that contributes to a reduction in back, neck, and knee pains when sitting and working for long hours. In addition, proper arrangement of the workstation for desk work is also an important consideration for assessing the factors leading to upper limb disorders. Facilities like arms rest, level of keyboard, placement of display screen equipment, and positioning of the seating chair all play a significant role in easing the risk to workers from experiencing ergonomic injuries. Though chair positioning and adjustment of display screen equipment distance depend on each worker's preference and convenience, insufficient space and lack of knowledge of workplace setup would potentially contribute to the risks. Our observations reveal that 66.7% of the employees are not able to make the right adjustment to keyboard, mouse, chair, and display screen equipment distance convenient to operate. Moreover, the employees have raised concerns of wrist pain, and eye strain occasionally because of improper ergonomic postures seen in 68.4% of employees working in corporate offices. Positioning the monitor at the right eye level is necessary to reduce eye glare, neck pain, and eye strains, around 65% of employees' monitors are positioned above the eye level while in other cases they are below eye level, which is a leading cause among employees experiencing neck pain and eye strain.

Observations on the working environment of employees in field site offices show employees are working in poor ergonomic work conditions. The number of work hours, workstation setup, adjustable chairs, work desks, reach of keyboard, mouse, display screen equipment, monitor, etc. are all factors considered to assess the significance of health concerns and ergonomic injuries. The survey indicates the workplace ergonomic assessment and not thoroughly followed, and employees are likely to face risks from poor ergonomic hazards.

### 18.4.5 Strategic Measures to Reduce Ergonomic Hazards Among Employees in Corporate Offices

Measures taken by oil and gas companies to mitigate the ergonomic risks to employees are examined, and responses show that employees are practicing rest breaks at indefinite intervals of work to avoid continuous exposure to monitors. Employees are either shifting their eye view away from the monitor or away from the workstation. Around 63% of the employees are found to be looking away from the monitor as a means of relief from eye strain, while the rest of the employees do not consider taking rest breaks. As we know, taking breaks is necessary during the day, 63.2% responded that they take regular breaks and stretch when sitting for a long duration. Around 42% of employees are taking one rest break per day to get up from their desks, 35% are taking 2–3 breaks and 22% of employees taking more than 3 breaks per day. Strategic measures to mitigate the ergonomic risks include improvising seating arrangements, and workplace ergonomic assessment to assess the risks from poor workstation settings. Moreover, inspecting improper functioning of chairs, adjusting monitors' view and angle to avoid glaring, provisioning of wrist support, etc. are aspects that should be reviewed every day by employees themselves before the start of work. Moreover, practicing rest breaks, either being away from workstations or staring away from monitors can significantly reduce the risks of ergonomic hazards while working in office environments.

Working in the corporate office for long hours could cause many occupational diseases, such as MSDS, back pain, neck pain, trigger finger, and carpal tunnel syndrome. From the responses, it is evident that the females were more than the males in the office because working in the office is more suitable for females than males. Moreover, the majority of responses were from ages 18–25 because working in this job needs the youth more than those over 40. The discomfort that the employees feel while working in the office substantially interferes with their work because if the working environment is not comfortable, it will affect their productivity. On the other hand, providing the workers with adjustable desks, comfortable chair that supports the spine and neck, and larger screens free from glare and lower than eye level, reducing the working hours, and providing many breaks will help to reduce work-related injuries and will reduce the employee's discomfort. Though the analysis of the risk rating from petrogas risk management document indicates a low to medium level of risk, but the study reveals most employees are at high risk of ergonomic hazards at PGR office. As a result, the risk rating is high. Furthermore, the control measures that were mentioned are not effective in mitigating the ergonomic hazards. Many changes should be made in the future to promote a safety culture and prevent work-related ergonomic injuries in the PGR corporate office.

Working in the Rima field has many ergonomic hazards. As mentioned in the results section, the driller who is working in a hoist is working for 12 h per day for 2 weeks, which exceeds the normal working hours. Furthermore, they are working in high temperatures without being provided with the proper personal protective equipment. As a result, most workers are suffering from Carpal tunnel syndrome, chronic back and neck pain, muscle pain, and tendinitis. It is clear from the interview that working in the field is more suitable for males because of their capability to work in hard conditions in the desert.

## 18.5 CONCLUSION

The study started with an introduction to the ergonomic hazards in the oil and gas industries in Oman and globally, as well as in global offices. The study focused on four main objectives: investigating the potential ergonomic hazards, evaluating the risk and implementing risk rating, identifying and assessing ergonomic factors and conditions, and finally proposing a mitigation measure to reduce the ergonomic hazards. As we know, working in an office has many ergonomic hazards because of the long hours that the workers are in one posture without changing it. Most ergonomic hazards arise from working in an office. Implementing the study of ergonomic assessment and evaluation of the risk in the corporate office will help reduce the ergonomic hazards among the employees, create a safe workplace environment where employees are encouraged to work and promote a safety culture. Furthermore, it aims to identify the ergonomic hazards that the employees are suffering from and try to implement mitigation measures to prevent work-related injuries in the office. This study helps oil and gas companies improve their offices, and it is beneficial for other workers in the office to make sure that they are working, and sitting in a safe posture that will not affect their health and will not cause work-related diseases such as MSDS, carpal tunnel syndrome, muscle pain, and back and neck pain. It aims to encourage the other companies to develop and improve their offices. The improvement will start after implementing the ergonomic assessment and getting responses from the employees.

## REFERENCES


Asad, M. M., Hassan, R., Latif, K., & Sherwani, F. (2019, June). Identification of potential ergonomic risk factors and mitigating measures for Malaysian oil and gas drilling industries: A conceptual research proposition. In *IOP Conference Series: Materials Science and Engineering* (Vol. 530, No. 1, p. 012052). IOP Publishing.

Chaiklieng, S., Suggaravetsiri, P., & Boonprakob, Y. (2010). Work ergonomic hazards for musculoskeletal pain among university office workers. *Walailak Journal of Science and Technology (WJST)*, *7*(2), 169–176.

Chim, J. M. (2014). The FITS model office ergonomics program: A model for best practice. *Work*, *48*(4), 495–501.

Eyayo, F. (2014). Evaluation of occupational health hazards among oil industry workers: A case study of refinery workers. *Journal of Environmental Science, Toxicology and Food Technology*, *8*, 2319–2399.

Funminiyi, J., Akinlolu, A., & Agboola, P. (2014). Assessment of ergonomic hazards and techno-stress among the workers of Obafemi Awolowo University, Ile-Ife, Osun State, Nigeria. *Australian Journal of Business and Management Research*, *4*(1), 27–34.

Hernan, U. J., & Paola, R. M. (2012, September). Assessment and strategic approach for ergonomic issues in critical jobs in the oil and gas workforce. In *SPE International Conference and Exhibition on Health, Safety, Environment, and Sustainability?* (pp. SPE–157451). Paper presented at the International Conference on Health, Safety and Environment in Oil and Gas Exploration and Production, Perth, Australia, September 2012. SPE.

Karhu, O., Kansi, P., & Kuorinka, I. (1977). Correcting working postures in industry: A practical method for analysis. *Applied Ergonomics*, *8*(4), 199–201.

Koehl, B., & Childre, F. (2009). An effective office ergonomic assessment and intervention program. *AAOHN Journal*, *57*(12), 488–490.

Lima, T. M., & Coelho, D. A. (2011). Prevention of musculoskeletal disorders (MSDs) in office work: A case study. *Work*, *39*(4), 397–408.

Madhwani, K. P., & Nag, P. K. (2017). Web-based kap intervention on office ergonomics: A unique technique for prevention of musculoskeletal discomfort in global corporate offices. *Indian Journal of Occupational and Environmental Medicine*, *21*(1), 18.

Mohammadipour, F., Pourranjbar, M., Naderi, S., & Rafie, F. (2018). Work-related musculoskeletal disorders in Iranian office workers: prevalence and risk factors. *Journal of Medicine and Life*, *11*(4), 328.

Ortiz-Hernández, L., Tamez-González, S., Martínez-Alcántara, S., & Méndez-Ramírez, I. (2003). Computer use increases the risk of musculoskeletal disorders among newspaper office workers. *Archives of Medical Research*, *34*(4), 331–342.

OSHA. (n.d.). *Oil and Gas Extraction*. Department of Labor Logo United States department of Labor. https://www.osha.gov/oil-and-gas-extraction/hazards

Petroleum Development Oman L.L.C. - hsedocs.pdo.com.om (2020). *Workshops and Industrial Safety Specification Including Offices & Laboratories*. Available at: https://hsedocs.pdo.com.om/HSEDSVisioFiles/SP-1256.pdf (Accessed: 04 July 2023).

Quemelo, P. R. V., Gasparato, F. D. S., & Vieira, E. R. (2015). Prevalence, risks and severity of musculoskeletal disorder symptoms among administrative employees of a Brazilian company. *Work, 52*(3), 533–540.

Roth, C. (2006, July 13). Fueling ergonomics in the oil and gas industry. *EHS Today*. https://www.ehstoday.com/health/article/21913785/fueling-ergonomics-in-the-oil-and-gas-industry

Samaei, S. I., Tirgar, A., Khanjani, N., Mostafaee, M., Bagheri Hosseinabadi, M., & Amrollahi, M. (2015). Assessment of ergonomics risk factors influencing incidence of musculoskeletal disorders among office workers. *Journal of Health and Safety at Work, 5*(4), 1–12.

Sonne, M., Villalta, D. L., & Andrews, D. M. (2012). Development and evaluation of an office ergonomic risk checklist: ROSA–Rapid office strain assessment. *Applied Ergonomics, 43*(1), 98–108.

Stanch, P. M., Moore, A., & Mcglothlin, J. D. (2023, February). Ergonomics: Fitting the task to the worker. *The Synergist*. https://synergist.aiha.org/202302-ergonomics-task-worker

UNC. (2023, May 24). *Ergonomics*. https://ehs.unc.edu/topics/ergonomics/

# 19 Enhancing Organizational Success

## *Exploring Ergonomics, Employee Deviant Workplace Behavior and Occupational Burnout*

*Ujjal Bhuyan, Shad Ahmad Khan, Alina Costin, and Anuj Kumar*

## 19.1 INTRODUCTION & BACKGROUND OF THE STUDY

The concept of human factor engineering has gained importance in the fields of organizational success and enhancing employee productivity. It relates to incorporating physical and psychological wellness aspects to designing systems, processes and products. The underlying intent is to enhance employee productivity, reduce occupational stress and burnout, and restrict human error (Gomathi & Rajini, 2019). In addition to this, in today's knowledge-based economies, employee commitment and satisfaction have a pivotal role in an organization's success, as these economies put thrust on human capital and intellectual capital. Extensive research in the field of employee satisfaction and deviant workplace behavior (DWB) portrayed favorable impact of the working environment on employee productivity and job satisfaction (performance). Organizational stress and burnout are not separately inflicted upon by the employee or his/her work environment, but it is the result of incongruence between the two (Person-Environment Theory of Stress). These concerns have laid the foundation for the organization's deliberate inclination towards ergonomics or ergonomic planning in workplace. The term 'Ergonomics', originating from Greek literature ('Ergo' means work and 'Nomos' means Natural law) has a history that goes way back to 400 BC, when Archeologists had uncovered sculptures and designs, portraying chairs with contoured backs. It refers to the structuring of workplaces, tasks, tools, and systems in a way that fits an individual's physical and cognitive abilities and limitations, that has an importance bearing on their level of engagement and job performance (Gumasing et al., 2023). It refers to the scientific domain that caters towards comprehending engagement of people with other components of the system or arrangements with an intent to optimize people's comfort and system-wide effectiveness (Mohamed et al., 2020). This science of designing


DOI: 10.1201/9781003532293-22

people-centric workplace elements and developing an organizational culture has an important implication towards psychological well-being of people and workforce productivity. A conducive working environment has a significant impact on job satisfaction and overall productivity at workplaces, which are contributing factors in determining organizational success in terms of enhancing brand reputation, customer contentment, and perceived service value (Díaz-Carrión et al., 2020). It has an important implication towards psychological well-being of employees in an organization (Christy & Duraisamy, 2020). An ideal workstation design would not only invite arrangements that focus upon meeting the workers' needs and equip them with a setting that enables them to maximize their full potential. As per International Labour Organisation (ILO) statistics, approximately, 2.3 million men and women lose their precious lives owing to job-related hazards or illness every year (World Statistic, 2011). Moreover, the past two decades have seen a constant rise in work-related musculoskeletal disorders, one of the major factors influencing the prevalence of occupational diseases. This has led to significant loss in labor hours, dip in job productivity and quality of life (Chheda & Sreeraj, 2020). This has raised concerns among policy makers across the world to put due thrust on regulating workplace ergonomics (safety, hygiene, and working conditions). The Government of India has come up with the Occupational Safety, Health and Working Conditions Code 2020 to regulate workplace elements to make it more employee-friendly. Previous literature studies have highlighted various dimensions of ergonomic variables, physical, cognitive, organizational, and spiritual ergonomics, playing a crucial role in workplace. The relationship between ergonomics and employee commitment (or productivity) can be established with the 'Social Exchange Theory' framework. According to social exchange theory (SET), exchanges between participants in mutually dependent interactions result in an expectation of reciprocity and perhaps the practice of reciprocity (Huang et al., 2016). Employees find it obligatory to value and serve the organization well, if they believe the organizations to be supportive and invested in them. Understanding these workplace ergonomic principles and practices can bring about a major departure from the existing workplace designs and systems and contribute significantly towards human resource management. Two common issues faced by organizations with respect to human resource management are: Deviant behavior and burnout. Employees exhibiting undesirable or deviance behavior from acceptable organizational norms is both pervasive and an expensive affair to the company. It is voluntary in nature which might result from a lack of motivation arising from unsuitable physical, organizational, and cognitive hindrances at workplace. Furthermore, occupational burnout also seems to be rising at an alarming rate and would prove to be detrimental to the organization at large. The present study attempts to shed light on the role of workplace ergonomics in employees' experiencing burnout and exhibiting DWB among academic stuff. This scope was lacking in previous literatures and would add to the uniqueness to the study.

### 19.1.1 Ergonomics and Occupational Burnout

The disease epidemiological landscape in India has seen a major shift from communicable diseases to non-communicable diseases (NCD) over the previous two decades. As per WHO report, NCD fatalities across the globe account for 74%;

the percentage is 66% for India. One of the major contributors towards NCD is stress (Stress in India, 2023). Workplace arrangements and norms have a significant (either positive or negative) impact on the employees' physiological and psychological conditions. Adverse workplace conditions could invite unwanted consequences; occupational burnout syndrome is an important concern at workplace for the managers, as it has a detrimental effect on employee productivity and psychological well-being. Individuals encounter burnout syndrome, owing to their inability to handle emotional pressure, thereby experiencing psychological exhaustion. Occupational stress can be defined as an "emotional state arising from psychosocial weariness when one fails to cope with job-related stress" (Baquero, 2023, p. 2). It can be expressed as skepticism and mental weariness, which is prevalent among people (Maslach & Jackson, 1981). Ergonomic designs and systems can prove to be pivotal in addressing employee stress reduction and making workplace more comfortable (Ayanyinka & Chukwvemeka, 2022). Evidences show that adopting people-centric approaches in organizational workplace arrangements can influence psychological and social working conditions and prevent work-related stress (Rypicz et al., 2023). Based on this, the researchers propose the following hypothesis:

**H01**: Ergonomics (physical, cognitive, and organizational) play a positive role in addressing occupational burnout.

### 19.1.2 Burnout and DWB

Studies on deviant workplace behavior (DWB) have gained importance over the years among researchers as it has detrimental implications for an organization and its workforce (Banks et al., 2012; Lubbadeh, 2021). It has the potential to undermine internal functioning of an organization and result in excessive cost borne by it. Moreover, such behaviors might put employees' career and personal growth at stake. DWB can be defined as the "deliberate behavior by an organization member, which the organization deems to be contrary to its legitimate interests" (Lubbadeh, 2021, p. 50). These are voluntary acts on the part of employees that are necessarily against the organizational norms and endanger the well-being of the organization and its employees in the process (Robinson & Bennett, 1995, p. 349). Previous research has identified a captivating connection between DWB and occupational burnout. Employees who have experienced emotional exhaustion would exhibit lower level of satisfaction and commitment towards the organization, which results in counterproductive work behaviors (Mulki et al., 2006). Song et al. (2021) and Liang and Hsieh (2007) in their studies have also advocated for a positive between the aforesaid behaviors. Moreover, studies have found that emotional exhaustion or burnout would increase the likelihood towards depersonalization. Experiencing depersonalization has been found to be an important predictor of DWB (Chullen et al., 2010). Based on this, the researchers propose the following hypothesis:

**H02**: Experiencing organizational burnout influences employees' DWB.

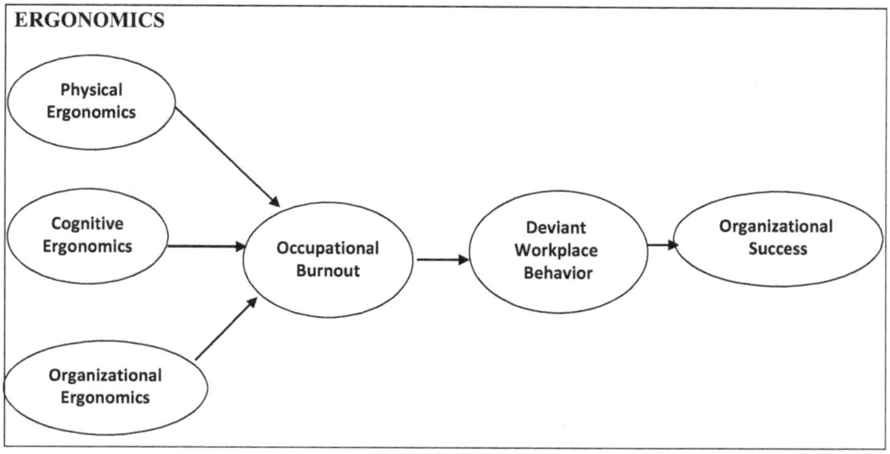


**FIGURE 19.1** Research model.

### 19.1.3 DWB AND ORGANIZATIONAL SUCCESS

Irrespective of an organization's products or services, the market they cater to or scale of operation, every organization strives towards survival and its sustainability in the long run. Organizational success has remained the cornerstone behind any organization's policy and strategy. To achieve this, investment in organization's human resource is of pivotal importance (Emeka et al., 2020). The significance of organizational success lies in its ability to measure the degree to which organizations meet their goals in spite of obstacles. Miller and Friesen (1978, p. 469), in their study, defined organizational success as "the degree to which firms are able to achieve their objectives subject to the constraints of long-term viability". Implementation of Ergonomic principles in organizational domain can bring about significant improvement in workforce productivity and performance, which is crucial for any organizational success. Apart from SET, affective event theory, as a theoretical framework, did support the linkage between DWB and organizational success. The theory posits that occurrences at workplace could trigger emotional reactions that might affect workers' attitude and behavior (Vroom, 1964). Employees' whose perceived organizational support and perceived supervisor support is not satisfactory is more likely to experience organizational burnout (stress) and eventually demonstrate counterproductive workplace behavior. Based on this, the researchers propose the following hypothesis (Figure 19.1):

**H03**: Employees' DWB negatively affects organizational success.

## 19.2 METHODOLOGY

A descriptive-analytical cross-sectional study design has been employed to carry out the present study. The data has been collected from a sample of 42 respondents working at varied institutions ranging from education, real estate, banking, social

**TABLE 19.1**
**Sources of Questionnaire Items**

| Sl. No. | Variables | Scale | References |
|---|---|---|---|
| 1. | Organizational ergonomics (OE) | 7-point Likert Scale | Tiongco et al. (2024) |
| 2. | Cognitive ergonomics (CE) | 7-point Likert Scale | Paajanen and Kalakoski (2017)<br>Tarvainen (2023) |
| 3. | Physical ergonomics (PE) | 7-point Likert Scale | Tiongco et al. (2024) |
| 4. | Organizational success (OS) | 7-point Likert Scale | Emeka et al. (2020) |
| 5. | Occupational burnout (OB) | 7-point Likert Scale | Gomathi and Rajini (2019) |
| 6. | Deviant Workplace Behavior (DWB) | 7-point Likert Scale | Chullen et al. (2010) |

*Source:* Authors' compilation.

works, etc. Out of the total sample size, 45.2% of the respondents were male, and 52.4% were female, whereas 1 preferred not to reveal gender identity. Majority of the respondents (45.2%) were in the age group of 20–30 years, 33.3% were in the age group of 30–40 years, and the remaining respondents were in the age group of 40 and above.

A well-structured questionnaire consisting of 58 items was administered to respondents to collect the required data via Google forms. Before administering the questionnaire to the respondents, the contents of it have been diligently discussed with subject experts for assessing the validity of the construct. The details of the concerned variables in the study have been stated below. All the items of the concerned variables in the study were drawn from the prior studies and a 7-point Likert scale has been adopted. The gathered data was analyzed through smart PLS, involving Partial Least Square Structural Equation modeling as per the recommendation of Ringle et al. (2015), Khan et al. (2021), Khan and Magd (2021), Shiva et al. (2020). This study being an indicative study, the sample size of 35 or more was acceptable and was feasible as performed by Purwanto (2021). The questionnaire items were inspired by the review of literature primarily by the sources mentioned in Table 19.1.

These constructs were assigned measuring items as question statements as provided in Table 19.2.

## 19.3 ANALYSIS AND INTERPRETATION

### 19.3.1 Model Quality Assessment

As a first step, the model assessment is performed where the factor load, reliability, and validity are checked.

As shown in Table 19.3, the factor loadings for many items were not satisfactory, as they were below the recommended threshold of 0.6 (considered liberal). However, items with values less than 0.6 were still retained primarily due to their performance in the model, as indicated by the reliability and validity values [with thresholds set

**TABLE 19.2**
**Items/Statements Used in the Questionnaire**

| Item Code | Item/Statement Used in the Questionnaire |
|---|---|
| PE1 | Safety in your workplace |
| PE2 | Hygiene in your workplace |
| PE3 | Repetitive movements in your workplace |
| PE4 | Carry or move objects |
| PE5 | Physical loads in terms of strength, neck, shoulder and back |
| PE6 | Workplace design |
| PE7 | Workloads are well distributed |
| PE8 | In general, how satisfied are you with the physical demands in your workplace? |
| OE1 | Communication with bosses |
| OE2 | Communication with colleagues |
| OE3 | Teamwork |
| OE4 | Your work schedule |
| OE5 | My roles and responsibilities are well defined |
| OE6 | The new paradigms in my work |
| OE7 | Suggestions and ideas are taken into account in my work team |
| OE8 | Quality improvement initiatives |
| OE9 | The way in which the institution is managed |
| OE10 | Resource management |
| OE11 | In my job, I can develop my skills |
| OE12 | I receive information on how I perform my work |
| OE13 | Medical services |
| CE1 | Is there some aspect of the working conditions that makes it difficult to see, hear or notice things? |
| CE2 | Is your memory load reasonable? |
| CE3 | Is communication clear and are instructions, memos and other documents clear and non-contradictory? |
| CE4 | The use of tools and equipment in my workplace causes mental fatigue |
| CE5 | Do you have enough information and support for problem-solving and decision-making? |
| CE6 | Do you have clear and concrete competence requirements? |
| CE7 | Do you have an appropriate amount of new things to learn? |
| CE8 | Is your working environment disturbance-free? |
| CE9 | Is there too much information to manage? |
| CE10 | Are your tools and systems functional and appropriate? |
| CE11 | I feel stressed due to the high work load assigned to me |
| CE12 | I feel burnout out because of the highly repetitive task assigned to me |
| DWB1 | Spent too much time fantasizing or daydreaming instead of working |
| DWB2 | Taken an additional or longer break than is acceptable at your workplace |
| DWB3 | Come in late to work without permission |
| DWB4 | Littered your work environment |
| DWB5 | Neglected to follow your boss's instructions |
| DWB6 | Intentionally worked slower than you could have worked |
| DWB7 | Discussed confidential company information with an unauthorized person |
| DWB8 | Put little effort into your work |
| DWB9 | Dragged out work in order to get overtime |

*(Continued)*

**TABLE 19.2 (*Continued*)**
**Items/Statements Used in the Questionnaire**

| Item Code | Item/Statement Used in the Questionnaire |
|---|---|
| DWB10 | Acted rudely toward someone at work |
| DWB11 | Said something hurtful to someone at work |
| OB1 | I feel emotionally drained from my work and have to face another day on the job |
| OB2 | I feel used up at the end of the workday |
| OB3 | I feel fatigued when I get up in the morning |
| OB4 | Working with people all day is really a strain for me |
| OB5 | I feel burned out from my work |
| OB6 | I feel I am working too hard on my job |
| OB7 | I feel frustrated by my job |
| OB8 | Working with people directly puts too much stress on me |
| OB9 | I feel like I am at the end of my rope |
| OS1 | There is efficiency in the organization as a result of its efficient workforce |
| OS2 | Efficient and productive workforce help an organization to be effective |
| OS3 | Employees work towards the corporate vision and mission as well as individual objectives |
| OS4 | Happy and productive employees are more confident and competent and are more likely to have a better impact on the organizational success |
| OS5 | Less stressed employees are more productive and innovative |

**TABLE 19.3**
**Model Assessment**

| Item | Load | Cronbach's Alpha | Composite Reliability (rho_a) | Composite Reliability (rho_c) | Average Variance Extracted (AVE) |
|---|---|---|---|---|---|
| **CE** | | 0.81 | 0.884 | 0.853 | 0.544 |
| **CE1** | 0.176 | | | | |
| **CE4** | 0.888 | | | | |
| **CE7** | 0.243 | | | | |
| **CE9** | 0.881 | | | | |
| **CE11** | 0.934 | | | | |
| **CE12** | 0.858 | | | | |
| **DWB** | | 0.939 | 0.956 | 0.948 | 0.627 |
| **DWB1** | 0.69 | | | | |
| **DWB2** | 0.645 | | | | |
| **DWB3** | 0.689 | | | | |
| **DWB4** | 0.594 | | | | |
| **DWB5** | 0.883 | | | | |
| **DWB6** | 0.826 | | | | |
| **DWB7** | 0.835 | | | | |
| **DWB8** | 0.798 | | | | |
| **DWB9** | 0.888 | | | | |
| **DWB10** | 0.923 | | | | |
| **DWB11** | 0.863 | | | | |

(*Continued*)

**TABLE 19.3 (*Continued*)**
**Model Assessment**

| Item | Load | Cronbach's Alpha | Composite Reliability (rho_a) | Composite Reliability (rho_c) | Average Variance Extracted (AVE) |
|---|---|---|---|---|---|
| **OB** | | 0.973 | 0.975 | 0.977 | 0.823 |
| **OB1** | 0.961 | | | | |
| **OB2** | 0.954 | | | | |
| **OB3** | 0.941 | | | | |
| **OB4** | 0.892 | | | | |
| **OB5** | 0.95 | | | | |
| **OB6** | 0.874 | | | | |
| **OB7** | 0.866 | | | | |
| **OB8** | 0.862 | | | | |
| **OB9** | 0.856 | | | | |
| **OE** | | 0.948 | 0.954 | 0.955 | 0.621 |
| **OE1** | 0.82 | | | | |
| **OE2** | 0.721 | | | | |
| **OE3** | 0.805 | | | | |
| **OE10** | 0.801 | | | | |
| **OE11** | 0.585 | | | | |
| **OE12** | 0.843 | | | | |
| **OE13** | 0.639 | | | | |
| **OE4** | 0.672 | | | | |
| **OE5** | 0.883 | | | | |
| **OE6** | 0.892 | | | | |
| **OE7** | 0.838 | | | | |
| **OE8** | 0.861 | | | | |
| **OE9** | 0.814 | | | | |
| **OS** | | 0.776 | 0.899 | 0.841 | 0.525 |
| **OS1** | 0.525 | | | | |
| **OS2** | 0.912 | | | | |
| **OS3** | 0.68 | | | | |
| **OS4** | 0.849 | | | | |
| **OS5** | 0.577 | | | | |
| | | 0.906 | 0.954 | 0.921 | 0.596 |
| **PE1** | 0.749 | | | | |
| **PE2** | 0.766 | | | | |
| **PE3** | 0.787 | | | | |
| **PE4** | 0.646 | | | | |
| **PE5** | 0.65 | | | | |
| **PE6** | 0.861 | | | | |
| **PE7** | 0.846 | | | | |
| **PE8** | 0.836 | | | | |

*Source:* Authors' Calculation.
*Note:* CE:2,3,5,6,8,10; OE: 4,5,6,7,8,9 were dropped due to poor performance in the model.

**TABLE 19.4**
**Discriminant Validity**

| | | CE | DWB | OB | OE | OS | PE |
|---|---|---|---|---|---|---|---|
| **HTMT** | **CE** | | | | | | |
| | **DWB** | 0.369 | | | | | |
| | **OB** | 0.739 | 0.369 | | | | |
| | **OE** | 0.386 | 0.215 | 0.546 | | | |
| | **OS** | 0.362 | 0.455 | 0.287 | 0.518 | | |
| | **PE** | 0.422 | 0.209 | 0.415 | 0.960 | 0.370 | |
| **Fornel larker** | **CE** | 0.738 | | | | | |
| | **DWB** | 0.18 | 0.792 | | | | |
| | **OB** | 0.75 | 0.373 | 0.907 | | | |
| | **OE** | −0.172 | −0.035 | −0.548 | 0.788 | | |
| | **OS** | −0.013 | −0.45 | −0.245 | 0.434 | 0.724 | |
| | **PE** | −0.108 | −0.004 | −0.448 | 0.900 | 0.285 | 0.772 |

*Source:* Authors' calculation.

**TABLE 19.5**
**Hypothesis Testing and Path Diagram**

| | Original Sample (O) | Sample Mean (M) | Standard Deviation (STDEV) | T Statistics (\|O/STDEV\|) | *P* Values |
|---|---|---|---|---|---|
| **H1-a: CE → OB** | 0.672 | 0.7 | 0.1 | 6.72 | 0 |
| **H1-b: OE → OB** | −0.502 | −0.435 | 0.311 | 1.615 | 0.106 |
| **H1-c: PE → OB** | 0.077 | 0.009 | 0.272 | 0.285 | 0.776 |
| **H2: OB → DWB** | 0.373 | 0.387 | 0.198 | 1.88 | 0.06 |
| **H3: DWB → OS** | −0.45 | −0.526 | 0.156 | 2.882 | 0.004 |

at 0.7 and 0.5, respectively (Hair et al., 2019)]. Given that this is an indicative study, the goal was to retain as many items as possible; thus, reliability and validity values were used as benchmarks. Further, the discriminant validity was checked through Fornell Larker Criteria and HTMT, both were found to be acceptable on the liberal side as presented in Table 19.4.

### 19.3.2 Hypotheses Testing and Path Analysis

Bootstrapping method was used to run the regression analysis to get the P-values to ascertain whether the hypotheses established were accepted or rejected as presented in Table 19.5.

Only one major hypothesis i.e., H3 was accepted ($p<0.05$), this means that Organizational success is significantly influenced by the DWB. One of the reasons for this type of outcome is the small sample size. To understand this relationship

**TABLE 19.6**
**Coefficient of Determination**

| | *R*-Square | *R*-Square Adjusted |
|---|---|---|
| **DWB** | 0.139 | 0.108 |
| **OB** | 0.745 | 0.716 |
| **OS** | 0.202 | 0.174 |

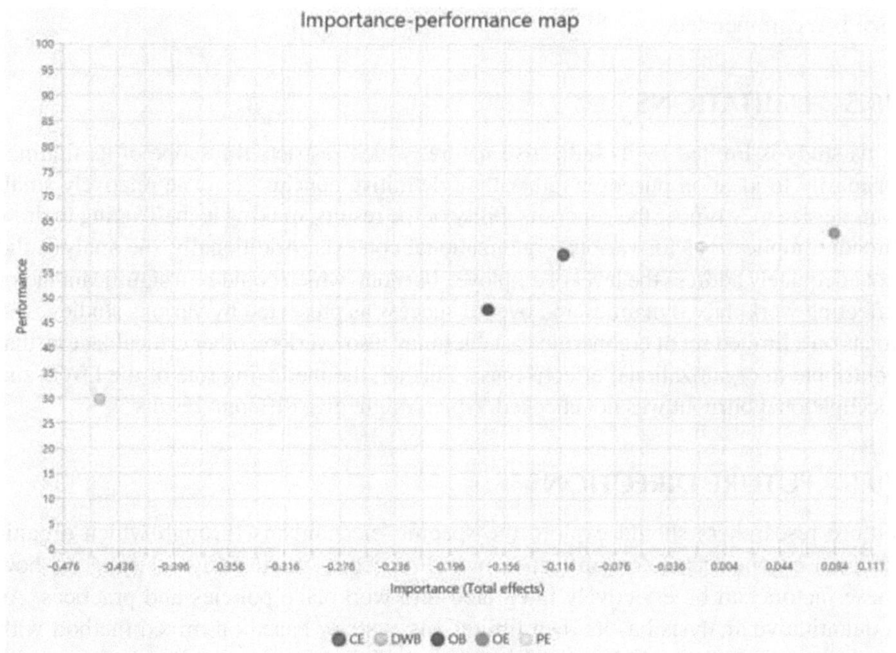


**FIGURE 19.2** Importance-performance map.

better, the R-Square (Coefficient of determination) values are to be referred as given in Table 19.6, the same was found highest for the OB at 0.745.

### 19.3.3 Importance-Performance Map Analysis (IPMA)

IPMA is performed to check the most important construct for the ultimate dependent variable i.e., OS in the present case. The top right variable is the most important and performing variable. As visible from Figure 19.2, the most important-performing variable is OE, i.e., organizational ergonomics, second best variable happened to be the PE, i.e., physical ergonomics, whereas the least important-performing variable is DWB (on the bottom left of the map).

## 19.4 CONCLUSION

The findings of this study conclude that ergonomic variables are important for an organization to achieve success. The most important variable that emerged is Organizational Ergonomic factors followed by Physical ergonomic factors. Although DWB of employees may significantly affect the organizational success, the same emerged to be the least important and performing variable. The level of burnout was not evident from the analysis of the existing responses. However, this study is only an indicative study and can be used for ideation purposes only. In order to get better reflections with a great scope of generalizability, a similar study on a larger sample size is recommended.

## 19.5 LIMITATIONS

This study is limited by its indicative nature, which restricts the scope of its findings primarily to ideation purposes rather than definitive conclusions. The relatively small sample size may hinder the generalizability of the results, making it challenging to draw broader implications for various organizational contexts. Additionally, the analysis did not adequately address the level of employee burnout, which could be a significant factor affecting workplace dynamics and overall success as presented by various studies. The focus on a limited set of ergonomic variables may also overlook other critical factors that contribute to organizational effectiveness. Further, the mediating role of the DWB and Occupational burnout was not checked, which might give different results.

## 19.6 FUTURE DIRECTIONS

Future researchers should explore the specific mechanisms through which organizational ergonomic factors influence overall success, particularly focusing on how these factors can be effectively integrated into workplace policies and practices. As a quantitative analysis has its own limitations, a study based on mixed method with larger sample size would be recommended. Additionally, studies should investigate the role of DWB in greater depth to understand its impact on organizational outcomes and employee morale. Investigating burnout more comprehensively could also provide valuable insights, as this variable was not adequately captured in the current study.

## REFERENCES

Ayanyinka, A. & Chukwvemeka, O. (2022). Ergonomics and stress reduction in organisation. *International Journal of Social Sciences and Management Review*, *5*(3), 12–21.

Banks, G. C., Whelpley, C. E., Oh, I., & Shin, K. (2012). (How) are emotionally exhausted employees harmful? *International Journal of Stress Management*, *19*(3), 198–216.

Baquero, A. (2023). Hotel employees' burnout and intention to quit: the role of psychological distress and financial well-being in a moderation mediation model. *Behavioral Sciences*, *13*(2), 84.

Chheda, P., & Sreeraj, S. R. (2020). Prevalence of musculoskeletal symptoms and quality of life in housekeeping workers of a tertiary care hospital in Navi Mumbai, India: A descriptive study. *MGM Journal of Medical Sciences*, *7*(3), 133–140.

Christy, D. V., & Duraisamy, D. S. (2020). Ergonomics and employee psychological well being. *International Journal of Management, 11*(3).

Chullen, L. C., Dunford, B. B., Angermeier, I., Boss, W. R., & Boss, A. D. (2010). Minimizing deviant behavior in healthcare organizations: The effects of supportive leadership and job design. *Journal of Healthcare Management, 55*(6), 381–398.

Díaz-Carrión, R., Navajas-Romero, V., & Casas-Rosal, J. C. (2020). Comparing working conditions and job satisfaction in hospitality workers across Europe. *International Journal of Hospitality Management, 90*, 102631.

Emeka, H., Ugwu, J. N., Okezie, B. N., Udeze, C. C., Azubuike, N. U., & Adama, L. (2020). Employee mentoring, career success and organizational success. *Humanities, 8*(4), 464–480.

Gomathi, K., & Rajini, G. (2019). Organizational ergonomics: Human engineering leading to employee well-being. *International Journal of Innovative Technology and Exploring Engineering, 8*(10), 3744–3749.

Gumasing, M. J. J., Rendon, E. R. A., & German, J. D. (2023). Sustainable ergonomic workplace: Fostering Job Satisfaction and productivity among business process outsourcing (BPO) workers. *Sustainability, 15*(18), 13516.

Hair, J. F., Sarstedt, M., & Ringle, C. M. (2019). Rethinking some of the rethinking of partial least squares. *European Journal of Marketing, 53*(4), 566–584.

Huang, Y. H., Lee, J., McFadden, A. C., Murphy, L. A., Robertson, M. M., Cheung, J. H., & Zohar, D. (2016). Beyond safety outcomes: An investigation of the impact of safety climate on job satisfaction, employee engagement and turnover using social exchange theory as the theoretical framework. *Applied Ergonomics, 55*, 248–257.

Khan, S. A., Epoc, F., Gangwar, V. P., Ligori T. A. A., Ansari, Z. A. (2021). Will Online banking sustain in Bhutan post Covid – 19? A quantitative analysis of the customer e-satisfaction and e-loyalty in the Kingdom of Bhutan, *Transnational Marketing Journal, 9*(3), 607–624. https://www.ceeol.com/search/article-detail?id=998408

Khan, S. A. & Magd, H. (2021). Empirical examination of MS teams in conducting webinar: Evidence from international online program conducted in Oman. *Journal of Content, Community and Communication, 14*, 159–175. https://doi.org/10.31620/JCCC.12.21/13

Liang, S. C., & Hsieh, A. -T. (2007). Burnout and workplace deviance among flight attendants in Taiwan. *Psychological Reports, 101*(2), 457–468.

Lubbadeh, T. (2021). Job burnout and counterproductive work behaviour of the Jordanian bank employees. *Organizacija, 54*(1), 49–62.

Maslach, C., & Jackson, S. E. (1981). The measurement of experienced burnout. *Journal of Organizational Behavior, 2*(2), 99–113.

Miller, D., & Friesen, P. H. (1978). Archetypes of strategy formulation. *Management Science, 24*(9), 921–933.

Mohamed, N. A., Tuah, S. N. A., & Othman, A. (2020). The dark sides of workstation: Ergonomics aspects and occupational diseases towards job performance at General Hospital in Malaysia. *Journal of Engineering and Health Sciences, 4*(1), 149–160.

Mulki, J. P., Jaramillo, F., & Locander, W. B. (2006). Emotional exhaustion and organizational deviance: Can the right job and a leader's style make a difference? *Journal of Business Research, 59*(12), 1222–1230.

Paajanen, T., & Kalakoski, V. (2017). Mitä työterveyslääkärin tulisi tietää kognitiivisesta ergonomiasta. *Työterveyslääkäri 2017, 35*(2), 16–21.

Purwanto, A. (2021). Education management research data analysis: comparison of results between Lisrel, Tetrad, GSCA, Amos, SmartPLS, WarpPLS, and SPSS for small samples. *Nidhomul Haq: Jurnal Manajemen Pendidikan Islam*, 6(2), 382–399.

Ringle, C. M., Wende, S, & Becker, J. (2015). *SmartPLS 3*. Bönningstedt: SmartPLS. Retrieved from https://www.smartpls.com

Robinson, S. L., & Bennett, R. J. (1995). A typology of deviant workplace behaviors: A multidimensional scaling study. *Academy of Management Journal*, *38*(2), 555–572.

Rypicz, Ł., Witczak, I., Gawłowski, P., Salehi, H. P., & Kołcz, A. (2023). Assessment of selected psychosocial risk factors: Stress, job burnout, and bullying in the case of medical staff as part of workplace ergonomics during the COVID-19 pandemic—A prospective pilot study. Frontiers in *Public Health*, *11*, 1169604.

Shiva, A., Narula, S., & Shahi, S. K. (2020). What drives retail investors'' investment decisions? Evidence from no mobile phone phobia (Nomophobia) and investor fear of missing out (I-FoMo). *Journal of Content, Community and Communication*, *10*(6), 2–20.

Song, Y. H., Skarlicki, D. P., Shao, R., & Park, J. (2021). Reducing customer-directed deviant behavior: The roles of psychological detachment and supervisory unfairness. *Journal of Management*, *47*(8), 2008–2036.

Stress in India. (2023, December 19). *Statista*. https://www.statista.com/topics/10075/stress-in-india/#topicOverview

Tarvainen, J. (2023). *The relevance of cognitive ergonomics for the work well-being of knowledge workers*. Metropolia University of Applied Sciences. https://urn.fi/URN:NBN:fi:amk-2023112130866

Tiongco, D. D. D., Vitug, P. Z., & Macindo, J. R. B. (2024). A home environment-moderated model of the influence of ergonomics on individual work performance among Filipino nurse educators working from home during the COVID-19 pandemic: An interaction moderation analysis using structural equation model. *Applied Nursing Research*, *76*, 151788.
https://doi.org/10.1016/j.apnr.2024.151788

Vroom, V. H. (1964). Affective Events Theory: A theoretical discussion of the structure, causes and consequences of affective experiences at work. *Research in Organizational Behavior: An Annual Series of Analytical Essays and Critical Reviews*, *18*(1), 34–74.

World Statistic. (2011, July 13). https://www.ilo.org/moscow/areas-of-work/occupational-safety-and-health/WCMS_249278/lang--en/index.htm#:~:text=The%20ILO%20estimates%20that%20some%202.3%20million%20women,and%20160%20million%20victims%20of%20work-related%20illnesses%20annually

# 20 Workplace Stress Reduction and Mental Health Promotion via Human Factors and Ergonomics

*Muhammad Zulkifl Hasan,*
*Muhammad Zunnurain Hussain,*
*Umair Waqas, and Sonia Umair*

## 20.1 INTRODUCTION

Work stress has emerged as a critical concern in today's workplaces and actually has broad effects on individual and organizational effectiveness as well as the health of the nation. With industries growing and competition increasing, employees are exposed to unparalleled hardships, from having to endure uncomfortable postures owing to designed workplace environment to the psychological pressures from within an organization. Mitigating these stressors becomes paramount in order to promote health and increased productivity for workers while also affecting the bottom line through decreased cases of absenteeism, high turnover rates, and overall healthcare costs.

There is something known as Human Factors and Ergonomics (HF&E) that can effectively problem-solve in relation to stress and adverse mental health at work. As a discipline that is distinct from the biomedical model of disability, HF&E involves the designing of systems, tools, and environments that complement the capacity of the human body; HF&E offers usable approaches to the actuality of working environments. This provides an opportunity to consider stress in physical, mental, and social incorporation as part of HF&E principles. Human-centric approaches to workplace stress reduction, as highlighted by Industry 4.0 paradigms, emphasize integrating sustainability into organizational culture to promote well-being and efficiency (Umair et al., 2025).

This chapter aims to identify the relationship between workplace stress, workers' mental health, and HF&E so as to establish what ergonomic and human-centered designs can do to reduce stress and promote mental health. Thus, presenting examples

DOI: 10.1201/9781003532293-23

of workplace stressors and their effects, effective means for HF&E solutions, and case studies, this work supports the idea of HF&E as a revolutionary approach to improving the state of workplaces and people's health.

The following sections develop a conceptual map to respond to workplace stress under the prism of HF&E alongside practical guidelines for organizations, managers, and policymakers.

### 20.1.1 Understanding Workplace Stress: Causes and Consequences

Workplace stress refers to the physical, emotional, and psychological strain employees experience due to challenges and demands that exceed their ability to cope effectively. This is a very prevalent problem in all fields and with people as well as businesses. Automated Information Systems surveys revealed that job stress remains a major factor that hampers employee satisfaction and performance, as well as has negative immediate and long-term impacts on psychological well-being and company performance.

### 20.1.2 Causes of Workplace Stress

Stress in the workplace results from different aspects that greatly affect the health of a person. Physical pressure factors such as visibility, noise, unfriendly workstation configurations, and thermal conditions are also disruptive stressors. For instance, noise-induced hearing loss and increased irritability may stem from exposure to high-intensity noise for extended duration and poor ergonomics cause musculoskeletal disorders (MSDs) in the employees apart from physical stress, they experience psychological stress as well.

One of the most frequent stressors is workload and time pressure was identified as another frequent stressor. Heavy working leads to a situation called burnout as a result of pressure to perform numerous tasks on a short timeline. Burn-out equally reduces productivity together with organization morale and motivation in the long run due to exhaustion. Likewise, role indistinctiveness and role contradiction result in role states of confusion, whenever there is confusion over job roles or conflicting demands. It also means employees have to think more to understand how to approach their tasks, making stress levels even higher.

They include organizational culture and management practices as well. Work culture has negative environments in the workplace, including poor leadership, lack of employee recognition, and inadequate support that contributes to employees' experiences of not being valued, whereas such environments may lead to the development of mental-related disorders such as anxiety and depression. Conflicts such as competition in assertiveness, misunderstanding, and competition between co-workers for control make tension higher and render working environment-unfriendly to collaboration.

Finally, technological stressors are likely to be the rise as more jobs involve the use of technology in the workplace. Technological advancements have introduced "techno-stress," where rapid digital transitions and excessive connectivity contribute to employee strain (Umair et al., 2025). Combined these all shows that stress in the workplace is complex and needs more attention in order to come up with better solutions for the betterment of the workplace experience.

### 20.1.3 Consequences of Workplace Stress

Organizational stress is significant, in a way that people and businesses suffer from its impact. Another is the impact on individuals' mental well-being, which is influenced by many forms of stress, including chronic ones that are associated with depression, anxiety, and sleep disturbances. In the long term, it can cause emotions to become depleted and some rational thinking capacity is lost so decisions and probable solutions can be compromised. Stress also impacts the well-being of the human body causing diseases like hypertension, cardiovascular diseases, and weakening immune systems. One example of stress-related diseases is migraines, and other gastrointestinal disorders more so among those who work under stressful conditions.

In regard to organizational implication, stress worsens productivity, as well as staff involvement. Stressed employees lose the ability to concentrate or be creative and have lower productivity as part of their symptoms. Performance error, delay and slippage, and poor quality of work are more likely to occur since the employees affected by burnout are.

With burnout affecting employees, organizational productivity is bound to suffer.

However, workplace stress results in high turnover rate and high rate of absenteeism as well. Quite a number of employees use their sick days or mental health days to relieve stress, and chronic stress makes them look for healthier workplaces. It creates huge financial implications for organizations, as organic stress turnovers are estimated to set back the American economy billions of dollars per year.

Last but not least, stress influences the reputation of an organization. The environments that trigger stress are often seen as negative, which discourages the best talent from joining an organization. Business organizations experiencing high employees' stress have enormous challenges in working on the employer branding, which in turn hampers their capacity to compete for employees in the job market. Imposing measures to address job stress is a viable way of improving the organizational health and consequently the quality of its production.

This chapter explains the causes and effects of workplace stress in order to design proper intervention plans for creating a healthy and less stressful working environment for employees.

### 20.1.4 The Role of Mental Health in Professional Environments

Stress and mental disorders remain one of the most influential factors that determine the efficiency and effectiveness of works and creativity of employees at workplaces. A mentally healthy workforce produces better outcomes and contributes more on target to the objectives of an organization. Mentally resilient workforces significantly enhance decision-making, innovation, and overall productivity, aligning organizational goals with employee well-being (Umair et al., 2025; Khan &

Ansari, 2024; Ansari et al., 2024; Khan & Naim, 2023). On the other hand, negative mental health brings not only the consequences at the individual level but also organizational consequences referring to the efficiency and organizational culture.

#### 20.1.4.1 Stress Matters in Employment

Stress or lack of stress is therefore one of the major determinants of productivity both at group level and corporate level. A healthy employee is always more productive as compared to an unhealthy individual because the former remains more attentive and focused, capable of organizing his/her tasks. It is also easier for them to cope with adversities and make valuable inputs to the achievements of organizational objectives. Besides, mental health leads to the enhancement of decision-making and innovation due to enhanced thought process and creative thinking. High levels of job stress have been proven to interfere with cognitive abilities including memory, problem-solving, and logical reasoning, whereas sound mental health enhances decision-making and provision of key ideas.

A healthy mind also gives the ability to build interpersonal relations and fosters both effective and efficient interpersonal communications. Psycho-social well-being in the workforce is hence more likely to enhance effective interpersonal relationships with co-workers that boost the morale of every unit. However, treating mental health issues contributes positively to a firm's attainment of reduced absenteeism and turnover. Where employees get support, they are in a position not to burnout, not take many sick leaves, or even quit their tasks, thus exhibiting more organizational commitment. Leadership strategies that prioritize employee mental health have proven to reduce turnover and absenteeism, fostering long-term organizational commitment (Umair et al., 2025).

### 20.1.5 The Issues of Mental Health in Workplace

However, several barriers affect the provision of support for mental health at the workplaces. There is a huge problem of mental illnesses remaining unrecognized and untackled because of the stigma associated with them which keeps employees from asking for help or speaking about their problems. It is usually the case that this lack of openness worsens conditions and or delays a correct intervention. The other problem is linked to high-stress rates inherent to contemporary workplaces and the need for individuals to cope with increased loads. Stress that is long-term often results in disorders such as anxiety, depression, as well as emotional depletion.

There is also limited in organization mental health support that exists to complicate the problem. Unfortunately, even today, many employers offer few or no resources for workers, including Employee Assistance Programs, counseling services and education about mental health. Besides, such factors as work-life imbalance due to enhanced productivity expectation and being always on will have detrimental impacts on the mental health. Finally, the economic instability of work insecurity, job loss, and economic problems puts pressure on the health of employees besides worsening their mental health. These challenges and that of mental health in workplace should be tackled to enable workplace to provide healthy, efficient, and sustainable environment.

## 20.2 OVERVIEW OF HF&E

HF&E is the study of ways in which people interact with environments, equipment, and systems that therefore seeks to create the best possible fit between people and their tools and workplace. Defined mostly in terms of work design, patient care, or the work people do and the physical, cognitive, or information working environment, HF&E aims at improving health, safety, comfort, and quality or efficiency. It is particularly helpful in working with stress at the workplace and mental health as it aims to offer effective interventions that can help to decrease the physical load and mental workload.

HF&E normally involves several aspects of the sciences with emphasis on psychology, engineering, design, and physiology to improve man/system relations. According to the International Ergonomics Association, HF&E is divided into three primary domains: The division of ergonomic can be physical ergonomics, cognitive ergonomics, and organizational ergonomics. Physical ergonomics is concerned with the task design especially the layout of working environment in order to eliminate or minimize postural hazards that lead to MSDs. Cognitive ergonomics, in contrast, is concerned with mental work processes that include decision-making, memory, and the load that should be carried by the CPU in order to enhance production effectiveness without overstraining cognition. Organizational ergonomics can be defined as sociotechnical systems, changes to communication, working processes, and the structure of an organization to support them. As a combined package, these domains give a clear structure for the places and conditions that directly affect the overall health of individuals at the workplace, and thus facilitating healthier places of work. Ergonomic designs, supported by Industry 4.0 innovations, facilitate the creation of adaptive workplaces that cater to diverse physical and cognitive needs, enhancing overall productivity (Umair et al., 2025).

From these principles, we can deduce that the prominence of HF&E incorporates the subject of Human Factors, which relate to people in relation to their ability, experience, and ailments needed when designing systems and surroundings. This approach means that workspaces or tasks will be easy for individuals to locate or perform /will ensure work environments and work functions are usable to everyone. The other is the adaptability of those systems to human differences in terms of physical and mental capabilities. HF&E fosters designs that work for such differences to ensure that workspace is easily accessible for such persons. Risk reduction is another of these principles which work aims at reducing different chances of danger that threaten to cause bodily harm or withdrawal. For the last, creating effective work processes or simple and efficient organizational structures together with removing as many avoidable difficulties and loads as possible ensures the highest rates of productivity together with the minimum of likely levels of stress and fatigue to reach the highest levels of performance by employees.

There are various advantages that have been accorded to the implementation of HF&E in the workplace and among those advantages consist of decreased stress and also the increased health of the employees. They include ergonomic workstations, as these can address issues resulting in discomfort and musculoskeletal injuries from poor arrangement of chairs, desks, and computer placements. Workflow efficiency makes a positive contribution towards decreasing cognitive load through the defragmentation of workloads and tasks. Specific modification of the physical environment involves controlling factors such as light penetration, temperature, and noise in the workplace so as to eliminate stress. Furthermore, the integration of technology has been made easy to embrace by incorporating User Interface, such software makes working environment to be less challenging and optimized systems minimize cognitive load hence making employees to work smart and hard.

The current trends in the development of the HF&E field reflect the trends that define the future of workplace design. Other trends being but are now possible through innovations in technology include an increase in the use of artificial intelligence (AI), especially in combination with wearables in the monitoring of working conditions and real-time adjustments that thus foster the development of personalized workspaces. Another area of interest is also mental comfort and this has been found out that in setting up physical comfort, consideration has to be given to mental comfort. Other trends include sustainability in design, the fact that many companies fit ergonomic designs into sustainable concepts using environment-friendly material as part of Corporate Social Responsibility. These changes draw attention to the need for technology, mental health, and sustainability to be infused into HF&E guidelines to best address the organizations and to assure that the designs of efficient and healthy workplaces are also supportive of environmental wellness. Sustainable ergonomic practices not only improve employee health but also align with corporate social responsibility goals, emphasizing environment-friendly workplace designs (Umair et al., 2025).

## 20.3 INTEGRATING HF&E IN WORKPLACE WELL-BEING STRATEGIES

The roles of HF&E in work health promotion programs are critical because they help to design workspaces that support health, safety, and efficiency of all workers. HF&E offer a structured framework through which stressors and strategies to eliminate regulatory obstacles to physical, cognitive, and organizational enhanced adaptation can be easily determined. Moreso, by adopting HF&E principles in the workplace, organizations should be in a much better position to drive improvements to staff prospects and their mental health.

Including HF&E with strategies for workplace well-being has several advantages centered around the concept of risk factor reduction that also involve physical and mental modifications. With regard to physical work environment and the organization of work processes, HF&E interventions mitigate physical discomfort, such as musculoskeletal stress, and cognitive overload, resulting in a decreased level of stress. Some of these are workstation interfaces, organization of workstation, and noise control that influence comfort, ergonomics, and thus interaction with the workstation. Thus, employees perform better and have less likelihood of being always off work and quitting, all of which make the organization more successful.

Furthermore, HF&E enables the management of stress at the organizational level rather than only the treatment of symptoms of stress in contingency and crisis situations HF&E. This broad view of health can be defined as a proper combination of physical, mental, and social health of the employee. Ergonomic furniture, changing work frequencies throughout the day, promoting exercise, and the right medications all combine to build a healthy workplace for the employees. Opportunities for reporting about mental health issues, providing access to counseling, and supporting social events also add to successful organizational culture with low stress and high collaboration. Incorporating wearable technologies and AI-driven solutions into HF&E strategies enhances employee well-being through real-time monitoring and personalized interventions (Umair et al., 2025).

These steps include, first, evaluating the workplace needs and, second, designing the ergonomic remedies in this case HF&E then applying policies that conform to HF&E. Several measures aimed at increasing well-being in the workplace must be accompanied by employee training to become effective; the use of technology, for example, the use of wearable and intelligent systems in the office also contributes to well-being strategies. But organizations may encounter problems like high startup cost, employee resistance to changes, and need for specialist in ergonomics and psychologists. Addressing these concerns falls under a broad gist that requires enough planning, assistance from experienced professionals, and constant assessment of success for HF&E strategies. There is a good rationale to incorporate HF&E into workplace well-being; current evidence suggests that such approaches invariably enhance well-being, engagement, and productivity among employees. Second, they benefit software organizations with an aim of enhancing satisfaction, customer retention as well as maintaining compliance with the occupational health and the safety laws. Organizations integrate HF&E into their setting, design for worker satisfaction, and mental well-being, resulting in enhanced and more effective workers. However, the integration process means a lot of investment of resources and efforts where the benefit to be derived is enormous more than one can imagine making it a worth to do for the welfare of the employees and success of the organization.

## 20.4 PART I: IDENTIFYING WORKPLACE STRESSORS

Workplace stress arises from a combination of physical, cognitive, organizational, and technological factors that challenge employees' ability to perform effectively. Identifying these stressors is a critical step in mitigating their impact through HF&E interventions.

### 20.4.1 Physical Stressors: Lighting, Noise, and Workstation Design

Lighting is important in workplace health because inadequate lighting conditions cause discomfort, eye strain, and headaches thus, adding to stress at workplace (Králiková et al., 2021). Light reflections from fluorescent lighting fixtures or screens that have not been mounted correctly, for example, discharges efficiency and raises stress. Conversely, specific levels of lighting or, in particular, the access to natural lighting have been identified as creating positive mood and decreasing stress levels thus creating a more appropriate environment for employees.

Another highly frequently identified physical stressor at many workplaces studies physical noise in the accomplishment of tasks as well as in the increase in physiological stress responses such as cortisol levels. Several studies using noise exposure levels above 55 decibels demonstrated that the noise affects performance and causes mental fatigue (Evans & Johnson, 2000). Correcting this stressor involves installing solutions like the acoustic panels or using a silent zone that counts as an ideal HF&E solution for a quieter and stressful-free workplace (Kamarulzaman et al., 2011).

Another area of ergonomics is also concerned with workstation design where means to reduce stress at workplace should take place. You get to have poor positions, MSDs, and even physical strain that all contribute to stress (Hämmig, 2020). For instance, a chair at the wrong tilt or a monitor placed at the wrong angle implies that it exerts some pressure on the neck and the back for quite some time. In order to resolve these challenges, ergonomic design principles support the usage of height adjustable furniture and tools and apparatus that can meet everyone's physical capabilities to work for long hours without suffering physical discomforts. This, we believe, leads to a healthier and therefore more productive workplace for all parties involved.

### 20.4.2 Cognitive Load and Mental Strain in the Workplace

Complexity refers to the ability demands that exceed the cognitive capacity of an employee; it occurs when employees are overloaded with chunks of information, or when they are pressed for time to perform a difficult work (Sweller, 1988). There are several causes of cognitive load at the workplace. One therefore realizes that factors such as multitasking, which actually splits attention and consequently leads to higher mental fatigue and a consequent reduction in performance and squash stress levels (Rubinstein et al., 2001). Further, unclear instructions or a failure in the clarity of a particular job may increase frustrations and pressure, which makes the chances of lower job performance among employees higher.

To reduce cognitive load potential, HF&E solutions can be applied. The following are some of the ways standardization is used: Minimizing, for instance, work-related requirements of tasks reduce the workload. Another strategy is to give people handouts that would contain written and pictorial explanations of carrying out a particular task, explaining that employee may find that he is able to work through a process faster this way, giving him the opportunity to focus on one thing at a time instead of having a dozen things running simultaneously through his mind. All these HF&E interventions improve efficiency of operations, ease of work, and thus better performance due to reduction in work-related mental load.

### 20.4.3 Organizational Factors: Culture, Workload, and Role Ambiguity

Poor industrial relations that include bad communication, lack of appreciation, and high levels of work-related conflict are major causes of stress. Cultural adaptability and leadership's role in fostering a collaborative and inclusive work environment are pivotal in mitigating organizational stress (Umair et al., 2025). A brief review of literature reveals that the organizational climates that stress cooperation and reception of all employees lower stress and increase well-being of the individual (Cartwright & Coope, 1994).

Overworking, provoking shifts, and frequent pressure from office supervisors, or setting incredibly tight deadlines to all their subordinates by investigations causes' burn out. The Job Demand-Resource model as developed by Bakker and Demerouti (2007) posits that high job demands that are not met by adequate resources result in employee burnout and withdrawal. One intervention related to the HF&E program is workload management in tasks and distribution with sufficient support systems put in place. The same can be said about the lack of clearly defined roles, it just leads to confusion, stress, and frustration. Literature reviews reveal that as employees know what is expected of them, they are more satisfied with the jobs and stress free (Beehr & Glazer, 2005).

### 20.4.4 Technology-Induced Stress: Balancing Innovation and Usability

Technology has continued to improve in the workplace and has therefore brought about certain benefits hence certain challenges to employees. For instance, despite advancing technology helping organizations work more efficiently, failure to use it appropriately or in the wrong approach results in stress referred to as technostress. This is majorly due to overload and complexity. Users often face the problem of an information overload because their email inboxes, notifications, and overall workloads are filled with digital work. This influx can potentially become a severe problem, especially when additional to intricate systems or user interfaces of lower design quality. Such factors are a source of increasing error rates and reducing the level of confidence, which, in turn, worsens stress (Tarafdar et al., 2015). The other emerging causes include inadequacies in training in new technologies. It gets worse when the employees never receive training because they end up becoming nervous and take time to settle down. To this, HF&E principles combine that easy-to-learn systems and intensive training strategies ought to be adopted to enable employees to operate in safe manner when handling new technologies (Salanova et al., 2000). Also, technology is almost always present, and that makes it more difficult to separate between the work and private life. Since the usage of digital communication is constant, the employees may possibly be at their work even later than working hours, which in effect means that they get less or no rest. Organizations can avoid this stressor by implementing career-related digital backstage techniques for instance avoiding the use of emails after working hours (Derks et al., 2015).

## 20.5 PART II: HF&E-BASED STRESS REDUCTION STRATEGIES

Organizational stress is a complex problem that needs a systematic approach in workplaces to address. Applying HF&E-based solutions being to reduce and eliminate physical and cognitive loads as well as environmental stressors at the workplace in order to improve employee health and effectiveness. Stress reduction and case studies embedded with relevant academic references form the basis of this section amid details on HF&E principles.

### 20.5.1 Ergonomic Considerations for Minimizing Stress at Workplace

Stress injury prevention at the workplace starts with ergonomic design of workstations. Lack of proper workstation design results in some level of discomfort in the workstation muscular skeletal discomforts, repetitive strain injuries, and fatigue, all, which cause stress (Karwowski & Marras, 2003). Recommendations about ergonomic interventions which involve; adjustable desks and chairs are important in mitigating these problems. Sit-stand desks encourage postural change, thus, counteracting the effects of pertaining health concerns, including cardiovascular diseases and obesity, that are related to extended sitting (Tudor-Locke et al., 2011). Lumbar support chairs provide an excellent back contour to encourage spinal posture, control lower back pain, and improve comfort.

Another important aspect related to stress is the location of monitors some of which are as follows: the use of uniform monitors situated at eye level reduces the chances of neck stress and sore eyes that may prove very unproductive in the long run. Anshel (2005) conducted a study that indicates that proper positioning of monitors increases the efficiency and work success as well as decreases on eye strain. Also important is the design of the displays and the keyboards and the mice for the fact that they minimize the strain on the wrists. Other employees' necessities are met with ergonomic keyboard and vertical mouse which reduces the effects such as carpal tunnel syndrome and RSIs so that employees are able to work throughout without a threat to their health.

Another source of stress is noise, especially where the working environment is organized the working environment is organized in an open space. It was revealed that high noise levels interfere with concentration and impose extra load on the information processing system (Banbury & Berry, 2005). Control of noise sources can reduce these stress factors and enhance organizational health. Some of them include the use of acoustic panels which afterwards lessen all the sounds that penetrate the entire room, thus fostering concentration. Free from noise disturbances, quiet zones give employees a chance to find isolation and help them to reduce stress. Also, the noise generated by white noise machines can overcome other distracting noises so that workers may stay on the task and cut down on stress-causing elements due to noises. These noise management interventions contribute significantly towards the provision of optimal working conditions as well as teaching and learning environment, on one hand, and promotion of optimal health, on the other hand.

### 20.5.2 Optimizing Workflows to Minimize Cognitive Overload

Cognitive overload is a situation where the quantitative and qualitative load of a task takes time to be processed by a person. HF&E interventions are intended to work on ways that simplify work procedures, lessen the bulk load on employees, and help them to concentrate, and do not leave them drained mentally. One of the ways to deal with the issue of cognitive overload is to minimize the state of the affairs that overwhelm and complicate the workers' tasks. In most organizations, complex work processes are a source of stress and mistakes. Kanban boards and agile frameworks in particular can prevent the employee from being overwhelmed as well as optimize the processes (Wright & McMahan, 2011). The last strategy of minimizing the cognitive load is to take advantage of robotic process automation where employees are required to perform repetitive and repetitive duties. Being able to methodologically and unambiguously state requirements, descriptions of methods as well as provide legal prescriptions and requirements for standard conditions of operation or SOPs, will alleviate confusion and mistakes and hence result in decreased mental load.

Task personalization is also another way of managing of cognitive load, including the encouragement of the prioritization of tasks. These categories assist in addressing organizational issues in that they assist the employees not to be exhausted on which project to tackle because it is urgent or important, something that Eisenhower Matrix tackles (Covey, 1989). This concept enables employees to prioritize their activities and avoid mental burnout; this means that employees will be more productive in their working activities.

Mainstreaming also has another attribute that is worth considering; that is the interruption should be minimal. They destroy productivity and increase stress levels from constantly interrupted emails, meetings among others. According to Newport (2016), organizations should introduce 'focus time' equity in which workers are allowed to work without interferences. Moreover, grouping activities that are similar in nature, for example outlining the response to frequently asked questions in emails, at certain periods of the day also minimizes the number of times employees switch from task to task which in turn increases their productivity. To top it all, dedicating spaces free of distractions for deep work enhances productivity, lowers the employee's mental load, and allows them to work less stressed. These interventions taken together promote cognitive health in the workplace.

### 20.5.3 Promoting Physical Activity and Proper Posture

Lack of physical activity is causative of workplace stress and chronic health disorders, something that may be addressed by integrating physical movement with other activities in the workplace (Tudor-Locke et al., 2011). The first way of encouraging the movement of people in the workplace is the famous walking meetings. Not only does such a flexible meeting encourage lateral thinking but it also gets rid of the evils of traditional meetings including sitting for extended periods. In addition, short periods of break with regard to stretches or a few physical exercises during work time can boost energy levels and help to release accumulated tension, and so on.

In addition to promoting physical movement, posture training and awareness are critical in preventing MSDs, which are a significant source of workplace absenteeism and stress (Punnett & Wegman, 2004). Posture-correcting devices, such as wearable technology that provides real-time feedback on posture, help employees maintain proper alignment and reduce the risk of discomfort. Ergonomic workshops are the fourth valuable approach; seminars and education of employees regarding adjustments of workstation or correct postural movements to prevent chronic pain and stress in body.

Supportive equipment also appears to be an important component in further improving comfort and decreasing stress. A footrest serves all employees but is particularly important for shorter persons because it helps to align the legs properly on a chair. Wrist rests help in reducing the stress that is usually caused by typing on keyboards for extended hours, hence it's promising to enhance the employees' productivity as they will not be tired easily. Special mats for floor, which decrease discomfort for people who use standing desks, are useful and help employees to feel they are at work. These interventions alongside the encouragement of movements and correct posture, serve to decrease physical stress and create conditions allowing people to feel better in the course of work.

### 20.5.4 The Role of Environmental Design in Mental Well-Being

Incorporation of nature into workplace settings has been explained to lower stress and improve psychological health can be attributed to biophilic design (Kellert et al., 2011). The first principle is a concealed or visible integration of natural elements, the most important of which is plants. Interior plants not only have functions in the process of air purification—but they also have a relaxing effect. People who come to work in space surrounded by trees, grass, and flowers have 30% less stress (Kellert et al., 2011). Other natural aspects of the ergonomic design include the water fountains and wooden accented furniture that work to complement and create a calmer work setting.

Other factors that are considered in the care of employees' include temperature and air quality come in handy in taking care of the employees. According to the study, employees of an organization are most effective in their tasks when the temperature ranges from 20°C to 24°C (Seppänen et al., 2006). But, air quality also has an impact where of comfort is concerned through lowering fatigue and number of headaches as well. New technological systems such as HVAC systems guarantee that the quality of air in the offices is the best so as to lead to productivity.

It is also important to take into regard the colors that can be used in the office space as they also meddle with mood and behavior. Bright and pastel colors make people more relaxed, and focused, that is why such colors as blue and green for example, are best suited to creating this environment. However, warm-toned colors such as yellow and orange increase creativity despite the fact that they should be used to some extent as they cause overstimulation (Elliot & Maier, 2014). With regard to the nature of the environment, temperature, and quality of air, the concept of using color has been shown to enhance productivity of the workers, create comfort levels at the workplace that are acceptable to the workers with minimum or no pressure.

Modern technologies present sophisticated possibilities of enhancing the state of Organization Health and decreasing stress. Wearable devices for instance measure fundamental stress parameters such as the heart rate variability and the galvanic skin response. These devices offer feedback in real time hence allowing the employees to always manage or lower their stress levels. Moreover, they encourage stress awareness and management techniques that perhaps can reduce stress and increase feasibility in dealing with it among employees (Hickey et al., 2021).

Virtual Reality (VR) training is one more output promising. Interactive virtual reality programs are used to expose the patient to this reality and learn how to relax these muscles and also learn how to carry out activities in a right postural alignment and ergonomics studies. Freina and Ott (2015) found that use of the VR training increases training retention and interaction as compared to other training approaches. Explaining concepts through actions makes it easier for the employees to replicate so with stress management techniques and ergonomic practices become appealing in a virtual platform. Smart ergonomic systems are also being applied in the current working places. Smart desks and chairs mostly use AI technologies that adjust to a user's needs, maintaining proper posture and comfortable seating. These systems are meant to reduce likelihood of acquiring stress-related injuries for instance through constant adjustments to meet the user's body needs hence creating healthier workforce and increased productivity. These technologies are effective in presenting a new, individualized strategy for promoting the health of employees in the post-industrial environment.

## 20.6 PART III: MENTAL HEALTH PROMOTION THROUGH HF&E

It is worth stating that HF&E is not limited to applying positive changes to the physical environment of the workplace, and includes the promotion of Mental Health. Thus, when applying HF&E principles to create top organizational structure, it is possible to develop basic levels of resistance to adversity, as well as positive workplace experience to improve levels of psychological well-being (Jain et al., 2019).

### 20.6.1 Building Resilience Through Workplace Design

That is why it is possible to ease a mental load at work with the help of an elegant design and an organization of the restoration environment. When the tools and systems are usable, this means that there will be less stress and certainly less frustration. For example, the use of interfaces that self-explain, and the use of tools to reduce options and provide smooth interaction enables reduction in cognitive demands, in such a way that the employee can work smartly without being overburdened a lot of system-related complexity (Norman, 2013). Third, employees are less stressed due to less ambiguity in their roles and clearer work processes, which provide more streamlined working as well.

As important as that is the making of opportunities to restore within the workplace. Common areas that include plant life and natural lighting are beneficial to reduce stress among the employees. Kaplan in his research shows that biophilic designs provide exposure associated with an enhanced mood and decrease in stress (Kaplan, 1995). It also addresses issues to do with quiet zones and relaxation areas in that it creates areas where employees can reduce stress and this can improve their well-being and thus productivity. These details of design of the work environment aim at promoting a healthy performance among the staff.

### 20.6.2 Encouraging Work-Life Balance via Flexible Ergonomic Policies

Particularly important is the provision of flexible work arrangements that support work-life balance, primary prevention of RTW, and primary prevention of MSD by mainstreaming HF&E into work arrangements. Through choice working hours and telework options, organizations must facilitate people to manage their time in order to raise productivity and to improve health and well-being (Allen et al., 2013). As for teleworkers, it is crucial to guarantee that environments with home offices are as comfortable as possible for avoiding muscle and joint problems.

Similar to them are the approaches to take frequent time-offs and create clear lines between working and non-working time. There should be timed signals to take breakfast, lunch, and coffee/tea or any form of break in order to avoid monotony of work for HF&E principles. Moreover, the availability of flexible working schedules helps with shared workload and a good number of employees can manage work without adversely affecting their well-being. All these strategies combined create a healthier workplace, work relocation, and increase the recovery time to enhance workplace performance.

### 20.6.3 Promoting Social Interaction and Team Collaboration

Work organization with reference to workspace increases team-supportive strategies and improves social interaction among co-workers. This way, through making large offices with partially opened spaces for interaction and cooperation a company can foster both active participation and efficiency of its employees. However, it is also necessary to provide openness and, in equal measure, provide 'freedom from' in the form of confined spaces where employees can avoid interruption. While HF&E intervention aims at creating these working environments comprehensively, it includes the needs of each employee in the aspect of disability.

Another wide determinant of workplace stress is Psychological safety. If employees are allowed to feel free and without worrying about being laughed at, or left out of a specific group they will engage more and work more efficiently. Establishing HF&E principles can ensure that psychological safety is achieved because spaces to which everyone has access are safe and inclusive. This strategy fosters an inclusion perspective and thereby provides relief from stress likely arising from perceptions of exclusion, hence improving psychological and psychological health in addition to team results (Edmondson, 1999).

### 20.6.4 Technology for Mental Health: Wearables and Feedback Systems

Health surveillance and supervision through the use of wearable technology are today popular in maintaining workforce health. Smart gadgets like wristbands and smart bands give real-time data with basic health indicators including level of physical activity or stress level of an individual. When connected with HF&E solutions these devices serve to help employees manage their mental health. Such a strategy enables people to identify the early warning signs of stress in their lives and spend their time productively and healthily (Sultan, 2015). In addition, digital mental health tools are becoming an essential part of workplace wellness strategies. Applications and platforms designed with HF&E principles offer employees convenient resources for managing stress and improving mental resilience. These tools include features like guided meditation sessions or cognitive-behavioral therapy (CBT) techniques, specifically tailored to address the unique stressors of the workplace. Such resources not only provide valuable coping mechanisms but also empower employees to take charge of their mental health (Richards, 2022).

Mental health promotion within the framework of HF&E is a complex process that encompasses both intervention and evaluation in relation to the physical, organizational/technological context of working. By promoting organizational and psychological capital, empowering the work-family interface, enabling social connectedness, and incorporating technology, organizations can promote employee well-being whilst promoting organizational performance.

## 20.7 PART IV: IMPLEMENTATION AND MEASUREMENT

If an organization wants to plan, direct, and control HF&E-based stress reduction interventions, it has to measure workplace stress levels, write organizational policies, educate workers, and measure the success levels of changes that have been made in the organization. In particular, all of these components should blend together in order to promote wellness both in the body and in mind.

### 20.7.1 Assessing Workplace Stress Levels: Tools and Techniques

Self-created questionnaires are broadly applied for the evaluation of stress at workplaces. Readily administered instruments, such as the Job Content Questionnaire (JCQ) or the Perceived Stress Scale (PSS), help organizations determine the extent to which employees perceive stress, and which stressors, are present. These tools describe load, demands, and relations with others, which are appropriate to identify directions and goals for change (Karasek et al., 1998).

Moreover, there is objective data including BMI, cholesterol level, blood pressure, etc., to make objective estimations of changes in the employees' health due to stress at work. Pulse, with pulse variability, blood pressure, and cortisol level gives an understanding of Stress on body (Chesney et al., 2005). This data can then be gathered by wearable devices in real time which will provide a fuller picture of stress levels across a single working day.

There are also the assessments of employee stress through focus groups and interviews as another effective method for further stress comprehension at the workplace. This qualitative data supplements survey results to open a deeper understanding of how design, organizational culture, and workload cause stress and impact the mental health of employees. By combining the two, an organization can be able design a better approach when dealing with workplace stress.

### 20.7.2 Developing and Implementing HF&E Policies

When initial sources of stress have been identified and evaluated, the organizations can adopt policies that conform to HF & E in order to reduce the stress factors. Examples of such policies include: ergonomic workstation provisions, flexible work schedules, and stress-free programs (Hämmig, 2020). They should be aimed at achieving the overall organizational objective, as well as satisfying employees' requirements by being realistic and realistic in the given setting.

Leadership commitment: it was also an important finding that for policy implementation, leadership should support HF&E principles. Such commitment includes ensuring that money for good ergonomic equipment is put in place, coming up with a way of training employees on how to tackle stress; and also coming up with a way of providing counseling services to members. If employees feel that their needs are valued by the company it is more likely that they will support such policies and engage in stress management programs more effectively (Kabat-Zinn, 1990). It helps make the workplace healthier for everyone and encourages them to exercise proper behavior that would lead to the prevention of stress over the long term.

### 20.7.3 Training Employees and Leaders on HF&E Approaches

While it is important to provide ergonomic tools, reduce workers' stress level, and integrate wellness exercise in the workplace, it is paramount that employees receive training that can enable them understand and practice all these elements. Such training can entail issues to do with body position, efficiency with time, and skills in calming down due to stress. Seminars or online meetings on such subjects are ways of familiarizing the workers with key aspects of ergonomics and mental health to apply them at work with efficiency for advancement.

As significant is the need to educate managers and team leaders about the signs of stress and the assistance that can be offered to their subordinates. Managers and leaders should be ready to apply "risk assessment" knowledge, which comes with HF&E for the purpose of applying stress reduction interventions in their group or organization, as well as design a work environment that fosters work-life balance. Stress management education training for leaders creates a positive working model at the organizational top, supporting a strong message of the importance of prevention and management of stress at the workplace (Cartwright & Cooper, 1994). This chapter points out that the rationale of empowering leaders who in turn support their staff is significant as it fosters more able and tenacious workforce.

### 20.7.4 Measuring Success: KPIs for Mental Health and Ergonomic Efficiency

It is also important for organizations to agree on certain specific goals or targets for the success of HF&E interventions where the concept of KPI may include mental health issues as well as ergonomic features of efficiency. Some of these KPIs may involve the usage of reported stress levels, from questionnaires or other self-complete forms, less absenteeism in relation to mental health issues, relative increase or decrease in productivity depending on output or quality control, and specific decline in injury rates, especially those stemming from ergonomic alterations. These related KPIs should be regularly reviewed to determine if the organizations initiative and expenditure on HF&E is delivering the required decrease in stress levels and increased employee well-being.

In addition to the basic concepts, which are measured in numbers, a survey to evaluate the level of employee happiness or even health is critical. Follow-up employee engagement surveys can measure perceived organizational stress reduction programs' effectiveness, reveal necessary enhancements as well as make sure that employees are not overwhelmed. Such results are quite helpful to understand what changes in organizational policies are useful in the future as well as to work on new solutions in the sphere of HF&E.

Thus, the successful application of HF&E-based stress reduction guidelines involves recognition of stress, formulation of stress-related policies, training of employees and leaders and, finally, evaluation of the effectiveness of attempts at stress diminution. Through the implementation of appropriate assessment methods, and through the modification of the results of the interventions applied, organizations can create conditions that decrease stress and increase mental well-being and overall productivity. By applying such practices, staff members are given a certain level of care for their health, as well as the overall business is benefitted in the long term.

## 20.8 PART V: CASE STUDIES AND FUTURE DIRECTIONS

With organizations understanding the value of HF&E in alleviating workplace stress and enhancing the mental health of employees it is imperative that the reader consider real examples of where such strategies have been applied. This section analyses examples of best practices and to some extent discusses the tendencies of HF&E regarding mental health promotion to predict further development of well-being trends in the workplace.

### 20.8.1 Success Stories: HF&E in Action for Stress Reduction

#### 20.8.1.1 Case Study 1: Google's Ergonomic Workspaces

Google is well known for its efforts in creating comfortable and healthy workplace environment especially through functional ergo friendly and creatively designed workplace. The company has followed open, flexible organization of services and desks provided with ergonomic furniture, which is height adjustable work desks and chairs enabling the employees to switch between sitting and standing (Sundstrom, 1986). Moreover, the company offers wellness policies that enable the employees to have on-site gym, meditation rooms and career counselors. It has been found that such programs help in lowering stress and making employees happier and more productive at workplaces (Garg, 2017). Hence, the Company has been adopting HF&E practices both in physical and mental fitness to become one of the most desirable places to work.

#### 20.8.1.2 Case Study 2: Steelcase's Research on Work Environment and Stress

Google has been at the forefront industry when it comes to maintaining healthy workplace policies particularly those touching on functional ergo friendly, creatively designed workplace. It has followed open and flexible organization of services and desks provided with ergonomic furniture that is; adjustable workstation and chairs allowing the employees to shift between sitting and standing (Sundstrom, 1986). However, the company provides policies on wellness for every employee to access gym services, meditation services, and career advisors on the company premises. Tremendously, such programs have been realized to ease stress in the working places and increase general well-being of the employees (Garg, 2017). Therefore, the Company has been implementing HF&E practices focusing on both physical and health well-being aspects to make it one of the best places to work at.

### 20.8.2 Emerging Trends in HF&E for Mental Health

Smart technologies are coming into the workplace through the application of HF&E principles. It is possible to use chairs and desks with motion sensors, providing real-time data and teaching to employees how to maintain the correct posture during the working day (Reinhold et al., 2008). These technologies are intended to self-learn and adapt to the changes in the movement of the user so that there is comfort and minimize the likelihood of developing MSDs which result from contracting improper postures. There are also advancements in applying ergonomically, digital platforms to the workplace environment that encourage mental health. Applications focused on reducing stress, and tracking one's mental health assist the employee by taking personal responsibility for their health. These platforms provide customized suggestions about how to minimize stress level, enhance mental hardness and enhance cognitive performance. With the help of such smart technologies, the organizations will be able to introduce more comprehensive approaches delivering the employers' health care not only in terms of their physical conditions but also mental health.

### 20.8.3 Biophilic Design and Nature Integration

Biophilia and the concept of bringing elements of the natural environment into the design of human-built interiors are fast becoming appreciated as an effective means of combating stress in the workplace. Studies have shown that things like greens, natural light, and water bodies can decrease cortisol levels and bring relaxation (Sadick & Kamardeen, 2020). One is in relation to the environment that causes relaxation leading to happiness and even better performance in cognitive exercises. Greenery, colored spaces, outdoor spaces, and elements related to nature can almost be seen in the work setting of many organizations today as many of these adapt change to make their setting include green areas. Such conditions not only help the working individuals to avoid stress coming from high-performance work environment but also help them to live healthy and happy lives. Biophilic design integration results in the enhancement of the employee's well-being by providing restorative effects in organizations that sectors entail strained psychological and physiological exertion for example the technology, finance, and healthcare fields. Making an effort to incorporate nature into the work environment, businesses can improve the employees' psychological state and efficiency of the labor process.

### 20.8.4 Physical Fitness and Employer Responsibility

It is clear that exercising and having good physical health contribute to psychological and physical health in the workplace setting. This is the employer's duty that employees should have access to the essentials in their workplace which will help them to practice an active lifestyle. Onsite gym, proactively providing workplace wellness programs or encouraging employees to take breaks and move around the workplace will ensure that your employees are healthy, low stressed, happier and more energetic. This means exercise increases the efficiency of the human brain, reduces anxiety levels and improves productivity. This way employers who encourage fitness enable their staffs to make right choices over their body, and in the process there will be a healthier, focused workforce. Promoting exercise not only helps with avoiding disease but also assists in retaining good concentration and ensures high-quality work is produced. Finally, incorporating physical health consciousness into workplace is an important feature of the physical health and satisfaction of an employee.

### 20.8.5 Hybrid and Remote Work Models

COVID-19 has accelerated the trend towards working remotely and has introduced many challenges, as well as new chances for HF&E. Despite the obvious advantages of working remotely, such as flexibility and comfort, the new situation adds new forms of pressure on workers. These are as follows: loneliness, insufficient ergonomic arrangements, and flexible boundaries between one's work and home life (Bocar et al., 2022; Magd & Khan, 2022; Grawitch et al., 2016; Khatri et al., 2023; Khatri et al., 2024). Unfortunately, many companies face these challenges, and that is why today a significant number of organizations have introduced a flexible working mode, which includes work from the office and remotely (Khatri et al., 2023; Khan & Naim, 2023). Due to the ergonomic difficulties associated with working during the lockdown, companies are providing support regarding creating comfortable home-office work environment. This comprises offering ergonomic chairs, sit-stand desks, and technological aids for use in home-based worker arrangements. Providing choices in work schedules and providing organizational workforce with the equipment required for home-based work are a number of measures that can be taken by an organization to reduce stress, wellness promotion, and productivity improvement.

## 20.9 THE FUTURE OF WORK: ADAPTING HF&E IN A POST-PANDEMIC WORLD

The coronavirus outbreak has become a new reality and has transformed the perception of the working process providing awareness of flexibility, mental health, and ergonomics. Looking ahead towards an even post-COVID-19 WFH/Remote workplace scenario, HF&E has the potential to grow further and rise to the challenges of a changed perspective towards work and productivity, new faces of workplaces as well as human, organizational, mechanical, and virtual well-being.

### 20.9.1 The Role of AI and Automation

There will also be more opportunities to lessen stress through workload and delegation as AI and automation increasingly become workplace fixtures (Zareen & Khan, 2023). AI can simplify and track the stress levels of employees by for example analyzing stress biomarkers or even re-arrange working hours to fit within an optimum cognitive workload. But the question will be how much of it would be possible to automate and how much of it will have to remain human interface so as to not stress the employees with rapid appeals to innovation (Khan & Naim, 2023; Khatri et al., 2023). This will be important in making AI and automation more of enablers for reducing stress than a way of minimizing human interfaces.

### 20.9.2 Personalized and Data-Driven Approaches

Using the identified trends, the further development of HF&E will possibly be connected with the aspects of individualization of stress management and possibilities using the data collection for them. Smart clothes and health applications will monitor the condition of employees' mental and physical health and show managers what kind of interventions are needed. The strategies can also be sensitive to employees and can be used in addressing various employees better and reducing burnout.

### 20.9.3 Workplace Design and Mobility

Another factor that has become a principle during the pandemic is mobility and flexible office designs, which will only grow in importance in the post-pandemic world. Therefore, organizations will have to begin contemplating structures of work that urge togetherness as more and more accept hybrid work settings by designing work environments that foster collaboration, creativity, and well-being while at the same time offering work-from-home schedules. Blended work environments are going to have to use technology and workplace design to deliver cohesive and effective experiences for in-building and at-home employees.

## 20.10 CONCLUSION AND RECOMMENDATIONS

Job stress is an important area that has a huge impact on employees' health and impact on the organizational performance. As shown in this chapter, workplace stress originates from numerous physical, cognitive, and organizational stressors, all of which are capable of leading to mental and physical health deterioration. Resultant mental health issues not only bring a problem for the person concerned but also for organization through cost terms like absenteeism, reduced productivity, high turnover.

HF&E is a suitable approach for solving the problem of stress and ensuring mental well-being at work. The application of ergonomic principles at the workplace can minimize physical strain and achieve reduced levels of cognitive load that can foster well-being at work. Reflection of this fact is provided by HF&E-based strategies that primarily aim at redesigning workplaces, working processes, and social settings.

Such measures can improve efficiency, prevent cases of staff burnout, and contribute positively to the quality of lives of the employee's outside workplace. The successful integration of HF&E principles into workplace design underscores the need for innovative leadership that champions employee-centric and sustainable strategies (Umair et al., 2025).

As highlighted by different stress-reducing measures described in this chapter, organizations stand a great chance of reducing stress through ergonomic work environment, physical activities at the workplace as well as policies that support employee's mental well-being. Flexible work schedules and the use of technology in monitoring and fashioning mental health are new strategies for prevailing stress. Moreover, since the work environment is becoming more dynamic with new technology and change in workplace flexibility, it becomes important that HR policies and leadership training should also work for mental health and ergonomics to sustain and remain successful in future.

## REFERENCES


Allen, T. D., Johnson, R. C., Kiburz, K. M., & Shockley, K. M. (2013). Work–family conflict and flexible work arrangements: Deconstructing flexibility. *Personnel Psychology, 66*(2), 345–376. https://doi.org/10.1111/peps.12012

Ansari, M. S. A., Khan, S. A., & Bhuyan, U. (2024). Exploring the strategic orientation factors influencing the organizational performance through bibliometric analysis. In Umair, S., Waqas, U., Mrugalska, B., & Al Shamsi, I. R. (Eds.), *Human Perspectives of Industry 4.0 Organizations* (pp. 68–88). CRC Press. https://doi.org/10.1201/9781032616810-4

Anshel, M. H. (2005). *Applied Exercise Psychology: A Practitioner's Guide to Improving Client Health and Fitness*. Springer Publishing Company.

Bakker, A. B., & Demerouti, E. (2007). The job demands-resources model: State of the art. *Journal of Managerial Psychology, 22*(3), 309–328. https://doi.org/10.1108/02683940710733115

Banbury, S. P., & Berry, D. C. (2005). Office noise and employee concentration: Identifying causes of disruption and potential improvements. *Ergonomics, 48*(1), 25–37. https://doi.org/10.1080/00140130412331311390

Beehr, T. A., & Glazer, S. (2005). Organizational role stress. In Edited by: Barling, J. Kelloway, E. K. Frone, M. R. Beehr T. A., & Glazer, S. (Eds.), *Handbook of Work Stress* (pp. 7–34). SAGE Publications, Inc. https://doi.org/10.4135/9781412975995.n2

Bocar, A. C, Khan, S. A., Epoc, F. (2022). COVID-19 Work from home stressors and the degree of its impact: Employers and employees actions. *International Journal of Technology Transfer and Commercialisation, 19*(2), 270–291. https://dx.doi.org/10.1504/ijttc.2022.124349

Cartwright, S., & Cooper, C. L. (1994). The human effects of mergers and acquisitions. *Journal of Organizational Behavior (1986-1998), 47*. https://www.proquest.com/openview/944cbd9104364208100b446af2b051f8/1?cbl=47893&pq-origsite=gscholar

Chesney, M., Lindeke, L., Johnson, L., Jukkala, A., Lynch, S., Disch, J., & Densford, K. J. (2005). Comparison of child and parent satisfaction ratings of ambulatory pediatric subspecialty care. *Journal of Pediatric Health Care, 19*(4), 221–229. https://doi.org/10.1016/j.pedhc.2005.02.003

Covey, S. R. (1989). *The 7 Habits of Highly Effective People: Powerful Lessons in Personal Change*. Free Press.

Derks, D., Van Duin, D., Tims, M., & Bakker, A. B. (2015). Smartphone use and work–home interference: The moderating role of social norms and employee work engagement. *Journal of Occupational and Organizational Psychology*, *88*(1), 155–177. https://doi.org/10.1111/joop.12083

Edmondson, A. (1999). Psychological safety and learning behavior in work teams. *Administrative Science Quarterly*, *44*(2), 350–383. https://doi.org/10.2307/2666999

Elliot, A. J., & Maier, M. A. (2014). Color psychology: Effects of perceiving color on psychological functioning in humans. *Annual Review of Psychology, 65*, 95–120. https://doi.org/10.1146/annurev-psych-010213-115035

Evans, G. W., & Johnson, D. (2000). Stress and open-office noise. *Journal of Applied Psychology*, *85*(5), 779. https://psycnet.apa.org/doi/10.1037/0021-9010.85.5.779

Freina, L., & Ott, M. (2015). A literature review on immersive virtual reality in education: State of the art and perspectives. *eLearning and Software for Education, 1*, 133–141.

Garg, N. (2017). Workplace spirituality and employee well-being: An empirical exploration. *Journal of Human Values*, *23*(2), 129–147. https://doi.org/10.1177/0971685816689741

Grawitch, M. J., Gottschalk, M., & Munz, D. C. (2016). The path to a healthy workplace: A critical review linking healthy workplace practices, employee well-being, and organizational improvements. *Consulting Psychology Journal: Practice and Research, 58*(3), 129–147. https://doi.org/10.1037/1065-9293.58.3.129

Hämmig, O. (2020). Work-and stress-related musculoskeletal and sleep disorders among health professionals: a cross-sectional study in a hospital setting in Switzerland. *BMC Musculoskeletal Disorders*, *21*, 1–11.

Hickey, B. A., Chalmers, T., Newton, P., Lin, C. T., Sibbritt, D., McLachlan, C. S., Clifton-Bligh, R., Morley, J., & Lal, S. (2021). Smart devices and wearable technologies to detect and monitor mental health conditions and stress: A systematic review. *Sensors, 21*(10), 3461.

Jain, A., Fernandez, R., & Goh, C. (2019). A holistic approach to mental health and ergonomics: Integrating mental wellness into workplace design. *Work*, *63*(1), 45–59. https://doi.org/10.3233/WOR-192903

Kabat-Zinn, J. (1990). *Full eCatastrophe Living: Using the Wisdom of Your Body and Mind to Face Stress, Pain and Illnss*. Delacorte.

Kamarulzaman, N., Saleh, A. A., Hashim, S. Z., Hashim, H., & Abdul-Ghani, A. A. (2011). An overview of the influence of physical office environments towards employee. *Procedia Engineering*, *20*, 262–268.

Kaplan, S. (1995). The restorative benefits of nature: Toward an integrative framework. *Journal of Environmental Psychology*, *15*(3), 169–182. https://doi.org/10.1016/0272-4944(95)90001-2

Karasek, R., Brisson, C., Kawakami, N., Houtman, I., Bongers, P., & Amick, B. (1998). The Job Content Questionnaire (JCQ): An instrument for internationally comparative assessments of psychosocial job characteristics. *Journal of Occupational Health Psychology*, *3*(4), 322. https://psycnet.apa.org/buy/1998-12418-004

Karwowski, W., & Marras, W. S. (2003). *Occupational Ergonomics: Principles of Work Design*. CRC Press.

Kellert, S. R., Heerwagen, J. H., & Mador, M. (2011). *Biophilic Design: The Theory, Science, and Practice of Bringing Buildings to Life*. Wiley.

Khan, S. A., & Ansari, M. S. A. (2024). Critical factors for relationship between strategic orientation, corporate success, and innovation. In Umair, S., Waqas, U., Mrugalska, B., & Al Shamsi, I. R. (Eds.), *Human Perspectives of Industry 4.0 Organizations* (pp. 89–110). CRC Press. https://doi.org/10.1201/9781032616810-5

Khan, S.A., & Naim, A. (2023). XAI in Society 5.0 through the lens of marketing and HRM. In Al-Turjman, F., Nayyar, A., Naved, M., Singh, A. K., & Bilal, M. (Eds.), *XAI Based Intelligent Systems for Society 5.0* (pp. 327–363). Elsevier. https://doi.org/10.1016/B978-0-323-95315-3.00004-8

Khatri, B., Sharma, N., Khan, S. A., & Nandini, G. (2024). Navigating employer branding in the digital age: Exploring the impact of social media networking during the COVID-19 pandemic. In Christiansen, B., Aziz, M., & O'Keeffe, E. (Eds.), *Global Practices on Effective Talent Acquisition and Retention* (pp. 313–332). IGI Global. https://doi.org/10.4018/979-8-3693-1938-3.ch017

Khatri, B., Shrimali, H., Khan, S. A., & Naim, A. (2023). Role of HR analytics in ensuring psychological wellbeing and job security: Learnings from COVID-19. In Yadav, R., Sinha, M., & Kureethara, J. (Eds.), *HR Analytics in an Era of Rapid Automation* (pp. 36–53). IGI Global. https://doi.org/10.4018/978-1-6684-8942-0.ch003

Králiková, R., Lumnitzer, E., Džuňová, L., & Yehorova, A. (2021). Analysis of the impact of working environment factors on employee's health and wellbeing; Workplace lighting design evaluation and improvement. *Sustainability*, *13*(16), 8816.

Magd, H. & Khan, S. A. (2022). Strategic framework for entrepreneurship education in promoting social entrepreneurship in GCC countries during and post COVID-19. In Magd, H., Singh, D., Syed, R., & Spicer, D. (Eds.), *International Perspectives on Value Creation and Sustainability Through Social Entrepreneurship* (pp. 61–75). IGI Global. https://doi.org/10.4018/978-1-6684-4666-9.ch004

Newport, C. (2016). *Deep Work: Rules for Focused Success in a Distracted World*. Grand Central Publishing.

Norman, R. (2013). *On Humanism*. Routledge. https://doi.org/10.4324/9780203814352

Punnett, L., & Wegman, D. H. (2004). Work-related musculoskeletal disorders: The epidemiologic evidence and the debate. *Journal of Electromyography and Kinesiology, 14*(1), 13–23. https://doi.org/10.1016/j.jelekin.2003.09.015

Reinhold, K., Tint, P., Tuulik, V., & Saarik, S. (2008). Innovations at workplace: Improvement of ergonomics. *Engineering Economics*, *60*(5), 85–94.

Richards, J. (2022). Putting employees at the centre of sustainable HRM: A review, map and research agenda. *Employee Relations: The International Journal*, *44*(3), 533–554. https://doi.org/10.1108/ER-01-2019-0037

Rubinstein, J. S., Meyer, D. E., & Evans, J. E. (2001). Executive control of cognitive processes in task switching. *Journal of Experimental Psychology: Human Perception and Performance*, *27*(4), 763. https://psycnet.apa.org/doi/10.1037/0096-1523.27.4.763

Sadick, A. M., & Kamardeen, I. (2020). Enhancing employees' performance and well-being with nature exposure embedded office workplace design. *Journal of Building Engineering*, *32*, 101789.

Salanova, M., Grau, R. M., Cifre, E., & Llorens, S. (2000). Computer training, frequency of usage and burnout: the moderating role of computer self-efficacy. *Computers in Human Behavior*, *16*(6), 575–590. https://doi.org/10.1016/S0747-5632(00)00028-5

Seppänen, O., Fisk, W. J., & Lei, Q. H. (2006). Ventilation and performance in office work. *Indoor Air, 16*(1), 28–36. https://doi.org/10.1111/j.1600-0668.2005.00394.x

Sultan, S. E. (2015). *Organism and Environment: Ecological Development, Niche Construction, and Adaptation*. Oxford University Press.

Sundstrom, E. (1986). *Workplaces: The Psychology of the Physical Environment in Offices and Factories*. Cambridge University Press.

Sweller, J. (1988). Cognitive load during problem solving: Effects on learning. *Cognitive Science*, *12*(2), 257-285. https://doi.org/10.1016/0364-0213(88)90023-7

Tarafdar, M., Pullins, E. B., & Ragu-Nathan, T. S. (2015). Technostress: Negative effect on performance and possible mitigations. *Information Systems Journal*, *25*(2), 103–132. https://doi.org/10.1111/isj.12042

Tudor-Locke, C., Craig, C. L., Brown, W. J., Clemes, S. A., De Cocker, K., Giles-Corti, B., Hatano, Y., Inoue, S., Matsudo, S. M., Mutrie, N., & Oppert, J. M. (2011). How many steps/day are enough? *International Journal of Behavioral Nutrition and Physical Activity, 8*(79), 1–17. https://doi.org/10.1186/1479-5868-8-79

Umair, S., Waqas, U., Mrugalska, B., & Al Shamsi, I. R. (2025). Human *Perspectives* of Industry 4.0 *Organizations*: Reviewing *Sustainable Performance*. Routledge.

Wright, P. M., & McMahan, G. C. (2011). Exploring human capital: Putting human back into strategic human resource management. *Human Resource Management Journal, 21*(2), 93–104. https://doi.org/10.1111/j.1748-8583.2010.00165.x

Zareen, S. & Khan, S. A. (2023). Exploring dependence of human resource management (HRM) on internet of things (IoT) and digital marketing in the digital era. In Naim, A. & Devi, V. (Eds.), *Global Applications of the Internet of Things in Digital Marketing* (pp. 51–66). IGI Global. https://doi.org/10.4018/978-1-6684-8166-0.ch004

# 21 Human-Centric Nutrition and the Microbiome for the Development of Business Strategies in Transforming Health

## *Special Focus on Arab World*

*Arshi Naim, Mohammad Ibrahim Khan, and Mohammad Faiz Khan*

## 21.1 INTRODUCTION

The rise of human-centric or personalized nutrition represents a change and a shift in how we approach health and wellness. On the contrary to traditional dietary, recommendations are given to follow the "one-size-fits-all" model. The personalized nutrition tailor's dietary interventions to individual needs, accounting for factors such as genetics, lifestyle, and, more recently, the gut microbiome and other stress levels present in an individual. The microbiome represents a diverse community of microorganisms residing in the human body, particularly in the gut, and has emerged as a critical determinant of health (Naim et al., 2023a,b). Several research indicates that the microbiome influences digestion, immunity, metabolic health, and even mental well-being. This growing body of evidence has positioned microbiome science at the forefront of innovation in health and nutrition. Arab Region has witnessed special care in maintaining good care of health because of their lifestyle (Bendriss et al., 2020).

The business world has started recognizing the potential of microbiome-based personalized nutrition for better health (Heniedy et al., 2024). Many firms in the food, health, and technology sectors are exploring ways to integrate scientific advancements into their products and services in the Arab region (Khan et al., 2025a,b). From microbiome-friendly foods and probiotics to advanced AI-driven nutrition apps, businesses are finding innovative ways to cater to the demand for tailored health solutions (Xiao et al., 2024). Personalized nutrition not only compels the implementation of a fixed plan but also gives significant challenges. Barriers such

DOI: 10.1201/9781003532293-24

as consumer skepticism, concerns about data privacy, and the high cost of personalized solutions limit the accessibility and acceptance of microbiome-based products. However, Gut Health is directly related to and dependent on mental health which works in a healthy way if metabolic and immune systems are strong and healthy. Gut health in a human body depends on mental health conditions which take control from metabolic activities and Immune systems' standards.

This research has identified a critical gap, despite the scientific progress in understanding the microbiome, there is a lack of clear strategies for translating these theoretical insights into actionable, consumer-friendly, and commercially viable solutions. Businesses often struggle to align their offerings with consumer needs and expectations, leading to missed opportunities in an otherwise promising market (Zyoud et al., 2024).

This study focuses on three key concepts, which are personalized nutrition, the microbiome, and their intersection with business strategies. Personalized nutrition involves tailoring dietary recommendations to individual needs, accounting for genetic, environmental, and lifestyle factors. Advances in microbiome research have uncovered the significant role of gut microbes in digestion, immunity, and overall health, making the microbiome a central player in personalized nutrition. However, these scientific advancements are not limited to healthcare and they open up to the opportunities for businesses in the food, health, and technology sectors (Al-Worafi, 2024). Businesses can leverage microbiome insights to develop innovative products and services, from probiotics and prebiotics to AI-driven dietary tools. Understanding these concepts in a business context aid to bridge the gap between consumer needs and scientific advancements, creating a foundation for sustainable and scalable business models (Escalante-Aburto et al., 2021).

The concept of personalized nutrition, particularly in relation to the microbiome, is gaining significant traction worldwide, but its application and adoption in the Arab region present unique challenges and opportunities (Kouarfaté & Durif, 2023). The Arab world, with its diverse cultural traditions, rapidly evolving healthcare landscape, and growing interest in health and wellness, is at a pivotal point in exploring how microbiome-based personalized nutrition can shape public health and business strategies. While the region has a long history of using natural, food-based remedies for maintaining health such as dates, yogurt, and herbal tea. The integration of modern scientific research on the microbiome into everyday dietary practices is still in the introductory stage (Alqahtani & Naim, 2022). The microbiome, the vast community of microorganisms in the human gut, plays a crucial role in digestion, immunity, and overall well-being, offering an exciting frontier for personalized health solutions (Hussain et al., 2019). However, barriers such as cultural perceptions of new health technologies, affordability issues, and a lack of widespread scientific education on the microbiome may hinder its mainstream acceptance. Despite these challenges, there is considerable potential for businesses in the Arab region to harness the benefits of personalized nutrition, especially as consumers become more health-conscious and open to digital health innovations (Reyed, 2023). Human-centric or Personalized Nutrition and Gut Microbiome work on six levels of hierarchy. These levels include a broad understanding of nutrition and its impact on gut health, role of multivitamins, microbiome analysis, tailored supplements based on the health requirements,

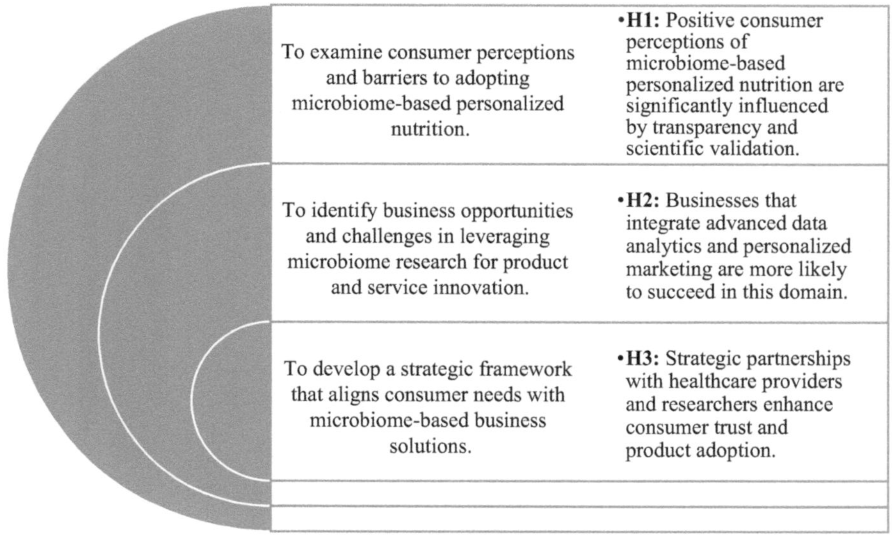


**FIGURE 21.1** Integration of research objectives and research hypothesis.

monitoring impact of multivitamins and measuring the health conditions post intake of these supplements and finally achieving health optimization.

This research aims to explore how businesses can navigate these cultural and economic complexities, provide insights into consumer perceptions, and identify strategies for successful market penetration, ultimately contributing to the growth of personalized nutrition in the Arab world. Figure 21.1 explains the hypotheses and the relationship between the research objectives.

- **RQ1**: To examine consumer perceptions and barriers to adopting microbiome-based personalized nutrition.
- **RQ2**: To identify business opportunities and challenges in leveraging microbiome research for product and service innovation.
- **RQ3**: To develop a strategic framework that aligns consumer needs with microbiome-based business solutions.

The current study has several significant benefits at multiple levels like for consumers, businesses, researchers, and policymakers. For consumers, the research benefits by addressing barriers like skepticism and cost, the study provides insights that could lead to more accessible and effective personalized nutrition solutions. For businesses, this study offers practical strategies for leveraging microbiome research to develop consumer-centric products, ensuring better market penetration and profitability. For researchers, the current study highlights a gap in microbiome and personalized nutrition studies, paving the way for further exploration of the link between science, consumer behavior, and business models. And finally, for policymakers the study gives insight to formulate guidelines and policies around data privacy and ethical business practices in the field of personalized nutrition.

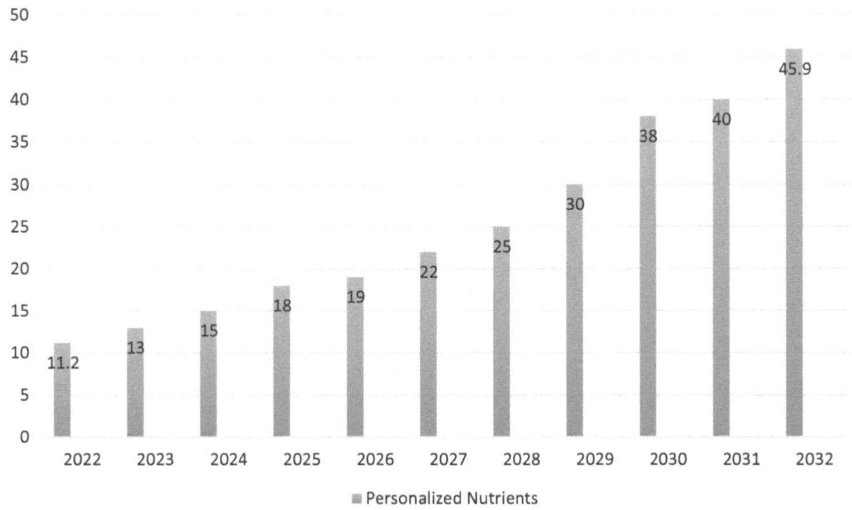


**FIGURE 21.2** Trend of personalized nutrition market in the world. (Source: https://www.sphericalinsights.com/reports/personalized-nutrition-market.)

The research has contributed from both theoretical and practical perspectives. Theoretically, it advances understanding of the interplay between microbiome science, consumer behavior, and business strategies, offering a new lens for exploring personalized nutrition. Practically, the study provides a strategic framework for businesses to design, market, and implement microbiome-based nutritional products effectively. It also identifies critical success factors such as transparency, collaborative ecosystems, and the use of cutting-edge technologies like artificial intelligence and big data analytics. These contributions aim to not only help businesses innovate but also promote the adoption of personalized nutrition as a mainstream approach, ultimately improving public health and fostering sustainable industry growth. The current study has extracted the secondary data to show the global personalized nutrition market in the world. Figure 21.2 gives a clear picture of growth and trend of personalized nutrition market in the world from the year 2022 to future years ending in 2032.

The study aims to bridge the gap by investigating the intersection and integration of personalized nutrition, microbiome research, and business innovation. It analyzes the challenges and opportunities faced by businesses in leveraging microbiome-driven personalized nutrition and identifies strategies and frameworks to create value for consumers while achieving sustainable growth. By focusing on consumer perceptions, barriers to adoption, and successful business practices, this research seeks to develop a comprehensive framework that can guide businesses toward effective integration of microbiome science into their strategies. Ultimately, the study contributes to both academic knowledge and practical applications, paving the way for innovation in the rapidly evolving field of personalized nutrition.

## 21.2 LITERATURE REVIEW

The concept of personalized nutrition has evolved significantly over the past two decades, driven by advancements in genetics, microbiome research, and digital health technologies (Murshid, 2024). Early approaches to personalized nutrition focused on genetic predispositions to certain diseases, using DNA-based insights to tailor dietary recommendations (Bouchard & Ordovas, 2012). However, this approach was limited by its cost and the inability to fully account for environmental and lifestyle factors (Mahmod et al., 2022). The discovery of the microbiome's pivotal role in human health marked a paradigm shift, emphasizing the gut as a dynamic interface between diet, metabolism, and overall wellness (Stewart & Kremer, 2023). Research such as the Human Microbiome Project (HMP) laid the foundation for understanding the microbiome's complexity, leading to an explosion of interest in microbiome-targeted therapies, including personalized nutrition. There are many tools used in the past to analyze the nutrition such as blood tests, DNA tests, Microbiome indicators, tools to test digestive nutrients and surveys to measure the lifestyle, habits, and food intake. These tools have helped in analyzing the plan and outcomes of personalized nutrition and measure the effectiveness.

In the context of the Arab world, limited literature exists on the integration of microbiome science into nutrition and business. Traditional dietary practices in the region, rooted in cultural and religious norms, have long emphasized the therapeutic potential of foods such as dates, fermented dairy products, and herbal teas (Namkhah et al., 2023). These foods naturally support gut health, providing a cultural advantage for introducing microbiome-based solutions. However, there is little documented research on how these practices can be scientifically aligned with modern personalized nutrition strategies, creating a gap this study seeks to address (Stewart & Kremer, 2023).

A competitive advantage of this research lies in its interdisciplinary focus, which combines insights from microbiome science, consumer behavior, and business strategy. Unlike most studies that examine personalized nutrition from either a biomedical or consumer perspective, this research uniquely investigates how businesses can leverage microbiome insights to create consumer-centric, culturally sensitive solutions. The inclusion of the Arab world as a focal point further strengthens the study's relevance, as it addresses a market that has been underexplored in global literature.

Moreover, this research highlights the role of technology and innovation in advancing the field. The rise of wearable devices, AI-driven nutrition apps, and at-home microbiome testing kits offer a competitive edge to businesses that can integrate these tools into their offerings. For example, global brands like Viome and ZOE have successfully capitalized on these trends in Western markets, yet there is significant potential for localized adaptations of these models in the Arab region (Namkhah et al., 2023).

By bridging gaps in existing literature and addressing the unique cultural and economic dynamics of the Arab world, this study contributes to both academic discourse and practical applications. It underscores the importance of culturally aligned business strategies, affordable innovations, and trust-building measures to foster the adoption of microbiome-based personalized nutrition in this diverse and rapidly evolving market.

## 21.3 RESEARCH METHODOLOGY

The study adopted a qualitative research methodology to explore the potential for microbiome-based personalized nutrition in the Arab world, focusing on consumer behavior, business opportunities, and cultural dynamics. Data collection involved semi-structured interviews with key stakeholders, including nutritionists, healthcare providers, business leaders in the nutrition and wellness industries, and consumers from various Arab countries. To ensure diverse perspectives, participants were selected from both affluent Gulf Cooperation Council (GCC) nations such as Saudi Arabia and the UAE, and economically diverse countries like Egypt, Jordan, and Morocco. This approach captured the unique socio-economic and cultural factors shaping the adoption of personalized nutrition across the region.

Additionally, content analysis of local marketing campaigns, health initiatives, and business case studies was conducted to understand how personalized nutrition aligns with regional trends, such as the growing emphasis on preventive healthcare and the integration of technology into wellness solutions. Coding and thematic analysis were used to identify recurring patterns, such as trust in healthcare professionals, affordability concerns, and the influence of traditional diets.

The methodology also included a review of regional health data and policy documents to contextualize findings within the broader public health priorities of the Arab world. By combining insights from interviews, secondary data, and cultural analysis, the study provided a comprehensive understanding of the opportunities and challenges for businesses operating in this emerging field.

## 21.4 RESULTS

The Arab region presents a unique landscape for the adoption of microbiome-based personalized nutrition due to its cultural, economic, and health-related dynamics. The results of the study reveal a mixed level of awareness and acceptance of personalized nutrition, largely influenced by socio-economic factors, traditional dietary practices, and the region's evolving healthcare priorities. Secondary data is collected from Frost & Sullivan, 2023 Report on Wellness-Driven Innovations in the Middle East and North Africa (MENA) Region, Regional Health Research Organization Survey, 2022, and Market Analysis of GCC Nutrition Sector, 2021. Figure 21.3 illustrates key indicators related to the adoption of microbiome-based personalized nutrition in the Arab region. These indicators are consumer interest, Awareness of Microbiome Solutions, Cost Concerns, and Market Growth Rate.

Consumer interest is 62%, which is a significant proportion of consumers in the region who express interest in personalized nutrition solutions. Awareness of Microbiome Solutions is 35%, which measures the awareness levels. The outcomes show a relatively lower level of awareness, indicating a need for educational initiatives for health-related knowledge. Cost Concerns are at 70%, which shows that the majority of potential consumers find such solutions expensive, and cause key barriers in adopting good nutrition. Market Growth Rate is at 8.5%, which is evidently proven that the region's nutrition sector is growing steadily, driven by health and wellness trends.

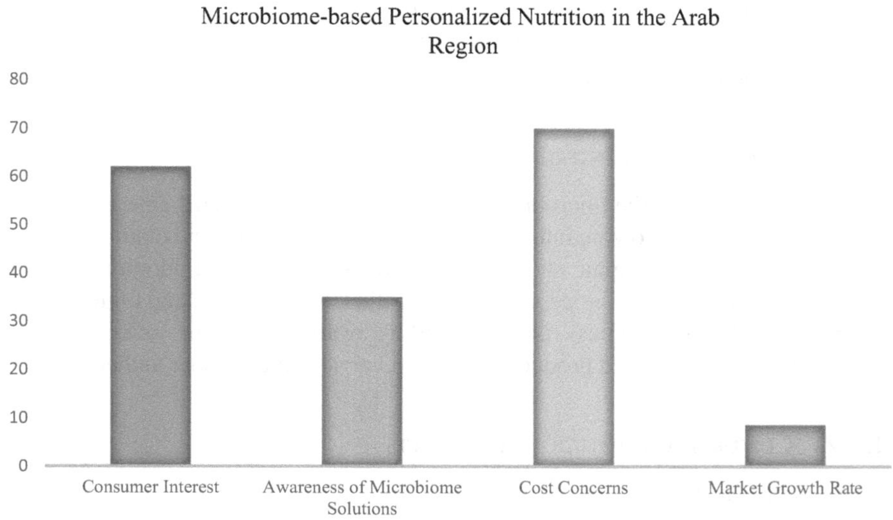


**FIGURE 21.3** Indicators for microbiome-based personalized nutrition in the Arab region.

These data points explain the necessity for businesses to focus on affordability, cultural alignment, and consumer education to successfully integrate microbiome-based personalized nutrition into the regional market.

To support the findings of this study, secondary data reveals a growing but cautious interest in microbiome-based personalized nutrition within the Arab region. Reports indicate that the global microbiome market is valued at approximately $1.8 billion as of 2023, with the MENA contributing a modest but expanding share. A 2022 survey by a regional health research organization showed that 62% of Arab consumers expressed interest in personalized nutrition solutions, though only 35% were aware of microbiome-based interventions. Skepticism, primarily driven by cultural norms and limited understanding of microbiome science, remains a significant barrier.

Cultural taboos and concerns about the origin of microbiome-derived products also play a role, with consumers preferring solutions aligned with Halal and traditional dietary practices. Furthermore, high costs were highlighted in a 2021 market analysis, showing that over 70% of potential consumers in GCC countries consider personalized nutrition too expensive, despite their interest. Data from healthcare providers show a lag in integrating microbiome solutions into mainstream practices due to limited awareness and training.

Despite these barriers, there is significant optimism. A 2023 report by Frost & Sullivan identified the Arab region as a high-growth market for wellness-driven innovations, with a compound annual growth rate of 8.5% expected in the nutrition sector by 2027. Investment in digital health, particularly mobile apps and wearable devices, has been accelerating in countries like Saudi Arabia and the UAE, offering a promising avenue for scalable, affordable solutions.

These data points underscore the need for businesses to address cultural and economic barriers while leveraging the region's digital transformation to drive consumer engagement and healthcare integration.

### 21.4.1 Awareness and Interest

While awareness about the microbiome and its role in health is still emerging in the Arab region, there is a growing interest, particularly among younger, health-conscious individuals. This demographic shows a strong inclination toward adopting innovative health solutions, including personalized nutrition. Social media and digital health platforms are playing a critical role in spreading awareness, but the scientific complexity of microbiome-based products remains a barrier for a broader audience.

### 21.4.2 Cultural and Religious Influences

Dietary habits in the Arab region are deeply rooted in cultural and religious practices. The acceptance of microbiome-based nutrition is higher when solutions are tailored to align with traditional diets and halal standards. For example, incorporating personalized nutrition into regional staples, such as dates, lentils, and whole grains, is more likely to resonate with consumers. Additionally, the communal nature of food in Arab culture requires businesses to consider family-based dietary solutions rather than individualistic approaches.

### 21.4.3 Barriers to Adoption

Several barriers were identified in the Arab region, which affect the widespread adoption of microbiome-driven personalized nutrition.

- **Cost:** The high cost of microbiome testing and personalized dietary plans is a significant deterrent, particularly for middle- and lower-income groups.
- **Data Privacy Concerns:** There is skepticism about sharing genetic or microbiome-related data, with consumers expressing fears about misuse or lack of transparency from businesses.
- **Trust in Science:** Many consumers are cautious about new health technologies, emphasizing the need for clear, evidence-based communication and endorsements by trusted healthcare professionals.

The adoption of microbiome-driven personalized nutrition in the Arab region faces several notable barriers that hinder its widespread acceptance. Among these, cost stands out as a significant deterrent. Microbiome testing and tailored dietary plans remain expensive, often costing hundreds to thousands of dollars, which is prohibitive for middle- and lower-income groups. For instance, while residents in affluent areas like Dubai and Riyadh have the disposable income to invest in such services, the broader population across countries like Egypt, Jordan, and Morocco faces financial constraints. This disparity makes personalized nutrition solutions largely inaccessible to a significant segment of the population. Businesses must address these gaps by creating scalable and affordable options to ensure inclusivity.

Another critical barrier is data privacy concerns, which resonate deeply in the Arab region. Consumers are hesitant to share sensitive microbiome or genetic data due to fears of misuse or unauthorized access. This concern is amplified by a lack of stringent data protection regulations in some Arab countries compared to global standards. For instance, individuals in Saudi Arabia and the UAE, despite being technologically advanced markets, often question how companies handle and store their personal data. This skepticism reflects a broader cultural priority on privacy and security. Companies looking to enter the market must establish clear, transparent policies and build trust by adhering to high standards of data protection.

Lastly, trust in science poses a significant hurdle. While there is growing awareness about health and wellness, many consumers remain cautious about embracing new health technologies. The idea of manipulating gut bacteria to improve health, for example, can seem abstract and unproven to many in the region. In countries like Oman and Kuwait, where traditional medicine and dietary practices hold strong cultural significance, people often seek advice from trusted healthcare providers or religious leaders. Without endorsements from these authorities or clear, evidence-based communication, microbiome-based solutions are viewed with skepticism.

Addressing these barriers requires a localized approach that respects cultural values and economic realities. For example, partnering with well-known healthcare providers, offering subsidized options, and launching educational campaigns tailored to the region's diverse population can foster greater acceptance and trust.

### 21.4.4 Opportunities for Businesses

Despite these challenges, the Arab region offers considerable opportunities for businesses willing to adapt their strategies.

- **Education and Awareness Campaigns:** Increasing consumer understanding of the microbiome's impact on health can foster trust and interest in personalized nutrition. Collaborations with healthcare providers, universities, and influencers can help disseminate this knowledge.
- **Affordability and Accessibility:** Developing cost-effective microbiome-based products and offering flexible payment options can expand access to a wider audience.
- **Technology Integration:** The region's high smartphone penetration and growing interest in digital health tools provide an opportunity for businesses to offer AI-driven nutrition apps and online consultations tailored to Arab dietary preferences.
- **Partnerships with Healthcare Systems:** Integrating personalized nutrition into public health initiatives or corporate wellness programs can enhance credibility and reach.

Addressing the barriers to adopting microbiome-driven personalized nutrition in the Arab region requires targeted strategies that align with the region's unique cultural, economic, and technological landscape.

### 21.4.5 Education and Awareness Campaigns

Education is the cornerstone for building trust and interest in microbiome-based personalized nutrition. Many individuals in the Arab region remain unaware of how gut health influences overall well-being. Launching collaborative campaigns with healthcare providers, universities, and influencers can bridge this gap. For instance, universities in Saudi Arabia and the UAE, known for their advanced research in healthcare, could host public seminars or webinars on the science behind the microbiome. Collaborating with trusted medical professionals and regional influencers, who already resonate with local audiences, can further amplify these efforts. For example, a partnership with popular fitness influencers in the GCC, who often promote healthy lifestyles, could make microbiome-based nutrition more relatable to younger generations.

### 21.4.6 Affordability and Accessibility

Affordability is a key driver for broader adoption. By developing cost-effective microbiome-based products, companies can cater to middle- and lower-income groups. One potential approach is to introduce tiered pricing models or flexible payment options, such as monthly subscription plans. In markets like Egypt and Jordan, where disposable income is lower compared to the Gulf, such initiatives can make personalized nutrition accessible to a wider audience. Additionally, businesses could leverage economies of scale by collaborating with local manufacturers to produce affordable testing kits or supplements tailored to regional dietary patterns.

### 21.4.7 Technology Integration

The Arab region's high smartphone penetration and growing interest in digital health offer immense potential for innovation. Countries like the UAE and Saudi Arabia have robust digital ecosystems, with a significant percentage of the population using apps for fitness and health tracking. Businesses could introduce AI-driven nutrition apps that provide personalized dietary recommendations based on microbiome data. These apps can be tailored to local tastes by incorporating traditional Arab dishes and dietary habits into the recommendations. For instance, an app suggests gut-friendly modifications to popular meals like Khabsa or couscous. Coupled with online consultations, such digital tools make it easier for consumers to access expert advice without visiting clinics.

### 21.4.8 Partnerships with Healthcare Systems

Collaborating with healthcare systems is essential for credibility and scalability. By integrating microbiome-based nutrition into public health initiatives or corporate wellness programs, businesses can reach a broader audience while gaining the trust of institutions. For example, in Saudi Arabia, the government's Vision 2030 initiative emphasizes preventive healthcare, presenting an opportunity to incorporate microbiome-focused programs into existing wellness initiatives.

Similarly, corporate wellness programs in the UAE could include microbiome testing as part of employee benefits, encouraging adoption among professionals who are already health-conscious.

These strategies, when implemented effectively, can create a robust ecosystem where microbiome-based personalized nutrition is both accessible and trusted, ultimately driving innovation and improving health outcomes across the Arab region.

### 21.4.9 Regional Health Priorities

With rising rates of obesity, diabetes, and cardiovascular diseases in the Arab region, governments and healthcare providers are increasingly prioritizing preventive health measures. Personalized nutrition, driven by microbiome insights, aligns well with these priorities by offering targeted, preventive solutions that address these chronic health challenges. Businesses that position their offerings within this preventive healthcare framework stand to gain both consumer trust and policy support.

The study further highlights that while the Arab region is gradually embracing the concept of personalized nutrition, practical implementation remains uneven across different countries. For instance, in the GCC countries like Saudi Arabia and the UAE, there is a higher adoption rate due to stronger economic conditions, advanced healthcare systems, and a tech-savvy population. Here, premium services such as microbiome testing kits, AI-powered nutrition apps, and personalized diet plans are gaining traction among urban, affluent consumers. Companies like The Food Doctor UAE and BiOptimizers KSA are already marketing microbiome-friendly supplements and dietary programs tailored to local needs, showing how businesses are aligning their offerings with regional preferences.

In contrast, countries with more constrained economies, such as Jordan, Egypt, and Morocco, face challenges in scaling personalized nutrition. Affordability remains a key barrier, as the cost of microbiome testing and follow-up consultations is often prohibitive for middle-income households. However, opportunities lie in creating low-cost, scalable solutions, such as community-based nutrition programs or integrating microbiome-friendly products into existing food staples. For example, fortifying widely consumed items like flatbreads, yogurt, or lentil soups with probiotics or prebiotics can be an accessible entry point for businesses targeting these markets.

Additionally, the growing focus on lifestyle diseases presents a compelling case for personalized nutrition in the Arab region. With obesity rates exceeding 35% in countries like Kuwait and Qatar, and diabetes prevalence reaching alarming levels across the region, there is a clear demand for preventive and personalized healthcare solutions. Businesses that align their products with national health initiatives, such as Saudi Arabia's Vision 2030 health goals, can gain significant traction. For instance, introducing gut-health-focused meal plans in collaboration with fitness centers or launching educational campaigns through local clinics could bridge the gap between consumer skepticism and product adoption.

These examples demonstrate that while the adoption of microbiome-based personalized nutrition is in its early stages, businesses that tailor their approaches to the unique cultural, economic, and health landscapes of the Arab region are well-positioned to succeed.

The results indicate that the Arab region holds significant potential for the growth of microbiome-based personalized nutrition, provided businesses adopt culturally sensitive, affordable, and trustworthy strategies. By addressing barriers and leveraging regional opportunities, businesses can play a pivotal role in advancing personalized health solutions that cater to the unique needs of Arab consumers while contributing to the broader goals of public health and wellness.

## 21.5 CONCLUSION

The Arab region holds significant promise for the growth and adoption of microbiome-based personalized nutrition, but several factors must be addressed to unlock its full potential. While there is a growing interest in health and wellness, cultural, economic, and technological barriers continue to limit widespread acceptance. Businesses that wish to succeed in this market must develop solutions that are culturally relevant, affordable, and backed by transparent scientific evidence. By focusing on consumer education, leveraging existing dietary traditions, and integrating advanced technologies like AI and microbiome testing, companies can foster trust and drive demand for personalized nutrition products. Furthermore, aligning these innovations with the region's health priorities, such as the rising prevalence of lifestyle diseases, can help position personalized nutrition as a crucial element of preventive healthcare. The research highlights that while challenges exist, they are not insurmountable, and with the right strategies, the Arab region could emerge as a key player in the global personalized nutrition market.

## REFERENCES

Al-Worafi, Y. M. (2024). Clinical nutrition and diet education in developing countries. In *Handbook of Medical and Health Sciences in Developing Countries: Education, Practice, and Research* (pp. 1–19). Springer International Publishing.

Alqahtani, H., & Naim, A. (2022). Critical success factors for transforming CRM to SCRM for building E-CRM. In *Building a Brand Image Through Electronic Customer Relationship Management* (pp. 139–168). IGI Global.

Bendriss, G., Al-Ali, D., Shafiq, A., Laswi, I., Mhaimeed, N., Salameh, M., Burney, Z., Pillai, K., Chaari, A., Zakaria, D., & Yousri, N. A. (2020). Targeting the gut microbiome: A brief report on the awareness, practice, and readiness to engage in clinical interventions in Qatar. *Qatar Medical Journal*, 2020(3), 47.

Bouchard, C., & Ordovas, J. M. (2012). Fundamentals of nutrigenetics and nutrigenomics. *Progress in Molecular Biology and Translational Science*, *108*, 1–15.

Escalante-Aburto, A., Trujillo-de Santiago, G., Álvarez, M. M., & Chuck-Hernández, C. (2021). Advances and prospective applications of 3D food printing for health improvement and personalized nutrition. *Comprehensive Reviews in Food Science and Food Safety*, *20*(6), 5722–5741.

Heniedy, A. M., Mahdy, D. M., Elenien, W. I. A., Mourad, S., & El-Kadi, R. A. (2024). Postbiotics as a health-promoting technique: A review article on scientific and commercial interest. *Process Biochemistry*, 144, 6–19.

Hussain, M. R., Quadri, N. N., Ahmad, N., Qamar, S., Khan, N., Naim, A., & Hussain, M. E. (2019, April). Effective cost optimization approach in Healthcare to Minimize the treatment cost of Brain-tumor Patients. In *2019 International Conference on Computer and Information Sciences (ICCIS)* (pp. 1–5). IEEE, Aljouf, Kingdom of Saudi Arabia.

Khan, M. I., Khan, M. F., Gauthaman, S. A., & Naim, A. (2025a). Enhancing business applications in ENT diseases management through data visualization. In *Data Visualization Tools for Business Applications* (pp. 209–226). IGI Global.

Khan, M. I., Khan, M. F., Gauthaman, S. A., & Naim, A. (2025b). Enhancing medical sector efficiency: A comprehensive case study based on data visualization tools for business applications. In *Data Visualization Tools for Business Applications* (pp. 227–248). IGI Global.

Kouarfaté, B. B., & Durif, F. N. (2023). A systematic review of determinants of cultured meat adoption: Impacts and guiding insights. *British Food Journal, 125*(8), 2737–2763.

Mahmod, A. I., Haif, S. K., Kamal, A., Al-Ataby, I. A., & Talib, W. H. (2022). Chemoprevention effect of the Mediterranean diet on colorectal cancer: Current studies and future prospects. *Frontiers in Nutrition, 9*, 924192.

Murshid, G. A. (2024). The rise of healthy longevity and healthspan in Saudi Arabia: From funding geroscience research to precision medicine and personalized digital twins. *Discover Medicine, 1*(1), 1–12.

Naim, A., Alqahtani, H., Fatima, S., & Khan, M. F. (2023a). Applications of artificial intelligence of things in green supply chain management: Challenges and future directions. In *Blockchain, IoT, and AI Technologies for Supply Chain Management* (pp. 99–126). CRC Press.

Naim, A., Khan, S. A., Malik, P. K., Hussain, M. R., & Dildar, M. S. (2023b, November). Internet of things support for marketing sports and fitness products. In *2023 3rd International Conference on Advancement in Electronics & Communication Engineering (AECE)* (pp. 215–219). IEEE, Raj Kumar Goel Institute of Technology, Ghaziabad, India.

Namkhah, Z., Fatemi, S. F., Mansoori, A., Nosratabadi, S., Ghayour-Mobarhan, M., & Sobhani, S. R. (2023). Advancing sustainability in the food and nutrition system: A review of artificial intelligence applications. *Frontiers in Nutrition, 10*, 1295241.

Reyed, R. M. (2023). Focusing on individualized nutrition within the algorithmic diet: An in-depth look at recent advances in nutritional science, microbial diversity studies, and human health. *Food Health, 5*(1), 5.

Stewart, J. D., & Kremer, P. (2023). Urban atmospheric microbiome. In *The Palgrave Encyclopedia of Urban and Regional Futures* (pp. 1956–1961). Springer International Publishing.

Xiao, J., Tian, W., Abdullah, Wang, H., Chen, M., Huang, Q., Song, M., & Cao, Y. (2024). Updated design strategies for oral delivery systems: Maximized bioefficacy of dietary bioactive compounds achieved by inducing proper digestive fate and sensory attributes. *Critical Reviews in Food Science and Nutrition, 64*(3), 817–836.

Zyoud, S. E. H., Shakhshir, M., Barqawi, A., Abushanab, A. S., Koni, A., Khilfeh, S., Shahwan, M., Jairoun, A. A., Taha, A. A., Abushamma, F., & Al-Jabi, S. W. (2024). Comprehensive visualization of bariatric surgery and gut microbiota research: A global analysis. *Translational Medicine Communications, 9*(1), 13.

# 22 Exploring the Association between Ergonomics and Financial Peace

## *Investigating Financial Well-Being and Well-Being Expenses in Banking Professionals*

*Prakash Bhatnagar and Himani Sharma*

## 22.1 INTRODUCTION

Ergonomics is crucial for the physical and mental health, as well as the financial stability, of banking professionals. Due to the sedentary nature of banking, which requires lengthy hours of sitting, intensive computer use, and high-stress conditions, inadequate ergonomic standards can result in various physical and mental health problems. These concerns not only impact the welfare of employees but also have substantial financial consequences for both individuals and organisations. Implementing appropriate ergonomic measures might help reduce the occurrence of musculoskeletal disorders (MSDs), which are widespread among office workers, especially those in the banking industry (Punnett & Wegman, 2004). These problems are caused by repetitive strain, improper posture, and insufficient workstation arrangement, resulting in long-lasting pain, reduced efficiency, and higher rates of absenteeism (Buckle & Devereux, 2002). As a result, employees experience the financial strain of paying medical bills, earning less while they recuperate, and even facing long-term disability payments. Poor ergonomics have a significant financial impact from an organisational standpoint. Companies incur expenses related to medical claims, workers' compensation, and reduced productivity resulting from sick leave and presenteeism (Goetzel et al., 2004). Implementing ergonomic treatments is both a concern for health and a strategy for saving costs. Research has demonstrated that implementing ergonomic enhancements can boost employee efficiency and decrease the occurrence and intensity of work-related injuries, resulting in reduced healthcare and operational expenses (Hedge, 2003).

 DOI: 10.1201/9781003532293-25

Moreover, one's financial health is closely connected to one's overall job contentment and productivity (Bhuyan et al., 2024). Employees who experience a lack of pain and discomfort are more inclined to be actively involved and driven in their work, which leads to a favourable organisational culture and enhanced financial results (Carter et al., 2013). In this sense, financial peace (FP) denotes the state of stability and security employees enjoy when they are not encumbered by financial concerns connected to their health. This enables them to concentrate more effectively on their professional duties and opportunities for career progression. Essentially, incorporating ergonomic concepts in the workplace is crucial for protecting banking professionals' physical well-being and financial stability. Taking proactive measures to address ergonomic concerns can result in significant advantages, such as decreased healthcare expenses, increased productivity, and higher employee happiness. Ultimately, this contributes to individuals' and organisations' financial stability and overall welfare.

## 22.2 BACKGROUND OF THE STUDY

Professionals in the banking sector are required to possess a distinctive combination of cognitive and technical skills due to the dynamic and high-pressure nature of the industry. In recent decades, there has been an increasing acknowledgement of the substantial influence that workplace ergonomics has on the health and productivity of personnel in this particular industry. Due to the extended periods spent sitting at their computers, banking professionals are at a significantly elevated risk of acquiring work-related musculoskeletal disorders and other health problems.

The emphasis on ergonomics in the workplace originated from the necessity to improve efficiency and decrease injuries in industrial environments. As work changed over time, the principles of ergonomics started being used in a broader range of industries, including banking. The transition from physical to predominantly cognitive and sedentary activities has emphasised the need for ergonomic interventions to reduce harmful consequences associated with extended periods of sitting, repetitive tasks, and inadequate workstation layout (Karwowski, 2006). In the banking industry, the consequences of inadequate ergonomic measures go beyond physical well-being. The psychological welfare of workers is also at risk, as elevated levels of stress and bodily discomfort can result in exhaustion, reduced job contentment, and diminished overall efficiency (Schaufeli et al., 2009). The financial ramifications of these outcomes are substantial, impacting both individuals through medical bills and decreased earnings and businesses through diminished productivity and heightened healthcare expenses.

Considering these difficulties, numerous financial institutions have invested in ergonomic examinations and treatments. These strategies include reconfiguring workstations, incorporating sit-stand desks, and enforcing periodic breaks and exercise regimens to mitigate the likelihood of harm and enhance general health (Robertson et al., 2013). Implementing proactive measures enhances employees' well-being and satisfaction and fosters a more favourable and efficient work atmosphere. Furthermore, the fiscal stability of banking professionals is intricately linked to their physical and emotional well-being. Health-related concerns can result in substantial economic strain, impacting an individual's capacity to handle their resources and prepare for the future efficiently. Financial institutions can promote their personnel's physical well-being and stability by addressing ergonomic concerns (Kantor & Streitfeld, 2015).

Ergonomics, the scientific discipline concerned with the design and organisation of the work environment to suit the requirements of workers, plays a crucial role in improving productivity, well-being, and overall job contentment in many sectors. Implementing ergonomic concepts is particularly crucial in the banking sector, where staff frequently perform repetitive operations, spend long hours using computers, and interact with customers. Implementing ergonomic standards in the banking business helps alleviate physical discomfort, minimise the likelihood of MSDs, and foster a better work environment.

The widespread occurrence of occupational health problems emphasises the importance of ergonomics in the banking industry. Research has indicated that some bank employees experience work-related musculoskeletal issues due to inadequate ergonomic measures. Frequent problems encompass cervical and lumbar discomfort, carpal tunnel syndrome, and ocular fatigue, frequently arising from suboptimal workstation configuration, insufficient sitting arrangements, and prolonged sedentary positions.

Furthermore, incorporating ergonomics goes beyond just physical well-being and includes cognitive and organisational aspects of ergonomics.Itentails maximising the efficiency of mental effort, enhancing the simplicity of use of software interfaces, and refining the overall sequence of tasks to minimise cognitive stress and enhance productivity (Dul & Weerdmeester, 2008). Ergonomic adjustments can significantly improve operational performance in the fast-paced banking environment, where accuracy and speed are crucial.

Introducing ergonomic solutions in the banking sector requires a comprehensive strategy encompassing ergonomic evaluations, staff education, and ergonomic furniture and equipment integration. Ergonomic evaluations aid in identifying potential hazards and enable the customisation of treatments to meet the individual requirements of the workforce (Punnett & Wegman, 2004). Training programmes provide employees with knowledge and skills to effectively maintain ergonomic postures and make necessary workplace adjustments. In addition, investing in ergonomic equipment and tools, such as seats that can be adjusted, workstations that allow for sitting and standing, and keyboards designed for ergonomic use, can significantly enhance employees' comfort and productivity. Ergonomics is essential to workplace well-being and productivity in the banking sector. By attending to the ergonomic requirements of employees, banks may cultivate a safer, more efficient, and more gratifying work setting, ultimately resulting in improved organisational outcomes and employee welfare.

### 22.2.1 Research Objectives

**RO1**: Investigating the influence of ergonomics on financial well-being among banking professionals in Delhi NCR.

**RO2**: To evaluate the role of financial well-being in the connection between ergonomics and financial peace.

**RO3**: To investigate the impact of wellness expenses on the connections between ergonomics, financial well-being, and ergonomics and financial peace.

**RO4**: To Provide recommendations to enhance ergonomics and wellness strategies in banking institutions to improve financial peace and overall well-being.

### 22.2.2 Research Hypotheses

**H1**: Ergonomic practices significantly influence the financial well-being of banking professionals in Delhi NCR.

**H2**: Financial well-being significantly mediates the relationship between ergonomics and financial peace.

**H3a**: Wellness expenses significantly moderate the relationship between ergonomics and financial well-being, with the nature of the relationship varying depending on the level of wellness expenses.

**H3b**: Wellness expenses significantly moderate the relationship between ergonomics and financial peace, with the nature of the relationship varying depending on the level of wellness expenses.

#### 22.2.2.1 Literature Review

Ergonomics and well-being expenses (WE): Ergonomics can significantly improve staff health and organisational productivity. Ergonomics can improve wellness and productivity, according to Bowden (2017). Workplaces that support physical and mental health improve performance. Douphrate and Rosecrance (2004) argue that ergonomic solutions' economic benefits justify their cost. They emphasise that effective ergonomic techniques improve health and business finances.

Shorrock and Williams (2016) examine how human factors and ergonomics (HFE) improve system performance and well-being. Their edited book shows how HFE can improve operational effectiveness and employee well-being in real-world settings, arguing for ergonomic workplace design. Ergonomics can boost productivity and well-being, as Suokko and Reiman (2021) demonstrate from a holistic manufacturing facility construction process. They advise avoiding suboptimization, which can harm other aspects of the system, and recommend a holistic approach.

According to Koningsveld and Thé (2000), ergonomics and economics bring challenges and opportunities. They emphasise the necessity for preventive measures to increase workers' well-being and suggest that these measures are directly tied to prevention costs, making ergonomics an economically worthwhile investment. Good ergonomics can save money, according to Porter (2005). Poor ergonomics has a substantial financial impact, while successful ergonomic solutions can cut costs and improve well-being, resulting in lower legacy expenses and more productivity.

Hoffmeister et al. (2015) propose an Ergonomics Climate Assessment to measure operational success and staff well-being. They claim that ergonomic work design improves operational effectiveness and employee well-being, benefiting both employers and employees. Chim (2019) explores how ergonomic office design affects employee health and quality of life. Ergonomic designs can save money by lowering rental costs and improving employee satisfaction and productivity.

Dorman (2000) discusses the economics of safety, health, and welfare at work, emphasising the importance of wage stability and health insurance for financial stability. He believes better ergonomics can avert health conditions that considerably cost individuals and organisations. Ray et al. (2012) examine ergonomics and workplace safety. They emphasise that poorly designed work environments increase injury risk and that safe and productive workspaces improve well-being and economic benefits.

### 22.2.3 Financial Security and Ergonomics

Cognitive ergonomics' impact on financial company employees' well-being has been extensively studied. Sanil and Nair (2013) studied employees in public sector banks, private sector banks, and NBFCs and found that cognitive ergonomics improves employee welfare. Tuwai et al. (2015) examined how corporate well-being initiatives affect Kenyan commercial bank workers' performance. Their findings stressed the importance of financial wellness and ergonomic education in employee success. Maharjan et al. (2023) examined how cognitive ergonomics affects industrial employee welfare in Kathmandu Valley, demonstrating how ergonomic treatments improve economic well-being. In another study, Estember and Aguila (2020) evaluated musculoskeletal diseases in Philippine bank employees and stressed the importance of ergonomic assessments to reduce injury risks and promote employee health.

Kibret, Gebremeskel et al. (2020) examined Ethiopian bankers' work-related musculoskeletal illnesses and risk variables. Ergonomic training significantly reduced these problems. Musa (2016) examined self-reported work-related musculoskeletal diseases in Dhaka, Bangladesh, indicating ergonomic and work stress difficulties as significant banking industry concerns. Andrew and Kishokumar (2014) discovered that working conditions and workload significantly affected occupational stress in financial services workers. They suggested innovative workplace practices to decrease stress and boost productivity. Burbar (2021) found that a supportive work atmosphere improves employee performance in the Palestinian banking sector.

Ccanchi, Dragichevich et al. (2023) examined occupational health and safety in financial organisations, emphasising the significance of complete occupational health interventions to address ergonomic issues and job stress. Ha et al. (2019) found a significant incidence of psychosocial diseases in Singapore banking industry employees and suggested health practices to manage work-related stress.

Kenyan research shows that employee wellness programmes boost commercial bank performance. Stress management training, health education, nutrition and diet programmes, and ergonomic measures minimise healthcare and injury-related costs and improve employee health (Ng'eno, 2020). Such programmes demonstrate the value of comprehensive wellness measures in reducing health risks and improving workforce health. A study on work-based wellness programmes

in Nigeria found that they boost employees' well-being and awareness. Employee health improves with a wellness lifestyle, highlighting the need for workplace wellness programmes (Mohamed, 2020).

Global labour standards emphasise employee safety, welfare, well-being, and health. These practices reduce dangers and manage ergonomic, mechanical, and energy challenges to make the workplace safer and more productive (Gutterman, 2023). Wellness programmes for millennial retention in automobile manufacturing emphasise ergonomics, injury avoidance, exercise, and nutrition. These programmes help employees reach their health goals at low expense, improving workplace happiness and retention (Patterson, 2022).

A study on ageing industrial workers' thermal comfort implies that ergonomic research, particularly anthropometric studies, is crucial to worker health and safety. This study shows that workplace ergonomics should incorporate workers' temperature sensations and physical comfort (Caporale et al., 2022). Workplace stress affects employee behaviour and health. Wellness programmes that address physical fitness and stress management are needed to reduce stress's harmful effects on employees (Mundhe, 2019). Motivational and obstacle factors to workplace wellness programme engagement in Namibian public service were explored. The results show that health prevention and treatment programmes encourage healthy behaviours, improving workplace well-being (Maletzky, 2017).

Office buildings in developing countries like Malaysia have benefited from WELL adoption. WELL-integrated workstations improve employee health, productivity, and comfort (Tan et al., 2022).

Comparative studies on traditional and virtual office productivity show that ergonomics and wellness programmes boost productivity. Flexible work arrangements, including working from home, boost productivity and save money (Ntanjana, 2022).

Ergonomics improves employee well-being and productivity, according to the literature. Ergonomic practices can minimise health costs and improve financial outcomes for employees and organisations. These studies demonstrate the importance of ergonomics in banking personnel's health and performance. They recommend ergonomic modifications, corporate well-being, and supportive work settings to reduce health risks and boost job satisfaction and productivity.

Finally, management case studies emphasise employee rights and well-being. Wellness programmes like medical coverage and labour cost management make workers healthier and more productive (Mohamed & Vasudevan, 2022).

## 22.3 RESEARCH GAP

This study lacks data on how wellness expenses mediate and moderate ergonomics and FP. In addition, it studies how financial tranquillity affects financial well-being (FWB) among Delhi-NCR banking workers. Previous studies have shown that ergonomics improves workplace well-being and financial results, but little is known about how WE and FP affect this relationship. FP's impact on banking personnel's FWB inside ergonomic procedures is also under-studied. Thus, studying these links among Delhi-NCR banking professionals might fill this knowledge vacuum and provide valuable insights to academics and businesses.

## 22.4 RESEARCH METHODOLOGY

### 22.4.1 Research Design

This investigation examines the impact of financial tranquillity and decorum on the FWB of banking personnel in the Delhi-NCR region. In order to evaluate these effects, the investigation implements an exploratory and descriptive methodology. The cross-sectional method is employed to collect data, capturing information at a specific juncture in time to identify relationships among variables.

### 22.4.2 Participants of the Research

The sample is comprised of 249 bankers from the Delhi-NCR region, who were selected through systematic random sampling to ensure that they represented a variety of institutions and departments. This method guarantees that the sample accurately represents the population, with an emphasis on minimizing bias and enabling the findings to be generalizable to a broader range of banking personnel in the region.

### 22.4.3 Data Collection Procedure and Analysis

Participants were obligated to complete online surveys that were meticulously devised. The survey questionnaire was modified to accurately assess variables associated with ergonomics, wellness expenses, financial serenity, and FWB. The data was efficiently collected and managed using electronic systems, which guaranteed a high level of reliability and quality of response.

This investigation investigates the correlations between FWB, financial calm, wellness expenses, and ergonomics among banking employees in the Delhi-NCR region. The study guarantees a representative sample by employing a stratified random sampling method, thereby enhancing the generalizability of the findings to the broader population of banking personnel in this region.

## 22.5 RESULT AND ANALYSIS

The data provides demographic information from a sample of 248 individuals. Regarding gender distribution, the majority of the sample is represented by females at 66.6%, with males accounting for 33.4%. In terms of age groups, the largest segment comprises individuals between 30 and 40 years old, accounting for 37.7% of the total. It is closely followed by the 20%–30% age group, which makes up 22.3%. The income distribution data reveals that a significant percentage of individuals (48.6%) belong to the 3L–5L bracket. Additionally, 37.2% fall within the 5L–8L range, while 14.2% are in the 8L-12L bracket. The majority of respondents are from Delhi (39.3%), followed by Gurugram (34.0%) and Noida (26.7%). The data indicates that most of the population is female and middle-aged. Many individuals earn between 3L and 8L and primarily live in Delhi and Gurugram (Table 22.1).

Figure 22.1 visually illustrates the impact of "Wellbeing Expenses" and "Ergonomics" on "Financial Wellbeing," which in turn affects "Financial Peace."

**TABLE 22.1**
**Descriptive Statics of the Samples**

| Demographic Variable | | Frequency | Per cent |
|---|---|---|---|
| Gender | Male | 83 | 33.4 |
| | Female | 165 | 66.6 |
| Age | 20–30 | 55 | 22.3 |
| | 30–40 | 93 | 37.7 |
| | 40–50 | 39 | 15.8 |
| | 50–60 | 42 | 17.0 |
| | 50-above | 19 | 7.2 |
| Income | 3L–5L | 120 | 48.6 |
| | 5L–8L | 92 | 37.2 |
| | 8L–12L | 36 | 14.2 |
| Area | Delhi | 97 | 39.3 |
| | Gurugram | 84 | 34.0 |
| | Noida | 67 | 26.7 |

*Source:* Author's Calculation.
*Note:* $N = 248$.

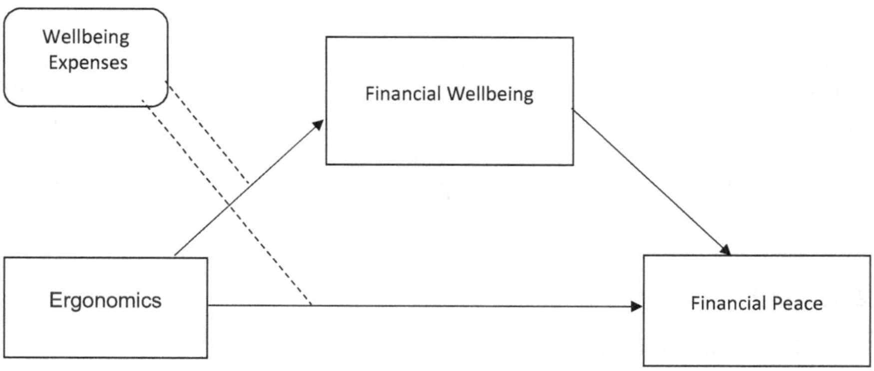


**FIGURE 22.1** Conceptual framework of the study.

## 22.5.1 Model Summary: Outcome Variable: FWB

### 22.5.1.1 Model Summary Interpretation

The regression Andrew F. Hyaes model summary reveals that the model accounts for about 22.96% of the variability in the dependent variable ($R$-squared$=0.2296$). The overall model demonstrates statistical significance, indicating that the predictors collectively exert a significant influence on the outcome ($F_{(3, 244)}=24.2333$, $p<0.001$) (Table 22.2).

Upon analysing the coefficients, it is evident that the constant term lacks significance ($p=0.3695$), suggesting that the intercept does not significantly deviate from

**TABLE 22.2**
**Model Summary**

| *R* | *R*-sq | MSE | *F* | df1 | df2 | *p* |
|---|---|---|---|---|---|---|
| 0.4791 | 0.2296 | 0.7621 | 24.2333 | 3.0000 | 244.0000 | 0.0000 |
| | | | **Model** | | | |
| | coeff | se | t | *p* | LLCI | ULCI |
| Constant | −5.4495 | 6.0608 | −0.8991 | 0.3695 | −17.3876 | 6.4886 |
| ER | 1.4711 | 0.4277 | 3.4398 | 0.0007 | 0.6287 | 2.3136 |
| WE | 2.7099 | 0.8925 | 3.0364 | 0.0027 | 0.9520 | 4.4678 |
| Int_1 | −0.2183 | 0.0630 | −3.4639 | 0.0006 | −0.3425 | −0.0942 |
| | | | **Product terms key:** | | | |
| Int_1 | ER | X | WE | | | |

zero. The variable ER demonstrates a noteworthy and meaningful impact on the outcome (coeff = 1.4711, $p = 0.0007$), indicating that an escalation in ER is linked to a rise in the dependent variable. In addition, our findings demonstrate a substantial and statistically significant effect (coeff = 2.7099, $p = 0.0027$), suggesting that an increase in the variable we examined is linked to a corresponding increase in the outcome.

The interaction term shows a significant and negative relationship (coeff = −0.2183, $p = 0.0006$), indicating that ER's impact on the outcome variable decreases as WE increase. This interaction demonstrates a moderating effect, where the relationship between ER and the dependent variable varies based on the level of WE. The model indicates that both ER and WE have a positive impact on the dependent variable. However, their combined effect is moderated by their interaction.

## 22.5.2 Outcome Variable: FP

### 22.5.2.1 Model Summary Interpretation

The regression analysis results suggest that the model can account for around 15.45% of the variability in the dependent variable, as indicated by an *R*-squared value of 0.1545. The model demonstrates a high level of statistical significance, as indicated by an *F*-statistic of 11.1009 ($p < 0.0001$). The study suggests that the predictors, when considered together, substantially influence the outcome (Table 22.3).

The constant term is 48.5555, indicating the expected value of the dependent variable when all predictors are zero. The constant demonstrates statistical significance ($p < 0.0001$), and the 95% confidence interval spans from 27.0546 to 70.0565.

The predictor ER has a coefficient of −3.2444, which suggests a negative relationship between ER and the dependent variable. As ER increases, the dependent variable tends to decrease, assuming all other variables remain constant. This relationship is statistically significant ($p = 0.0001$) with a 95% confidence interval from −4.7955 to −1.6934.

**TABLE 22.3**
**Model Summary**

| *R* | *R*-sq | MSE | *F* | df1 | df2 | *p* |
|---|---|---|---|---|---|---|
| 0.3931 | 0.1545 | 2.4638 | 11.1009 | 4.0000 | 243.0000 | 0.0000 |
| | | | **Model** | | | |
| | coeff | se | *t* | *p* | LLCI | ULCI |
| Constant | 48.5555 | 10.9154 | 4.4483 | 0.0000 | 27.0546 | 70.0565 |
| ER | −3.2444 | 0.7874 | −4.1204 | 0.0001 | −4.7955 | −1.6934 |
| FWB | 0.4996 | 0.1151 | 4.3405 | 0.0000 | 0.2729 | 0.7264 |
| WE | −6.1425 | 1.6347 | −3.7575 | 0.0002 | −9.3625 | −2.9224 |
| Int_1 | 0.4319 | 0.1161 | 3.7202 | 0.0002 | 0.2032 | 0.6605 |
| | | | **Product terms key:** | | | |
| Int_1 | ER | x | WE | | | |

The predictor FWB has a coefficient of 0.4996, indicating that a one-unit increase in FWB leads to a corresponding increase of 0.4996 units in the dependent variable while keeping other variables constant. This effect is highly significant ($p<0.0001$) and has a 95% confidence interval ranging from 0.2729 to 0.7264.

The predictor has a coefficient of −6.1425, indicating that an increase in the variable results in a decrease in the dependent variable while keeping other variables constant. This relationship is statistically significant ($p=0.0002$) with a 95% confidence interval ranging from −9.3625 to −2.9224.

The interaction term Int_1 (ER x WE) has a coefficient of 0.4319, suggesting that the level of WE influences the impact of ER on the dependent variable. For every unit increase in the interaction term, the dependent variable has a corresponding increase of 0.4319 units. This interaction is statistically significant, with a 95% confidence interval between 0.2032 and 0.6605.

The model demonstrates the noteworthy impact of ER, FWB, WE, and their interaction on the dependent variable. ER and WE exhibit adverse main effects, while FWB has a positive main effect. Additionally, the interaction between ER and WE helps alleviate ER's negative influence on the dependent variable.

## 22.6 FINDINGS AND DISCUSSION

Based on the data collected from a sample of 248 individuals, it is evident that there is a higher representation of females (66.6%) compared to males (33.4%). The largest age group comprises individuals between 30 and 40, making up 37.7% of the population. The age group of 20–30 years is closely behind, accounting for 22.3% of the population. The distribution of income reveals that a significant portion of individuals, approximately 48.6%, earn between 3L and 5L. Additionally, 37.2% of people fall within the income range of 5L-8L, while a smaller percentage, precisely 14.2%, earn between 8L and 12L. Delhi has the highest number of respondents at 39.3%, followed by Gurugram at 34.0% and Noida at 26.7%. Most of this sample comprises middle-aged women, with a significant number earning between 3L and 8L. They are primarily located in Delhi and Gurugram.

This study highlights ergonomics' important role in promoting FWB and FP. The impact of ergonomics on FWB varies depending on the level of WE. When WE are low, ergonomics has a positive effect on FWB. However, when WE are high, ergonomics harms FWB. On the other hand, when WE levels are low, it harms FP. As WE levels increase, this impact becomes less significant and insignificant at high WE levels. WE are crucial in managing the relationship between ergonomics and financial outcomes. Doing so helps minimize the negative impact on FP and improves FWB.

The literature provides ample evidence to support these findings, highlighting the importance of implementing ergonomic practices to improve employee well-being and boost overall organisational productivity. Bowden (2017) and Douphrate and Rosecrance (2004) emphasise the importance of creating work environments that promote physical and mental health, ultimately leading to improved well-being and productivity. They advocate for the use of ergonomic approaches to achieve these goals. Shorrock and Williams (2016) and Suokko and Reiman (2021) delve into the practical application of HFE to enhance system performance and promote employee well-being, cautioning against the pitfalls of suboptimization.

The importance of ergonomics in the workplace is emphasized by Koningsveld and Thé (2000), who advocate for preventive measures to enhance worker well-being. Porter (2005) also highlights the potential cost savings of effective ergonomic interventions. Hoffmeister et al. (2015) and Chim (2019) highlight the significant benefits of implementing ergonomic workplace design. This includes improvements in operational performance and increased employee satisfaction.

The research focused on the financial industry uncovers comparable patterns. According to Sanil and Nair (2013) and Tuwai et al. (2015), there is a strong emphasis on the significance of cognitive ergonomics and corporate well-being practices in improving employee performance. In their studies, Maharjan et al. (2023) and Estember and Aguila (2020) highlight the significant benefits of implementing ergonomic interventions. These interventions have been found to reduce MSDs and promote better employee health effectively.

The research suggests that implementing ergonomic practices is essential for improving the well-being and productivity of employees. Organizations can reap substantial economic benefits and enhance job satisfaction and performance by addressing health risks and fostering a supportive work environment.

## 22.7 CONCLUSION

This study highlights the significant influence of demographics and ergonomic practices on employees' FWB and FP. The sample primarily comprises middle-aged women with moderate incomes living in Delhi and Gurugram. The impact of ergonomics on financial outcomes is complex, with a positive influence on FWB when WE are low but a negative influence when WE are high. On the other hand, the impact of ergonomics on FP is less pronounced at low WE levels and becomes less significant as WE increases, eventually becoming insignificant at high levels.

The literature highlights the significance of ergonomic practices in improving employee well-being and organizational productivity. Implementing effective ergonomic interventions can help address health risks, minimize expenses related to inadequate ergonomic practices, and improve job satisfaction and performance. The significance of WE as a moderating factor highlights the importance of investing in employee well-being to fully capitalize on the positive impact of ergonomics on financial results.

Ultimately, this study highlights the importance of organizations implementing thorough ergonomic practices and initiatives focused on well-being. By fostering work environments that prioritize the well-being of their employees, companies can experience improved financial outcomes and overall peace. This, in turn, leads to increased productivity and economic advantages. Considering the importance of incorporating ergonomic principles in workplace design, it becomes evident that doing so can significantly contribute to cultivating a healthier and more productive workforce, ultimately enhancing overall organizational performance.

## REFERENCES

Andrew, A., & Kishokumar, R. (2014). Influence of working environment and workload on occupational stress among staff in the financial services industry. *International Journal on Global*, *7*, 2538.

Bhuyan, U., Khan, S. A., & Saleem, M. (2024). Gaining perspective on money attitude and financial planning of educators in Northeast India: Implications for sustainable finance. In S. Yıldırım, D. Yıldırım, I. Demirtaş, & V. Kandpal (Eds.), *Green Transition Impacts on the Economy, Society, and Environment* (pp. 244–262). IGI Global. https://doi.org/10.4018/979-8-3693-3985-5.ch014

Bowden, S. (2017). Optimising well-being and productivity through an ergonomics-based approach. In C. C. March & C. Steven (Eds.), *Creating the Productive Workplace* (pp. 160–172). Taylor & Francis.

Buckle, P., & Devereux, J. (2002). The nature of work-related neck and upper limb musculoskeletal disorders. *Applied Ergonomics*, *33*(3), 207217. https://doi.org/10.1016/S0003-6870(02)00014-5

Burbar, M. Y. (2021). The impact of work environment on employees' performance in the banking sector in Palestine. *International Business Research*, *14*(1), 112. https://doi.org/10.5539/ibr.v14n1p1

Caporale, A., Galizia, F. G., & Botti, L. (2022). Thermal comfort prediction of aged industrial workers based on occupants' basal metabolic rate. In *Occupational Ergonomics*. AHFE International. Retrieved from https://books.google.com

Carter, B., Danford, A., Howcroft, D., Richardson, H., Smith, A., & Taylor, P. (2013). 'Stressed out of my box': employee experience of lean working and occupational ill-health in clerical work in the UK public sector. *Work, Employment and Society*, *27*(5), 747–767. https://doi.org/10.1177/0950017012469064

Ccanchi, C. A. C., Dragichevich, C. O. Q., Claudio, B. A. M., & Ruiz, J. A. Z. (2023). Occupational health and safety in a financial company. *SCT Proceedings in Interdisciplinary Insights and Innovations*, *1*, 49–49.

Chim, J. M. Y. (2019). 6Ws in ergonomics workplace design. In *Proceedings of the International Ergonomics Association (IEA 2018)*. Florence: Springer.

Dorman, P. (2000). *The Economics of Safety, Health, and Well-Being at Work: An Overview*. ResearchGate. Retrieved from https://www.researchgate.net/publication/20000000

Douphrate, D. I., & Rosecrance, J. (2004). The economics and cost justification of ergonomics. In *Proceedings of the 2nd Annual International Symposium on Ergonomics*. ResearchGate. Retrieved from https://www.researchgate.net/publication/262949742

Dul, J., & Weerdmeester, B. (2008). *Ergonomics for Beginners: A Quick Reference Guide* (3rd ed.). CRC Press.

Estember, R. D., & Aguila, D. K. S. (2020). Effects of workplace design, work methods design, and health problems on the prevalence of musculoskeletal disorders among bank tellers in the Philippines. In *IEEE 7th International Conference on Industrial Engineering and Applications*, p. 2529. Bangkok: IEEE. https://doi.org/10.1109/ICIEA49774.2020.9101974

Goetzel, R. Z., Long, S. R., Ozminkowski, R. J., Hawkins, K., Wang, S., & Lynch, W. (2004). Health, absence, disability, and presenteeism cost estimates of certain physical and mental health conditions affecting US employers. *Journal of Occupational and Environmental Medicine*, *46*(4), 398412. https://doi.org/10.1097/01.jom.0000121151.40413.bd

Gutterman, A. S. (2023). *Sustainability and Human Rights Standards and Instruments Relating to Labor Practices*. Available at SSRN 4551042. Retrieved from https://papers.ssrn.com

Ha, H., Palaniappan, K., & Pidani, R. (2019). Psychosocial well-being of employees in the banking sector in Singapore. *Employment and Employment*, *25*(3), 200215. https://doi.org/10.1504/IJEWE.2019.103021

Hedge, A. (2003). Ergonomic workplace design for health, wellness, and productivity. *Ergonomics*, *46*(13), 316. https://doi.org/10.1080/00140130303533

Hoffmeister, K., Gibbons, A., & Schwatka, N. (2015). Ergonomics climate assessment: A measure of operational performance and employee well-being. *Applied Ergonomics*, *46*, 99107. https://doi.org/10.1016/j.apergo.2014.07.004

Kantor, J., & Streitfeld, D. (2015, February 22). Inside Amazon: Wrestling big ideas in a bruising workplace. *The New York Times*. https://www.nytimes.com/2015/08/16/technology/inside-amazon-wrestling-big-ideas-in-a-bruising-workplace.html

Karwowski, W. (2006). Ergonomics and human factors: The paradigms for science, engineering, design, technology, and management of humancompatible systems. *Ergonomics*, *49*(7), 671676. https://doi.org/10.1080/00140130400029167

Kibret, A. K., Gebremeskel, B. F., Gezae, K. E., & Tsegay, G. S. (2020). Workrelated musculoskeletal disorders and associated factors among bankers in Ethiopia. *Pain Research and Management*, *25*(2), 215225. https://doi.org/10.1155/2020/8735169

Koningsveld, E. A. P., & Thé, K. A. (2000). Ergonomics and economics, dilemmas and chances. *Human Factors and Ergonomics in Manufacturing & Service Industries*, *10*(2), 121132. https://doi.org/10.1177/154193120004403101

Maharjan, K., Dhakal, K., Koirala, R., Adhikari, K., Danuwar, R. K., & Poudel, D. K. (2023). Measure the effects of cognitive ergonomics on industrial employee well-being in Kathmandu Valley. *Journal of Economic*, *35*(1), 1525. https://doi.org/10.3126/tjec.v14i1-2.62317

Maletzky, M. M. (2017). *Employee Participation in Workplace Wellness Programmes in the Namibian Public Service: Motivational Factors and Barriers*. University of the Western Cape. Retrieved from https://etd.uwc.ac.za

Mohamed, D. R. Z., & Vasudevan, D. R. H. (2022). *Case Studies in Management Perspective*. ResearchGate. Retrieved from https://www.researchgate.net

Mohamed, J. M. (2020). *The Perceived Benefits of Workbased Wellness Programs: A Case Study of Nigerian Companies and Employees*. Academia.edu. Retrieved from https://www.academia.edu

Mundhe, E. S. (2019). *A Study of Workplace Stress and Work Stress*. Research Journey. Retrieved from https://www.researchgate.net

Musa, A. M. (2016). Selfreported workrelated musculoskeletal disorders (MSDs) among bank workers in Dhaka, Bangladesh. *Journal of Occupational Health, 20*(1), 4050. http://202.4.109.28:8080/xmlui/handle/123456789/233

Ng'eno, W. K. (2020). *Wellness Programs and Employee Performance in Commercial banks, Kenya.* Kenyatta University Institutional Repository. Retrieved from https://ir-library.ku.ac.ke/server/api/core/bitstreams/771b43c8-4b35-4da0-90e4-575d5be36352/content Ntanjana, S. (2022). *Comparison of Employee Productivity Levels between Traditional and Virtual Offices at Tertiary Institutions in the Western Cape.* Cape Peninsula University of Technology. Retrieved from https://etd.cput.ac.za/bitstream/20.500.11838/3786/1/Ntanjana_Siyasanga_216042488.pdf

Patterson, C. (2022). *Millennial Retention through Wellness in the Automotive Manufacturing Industry.* Scholar Commons. Retrieved from https://scholarcommons.sc.edu

Porter, M. (2005). *Poor Ergonomics Costs but Can Good Be Made to Pay?* Retrieved from https://nrl.northumbria.ac.uk/id/eprint/1787

Punnett, L., & Wegman, D. H. (2004). Workrelated musculoskeletal disorders: The epidemiologic evidence and the debate. *Journal of Electromyography and Kinesiology, 14*(1), 1323. https://doi.org/10.1016/j.jelekin.2003.09.015

Ray, T. K., Waters, T. R., & Hudock, S. (2012). Economics of ergonomics. *Occupational Ergonomics, 10*(3), 123134.

Robertson, M., Ciriello, V. M., & Garabet, A. M. (2013). Office ergonomics training and a sitstand workstation: Effects on musculoskeletal and visual symptoms and performance of office workers. *Applied Ergonomics, 44*(1), 7385. https://doi.org/10.1016/j.apergo.2012.05.001

Sanil, S. K., & Nair, V. K. (2013). Cognitive ergonomics and employee well-being in financial companies. *Journal of Strategic Management, 12*(3), 5869. https://www.proquest.com/openview/b723d547479f3c47d18a848f7171ac0f/1?cbl=2030931&pq-origsite=gscholar

Schaufeli, W. B., Bakker, A. B., & Van Rhenen, W. (2009). How changes in job demands and resources predict burnout, work engagement, and sickness absenteeism. *Journal of Organizational Behavior: The International Journal of Industrial, Occupational and Organizational Psychology and Behavior, 30*(7), 893917. https://doi.org/10.1002

Shorrock, S., & Williams, C. (2016). *Human Factors and Ergonomics in Practice: Improving System Performance and Human Well-Being in the Real World.* Taylor & Francis. https://doi.org/10.1201/9781315587332

Suokko, T., & Reiman, A. (2021). Ergonomics is profitable – Experiences from a holistic manufacturing plant level development process. In *Congress of the International Ergonomics Association (IEA 2021).* Springer.

Tan, C. Y. M., Rahman, R. A., Xia, B., & Chen, Q. (2022). *Streamlining WELL Concepts of Office Buildings for Developing Countries: The Case of Malaysia.* Construction Research Congress. Retrieved from https://ascelibrary.org

Tuwai, B. B., Kamau, C., & Kuria, S. (2015). Effect of corporate well-being practices on employees' performance among commercial banks in Kenya. *International Journal of Scientific, 8*(4), 100115. https://www.ijsrp.org/research-paper-0515/ijsrp-p4124.pdf

# Index


academic 4, 25, 29, 30, 47–60, 135, 275
accessibility 105, 152, 155, 158, 185, 213, 215, 223–225
accuracy 92, 141, 152, 155, 156, 163, 165, 170, 172–173, 178–179, 215, 218, 223, 326
adaptive 15, 148, 154, 190, 215, 224, 291
adoption 4, 15, 53, 66, 71, 73, 76, 111, 126, 163, 170, 190, 250, 256, 312, 314–318, 321–322
advancements 7, 30, 156, 162, 188, 190, 213, 219, 224, 288, 304, 311–312, 315
affective event theory 277
AI (artificial intelligence) 7, 17, 147, 162, 185, 198, 213, 274, 314
algorithms 151–155, 157–158, 163–165, 178–180, 187, 214, 217–218, 221, 223–226
applications 4, 77–79, 86, 98, 116, 149, 153, 154, 158, 164, 188, 192, 200, 204–207, 213, 216, 219, 223, 226, 229, 235, 303–304, 306, 312, 334
Arab world 311–316
areas 41, 72, 113, 127, 128, 156
assessment 7, 12–14, 16, 23, 38, 40, 45, 47, 52, 56, 67, 78, 84–86, 93, 106, 134, 151, 157, 164–166, 169, 175, 188, 200, 208, 214, 222, 256, 257, 263–272, 278, 301–303, 327, 328
assets 14, 16, 20, 24–26, 50, 215, 251
augmented reality (AR) 188, 192, 200–202, 209
automation 80, 101, 154, 160, 185, 189–191, 193–194, 201, 205–207, 209, 215, 218, 227, 297, 306
awareness 22, 35, 37–40, 44–45, 66, 71, 76, 78–79, 132, 149, 229, 239, 244, 263–265, 297–305, 316–320, 329

banking 214, 229, 277, 324–330
behavior 8, 21, 24, 25, 29, 41, 47–50, 52–53, 67, 69, 71, 91–92, 104, 108, 113, 120–121, 127, 132, 134, 137, 147–154, 179, 213, 215–216, 223–224, 228, 264–265, 274–278, 298, 300, 302, 313–316
Big Data 79, 85, 200–202, 205, 314
burnout 52, 94, 137, 235, 237, 251, 255, 257, 274–279, 284, 288–290, 295, 297, 306–307

carpal tunnel syndrome 326
cognitive 4–6, 21–23, 26, 35–36, 49, 67, 115, 134, 148–151, 154, 156–157, 204, 209, 215, 223–224, 237, 248–251, 258, 262, 274–278, 290–292, 294–297, 299–300, 304, 306, 325–326, 328, 334
collaboration 11, 15, 52, 69, 98, 125–126, 147, 157, 165, 190–191, 203, 205, 225, 226, 228–229, 257, 288, 293, 300, 306, 319
compliance 58, 293
conflicting organizational priorities 72–73
convergence 147
corporate success 3–4, 6, 8–9, 11–17, 76, 85, 198, 213, 248, 250, 252–254, 256, 258
cybersecurity 218–219

deviant behavior 275
digital age 147, 157
digital marketing 218
digital technologies 198, 205
digital transformation 150–153, 224, 318
digital twins 192, 198, 202, 227
deviant workplace behavior (DWB) 274–284

economic production 219
ecosystems 214, 320
educational outcomes 213
environmental sustainability principles 65
ethical considerations 148, 155, 178, 215, 221, 225, 226
ethical standards 58, 178, 221
ethical use 180
ethics 156, 216, 222, 228
evolution 4, 5, 149, 188, 190, 191, 198, 200, 205, 209
explainable AI (XAI) 163, 165, 226
extraversion 22, 48, 51, 53–59

facilities 81, 241, 270
fairness 151–155, 158, 162–163, 165–167, 170–174, 176, 178–180, 216, 221–222
fatigue 189, 190, 251, 291, 294, 298
financial performance 3, 6, 8, 9, 11, 13–16, 20, 126
financial tranquillity 329–330
fintech 214
fitness 59, 84, 150, 303, 305, 320, 321, 329

gadgets 201, 300
genetics 311, 315
Germany 126, 198
GHRM (Green Human Resource Management) 65–74
globalization x
governance 20, 219, 226, 228

hazards 133–135
healthcare 4, 6, 10, 77, 79–80, 84–85, 87, 91, 148, 154–156, 207, 214, 224, 227, 249, 258, 287, 304, 312–313, 316–321, 324–326, 328
health conditions 235, 312, 313, 328
health promotion 133, 287, 292, 299, 301, 303
heuristics 190
HF&E (Human Factors and Ergonomics) ix, 89–90, 93–95, 97–98, 150, 154, 158, 198–199, 213, 215–217, 220–228, 248, 287–288, 291–296, 327
hospitals 76–78, 80, 84–87
HR policies 60, 307
HR records 166, 170–171
human-centric 183, 311
human expertise 193
human intelligence 213
human resource management (HRM) 12, 65–66, 68–69, 71, 73–74, 90, 99, 133, 275
hygiene 81, 121, 275, 279

implementation 12, 15–16, 34–39, 45, 67, 72, 77–78, 94, 96, 104–105, 107, 111–113, 115, 152–153, 162, 167, 169–171, 188, 192–193, 199, 215, 238, 243–244, 250, 271, 303, 311, 321

India 19–20, 55, 58, 76, 78–79, 86–87, 131, 275
individual needs 15, 49, 94
industrial engineering 67
industries 9, 35, 148, 151–153, 158, 180, 188, 190, 198–199, 206, 229, 239, 244, 248, 250, 271, 287
Industry 5.0 183, 190, 192, 193
innovation 3–4, 6, 8–9, 11–12, 14–16, 34, 67–68, 93, 98, 112, 123–124, 126–128, 190, 192–194, 201, 204–205, 208, 229, 257, 290, 295, 306
intelligent tutoring systems 214
interdisciplinary 40, 69, 134, 157, 217, 219, 225, 228, 315
interrelationships 24
IoT (Internet of Things) 152, 190, 192–193, 198, 200–201

layout 82, 86, 149–151, 154, 157, 183–187, 244, 291, 325
leadership 12, 15–16, 21–24, 26–29, 66, 86, 91, 105, 108–111, 114, 122–128, 236, 242–243, 254, 290, 302, 307
learning 22, 47, 49–51, 55, 66–67, 107–108, 112, 115, 147, 151, 156, 189–191, 198, 213–214, 220, 223–224, 242, 298

maintenance 77, 98, 152, 157–158, 199, 205, 209, 214, 224, 227
managerial characteristics 25
managerial traits 24
manufacturing 4, 7, 10, 67–68, 152, 157, 188, 190, 198, 201–203, 205–206, 209, 223–224, 227–228, 263, 327, 329
medical 20, 50, 86, 98, 131, 148, 152, 201, 205, 223–224, 266, 279, 287, 320, 324–325, 329
mitigating risks ix, 162, 166–167
motivation 111, 113, 133–134, 136–137, 139–140, 239, 249, 251, 275, 288, 329
musculoskeletal conditions 263, 266–267
musculoskeletal disorders (MSDs) 6, 37, 40, 42, 78–79, 92, 96–97, 148, 151, 156, 187, 328–239, 241, 243, 263–267, 269, 271–272, 275, 291, 294, 297, 299, 304, 325–326, 334

neural networks 214, 227
neuroticism 22, 48, 52–58
nutrition 59, 311–322

Occupational Safety, Health and Working Conditions Code 2020 275
oil and gas 34–40, 45, 207, 262–265, 267–272
operational costs 186, 198
operational efficiency 262
operational performance 20, 190, 250, 326
order picking 183, 187, 190–191, 199
organizational culture x, 70, 90–91, 112, 165–166, 179, 275, 287–288
organizational dynamics 65
organizational outcomes 5, 120, 124, 132, 179, 251, 284
organizational performance 14, 67, 71, 94–95, 98, 121, 127, 132, 249
organizational success 4, 86, 94–95, 106, 110, 120, 122, 124, 128, 135, 251, 258, 274–275, 280, 282, 284
OWAS (Ovako Working Analysis System) 188–190, 265

pallet selection 189–190
personal values 140
personalized nutrition 311–322
picking speed 186
policy development 16

policy makers 275
promotion 133, 287, 292, 296, 299, 301, 303, 305
psychological factors 20, 22, 267
psychological well-being 131

qualitative data 8, 173, 301
qualitative methods 8, 70, 171–172
quantitative data 14, 148, 166
quantitative methods 73, 80, 85, 172, 258

repetitive movements 279
resource efficiency 186
retail 27, 152–153, 168–169, 215, 225, 227
risk management 223, 267, 269, 271
robotics 152–153, 158, 185, 190–191, 198–209
sensors 191–192, 200, 205, 208

SET (Social Exchange Theory) 69, 275
SLR (Systematic Literature Review) 19–20, 23–24, 28, 192, 208
smart factory 206, 208
strategic framework 313–314
strategic orientation ix, 3–6, 9, 16, 34, 48–49, 89–90, 93–94, 97–98, 198
strategic partnerships 145
strategic planning 313
structured approach 21
structured overview 12, 114–115
success factors 24, 314
supply chain 191, 193, 227
supportive environments 49
sustainability 4, 20, 65, 67–71, 73–74, 103, 193, 227, 253, 256, 287, 292
sustainable competitive advantage x
sustainable industry growth 314
sustainable management 65
symbiotic potential 65

techniques 11, 75, 112
technological advancements ix, 7, 103, 188, 190, 224, 235, 288
technological needs 229
telehealth platforms 227
terrain 66, 228
text analysis 79
thematic analysis 19–21, 24–26, 148, 172
theoretical frameworks ix, 65, 69, 103–106, 124, 136, 250–251, 277
therapies 315
transportation 34–35, 184, 214, 223–224, 227
travel distance 184, 186, 187
trust 22, 28, 69, 105, 107–109, 112, 155, 162–165, 167, 170–172, 176–180, 220, 313, 316, 318–321
trustworthy 209, 221, 226, 322
turing test 213
tutoring systems 214

unorganized sector 78
usability 147, 149–150, 158, 171, 192, 215, 228
usage 158, 191, 205–206, 250, 294–295, 302
user demands 215, 226
user expectations 215

virtual assistants 153, 158, 223, 227
virtual reality 17, 298
visual data 214
visualization 177, 202, 226
warehouse 152, 158, 183–189, 191–194

Warehousing 5.0 188, 190–194
well-being 131, 248, 262
WMSDs (work-related musculoskeletal disorders) 76, 85–86, 92, 96–97, 239
word cloud analysis 12
work environment 90, 92, 94–95, 108, 114–115, 133, 237–238, 242, 249, 255, 257, 262, 279, 292, 294, 299, 302–304, 307, 326
work commitment 60
work-life balance 91, 122, 128, 241, 254, 299
work satisfaction 60, 133
workplace design 5, 14, 16, 69, 73, 250, 255, 258, 279, 299, 306–307
workplace ergonomics 92, 149–150, 157–158, 275, 325, 329
workplace injuries 189–190
workstations 6, 80, 85–86, 92, 98, 187, 191, 193, 236, 243, 254, 266, 271, 326, 329
world 47, 120, 158, 206, 208, 213, 275, 305, 311, 314